Contents

SYSTEMS ANALYSIS AND DESIGN

A STRUCTURED APPROACH

WILLIAM S. DAVIS

MIAMI UNIVERSITY, OXFORD, OHIO

WITH CONTRIBUTIONS BY

Mrs. Laurena Burk
Dr. Teruo Fujii
Dr. David Haddad

ADDISON-WESLEY PUBLISHING COMPANY

Reading, Massachusetts • Menlo Park, California
London • Amsterdam • Don Mills, Ontario • Sydney

Sponsoring Editor: William S. Gruener

Production Editors: William S. Davis
Dr. S. Allison McCormack
Venis V. Torge

Designers: William S. Davis
Venis V. Torge

Art: William Seitz
William C. Davis

Cover Design: Richard Hannus

Library of Congress Cataloging in Publication Data

Davis, William S., 1943-
Systems analysis and design.

Includes index.
1. System design. 2. System analysis. I. Title.
QA76.9.S88D33 1983 001.6'1 82-11675
ISBN 0-201-10271-4

Reproduced by Addison-Wesley from camera-ready copy supplied by the author.

ISBN 0-201-10271-4
CDEFGHIJ-DO-89876543

to LARRY

Preface

The subject of this book is systems analysis and design. It assumes a reasonable understanding of computer concepts and terminology, and an ability to program in at least one compiler or assembler language. The material is designed to support a first course in analysis and design in an applied, computer-related college or university program. Additionally, it should prove interesting to the manager seeking an understanding of the analysis and design process, to the programmer contemplating a move into systems analysis, and to many others as well.

Recently, the structured approach to systems analysis and design has become widely accepted. Unfortunately, much of the material on this new approach is aimed at practicing professional analysts or technical managers. Experience in analyzing and designing systems is assumed; the structured techniques are presented as alternatives to traditional tools. While ideal for experienced analysts, this professional orientation creates a very real problem for the beginner, the student analyst. Before the value of a new tool can be appreciated, the student must have a sense of where that tool fits in the broader context of the analysis and design process. What does it mean to define a problem? What is a feasibility study? Why should a broad, general analysis preceed system design? Why is it so important that a program be designed *in the context* of a system plan? The experienced analyst knows the answers. The beginner may memorize sets of words and phrases that seem to answer the questions, but without active exposure to the process, cannot understand the answers. The beginner must start with the traditional system life cycle, and learn the analyst's tools and techniques within this context.

A number of textbooks cover the traditional system life cycle, and many are quite good; unfortunately, few incorporate the structured methodologies. There are many excellent books on the structured techniques, but few incorporate the underlying principles of analysis and design that are so essential to the beginning student. This text includes key elements of both the old and the new, introducing the student to structured methodologies in the context of the traditional system life cycle.

The text begins with an overview of the life cycle and then presents a series of three case studies: a payroll system, a professional office management system, and an on-line games and recreation system. In each case study, one chapter is devoted to each step in the system life cycle; thus the student sees how a system progresses from problem definition, through a feasibility study, analysis, system design, detailed design, implementation, and maintenance. Specific exit criteria are defined for each step; thus the student sees a structured methodology used in the context of the underlying process. Additionally, the cases expose the student to a variety of systems including a traditional batch processing application, a microcomputer-based turnkey system, and a large, on-line, real-time system.

A beginner's first task is to gain a sense of the underlying process of analyzing and designing systems. Too often, a detailed description of a specific tool can get in the way of this broad perspective; thus an unusual structure has been selected for the text. Instead of interrupting the flow of a case study to introduce a key tool or technique, the technical details have been removed from the cases and grouped, in the form of brief tutorials, in Part IV. With the details "out of the way," the student is able to concentrate on the process, turning to the appropriate tutorials only after the underlying concepts are understood. As the student gains confidence and experience, these same modules become a valuable reference. Finally, there is the question of flexibility. The tutorials of Part IV are written as independent modules; the student can read them in any order. Consequently, the instructor can "pick and choose," thus customizing the course.

Systems analysis is an exciting and challenging profession. A key objective of this text is to convey to the reader a sense of that excitement and challenge. There is more to it, however. Like any profession, systems analysis demands dedication and hard work, factors that are stressed again and again in the case studies that follow.

One final note: Some two decades ago, one of the first departments of systems analysis was founded at Miami University by perhaps the finest professional systems analyst the author has ever known, Professor Lawrence J. Prince. This book is dedicated to him.

Oxford Ohio

November, 1982

W.S.D.

PART I
The System Development Process

1

Structured Systems Analysis And Design

OVERVIEW

This chapter serves as an overview of the entire text. We begin by defining the term system. Then we consider the systems analysis and design process, emphasizing the work done by the systems analyst. An example from engineering is used to illustrate the need for a methodical approach to designing a large and complex project. For the analyst, the structured approach to systems analysis and design provides such a methodology. The key to structured systems analysis and design is the system life cycle, so we briefly describe how the structured approach guides the analyst through the steps of this life cycle. The chapter ends with an outline of the balance of the text.

WHAT IS A SYSTEM?

The fact that you are reading this text indicates that you already know something about the computer. Some readers will be professional programmers or managers looking for a better understanding of the systems analysis and design process. Others will be students, future managers or future programmers, who are just learning about systems analysis and design. An experienced systems analyst might be interested in a solid overview of some of the new techniques used in this discipline. You all have something in common: you know what a program is. You have probably written programs; if not, you have certainly used one.

Programmers and users tend to focus on a single program. The programmer sees a specific job to be done, while the user sees a specific problem to be solved. Try to take a somewhat broader view of that program. Ask yourself a few questions. Why was that particular program written? Programs do not happen by chance. Obviously someone wanted the program or it would not have been written, but why that program? And why was the program designed as it was? For example, some programs are interactive, while others require the user to prepare a complete set of commands and data and submit them to the computer in batch mode. Why? Finally, what about the support elements—hardware, software, procedures, and operators, for example—that surround the program? How do all these elements come together to support you as you write or execute your program?

Obviously, someone must have planned not merely the program, but the entire environment in which that program exists. We are dealing with something much broader than a program. We are dealing with a *system*.

What is a system? We hear about the solar system, the ecological system, the blood system, the grading system, and the educational system, but what does the term mean? The American National Standards Committee suggests the following definition:

> **SYSTEM: In data processing, a collection of men, machines, and methods organized to accomplish a set of specific functions.**

The programs you have used or written were part of a larger system that included other programs, hardware, data, human beings, and various procedures. The system defines the environment. The program exists in the context of that system.

Any activity that involves numerous components, be they men, machines, or methods, requires careful coordination; although it may not be clear from the definition, the components must work together if the specific function is to be accomplished. Coordination implies planning. Who plans the system? How is the system planned?

The system begins with a *user*. The user has a need for technical support, but doesn't know enough about the computer to actually do the job. In another part of the organization are the programmers. They know a great deal about the computer, but often do not have a clear understanding of the user's needs. The user knows the problem but can't solve it. The programmers might be able to solve the problem, if only they understood the problem. Complicating matters is a communication gap (Fig. 1.1): at times, programmers and users seem almost to speak a different language.

Enter the *systems analyst*, a professional whose basic responsibility is to translate user needs into the technical specifications needed by the programmers (Fig. 1.2). The systems analyst begins by developing a logical description of the user's needs. Using this logical description, the analyst designs a system that solves the problem. The system then serves as a reference point for developing the programmers' technical specifications. Management, of course, has a responsibility to control the system development process (Fig. 1.3). Computers and programmers are quite expensive. Management sees the system as an investment, and expects investment funds to be spent wisely. Providing management with a means to control the process is another key responsibility of the systems analyst.

Fig. 1.1: *A communication gap separates the user from the programmer.*

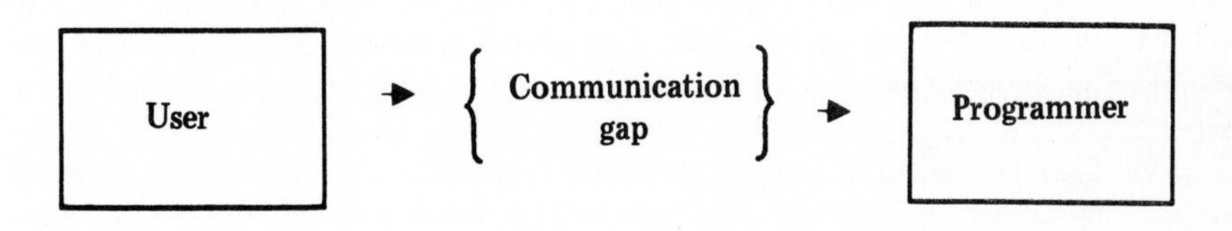

Fig. 1.2: *The systems analyst translates user needs into the technical specifications needed by the programmer.*

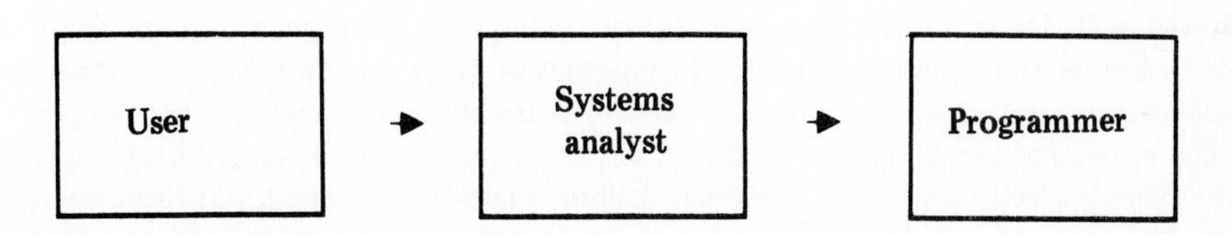

Fig. 1.3: *Management must control the system development process.*

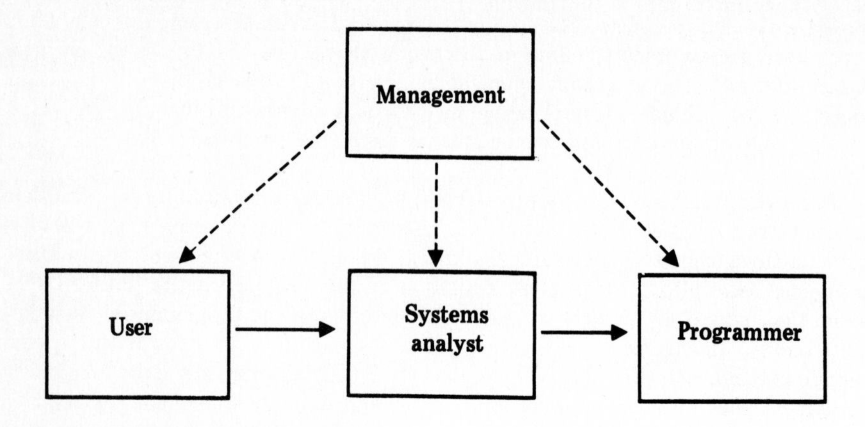

The analyst has a very difficult job. This individual must be able to deal concurrently with a user group, the technical professionals, and management. Users are concerned with such factors as ease of use and response. Programmers worry about bits, bytes, and file structures. Management wants to know about return on investment, the cost/benefit ratio, and the development schedule. Although our discussion has emphasized the development of software, we should point out that the analyst must consider the hardware, the people who operate that hardware, data entry, security, auditing—in short, every component of the system. Many systems do not even require a computer. People who can deal with such diverse criteria are relatively rare, which may be why professional systems analysts are well paid.

How does the systems analyst go about translating user needs into technical specifications under management control? A process, a *methodology*, is involved. Perhaps the best way to grasp the idea of a methodical process is to consider an example from a different profession: engineering.

THE ENGINEERING METHODOLOGY

An engineer is a technical professional who translates user needs into physical solutions; the engineer's job parallels that of the systems analyst in many ways. There is a touch of art involved in both professions. Inspiration may be a fine guide for projects such as laying out a garden or building a one-room log cabin, but a lack of careful, methodical planning can be a disaster on a large project. The Winchester Mystery House near San Jose, California is an excellent example of the impact of poor planning. Each room in the house is beautifully constructed, using the very best materials and workmanship, but the house makes no sense; stairways end at ceilings, doors open to nowhere, windows look out onto blank walls, and chimneys stop just short of the roof. The house evolved; it was not designed. A good engineer is creative, but this creativity must be exercised within the bounds of a well defined, well structured engineering methodology.

What are the characteristics of this engineering methodology? The engineer begins with a logical image of a project, often in the form of rough sketches. Models are built. Blueprints are drawn. Finally, the physical project (a bridge, a building) is constructed. The methodology begins with a high level, logical analysis of the task. Gradually, step by step, the physical details are introduced. At each step, specific documents or models must be completed before the next step can begin. These documents serve as aids to communication, clearly defining what is known about the project to date, and creating a base for subsequent work. Additionally, the methodology serves as a memory aid. If the engineer cannot complete the document or model required by a given step, then something is missing, and these missing details must be resolved before moving on to the next step.

Consider, for example, the problem of constructing a new office tower. The engineer may start by making a rough sketch of the building. This preliminary step can be quite creative. Such factors as room size, the nature of the land on which the tower will be constructed, and local building codes might be considered, but the engineer is basically dealing with ideas at this point, ignoring most physical construction details.

Once a logical design has been accepted, a model of the building is constructed. The model serves a variety of purposes. It is a miniature prototype of the building; many potential construction problems can be anticipated and corrected during the creation of the prototype. The model also serves as an excellent communication tool; engineers must communicate their designs to non-engineers, and there is no better way than with a physical model. (For example, you may be able to find a model of a new science building or sports complex on your own campus.)

Next, blueprints are developed. A blueprint is a graphic model of the building, drawn to exact scale. A non-technical user can look at a blueprint and envision the finished product. A contractor can read a blueprint and create the building. As a memory aid, the blueprint clearly defines what the building will contain and, by process of elimination, what it will not contain. Certain logical requirements were defined during the early stages of the project; have these requirements been met? A quick glance at the blueprints should answer this question.

A blueprint is, of course, relatively easy to change; it exists only on paper. The blueprint is a logical representation of the building. It is much closer to the physical structure than the engineer's rough sketches were, but it is still a logical model. The blueprints can, however, be used to estimate the material requirements and the final cost of the building. If these numbers are not acceptable, the blueprints can be changed, and costs reestimated. Finally, the materials can be ordered and construction can commence. The contractor will, of course, use the blueprints as a primary reference document, carefully building the physical structure to the logical specifications.

Note the methodical, step-by-step approach of the engineer. Note the documentation created at each step, and note how this documentation is used as a communication tool, as a memory aid, and as the foundation for the next phase. Creativity is involved, but only within the bounds imposed by the methodology. The systems analysis and design process must be subjected to a similar discipline.

STRUCTURED SYSTEMS ANALYSIS AND DESIGN

There is a touch of art involved in systems analysis, too. While the purely creative, "design as you go" approach may work for small or relatively simple projects, it can be a disaster on a large, complex system. Engineering is an old profession; its methodology has developed over many many years. Systems analysis is a relatively new profession; its methodology is still evolving, and many different versions of the "correct" approach exist. We can, however, begin to define a widely accepted process known as *structured systems analysis and design*. It is a step-by-step approach to system development, beginning with logical design, and gradually moving to physical design. Specific documentation requirements are associated with each step in the process. This documentation can be used for communication. It also serves as a memory aid, and the output documentation for one step serves as input to the next step. It is this structured approach to systems analysis and design that will be presented here.

THE SYSTEM LIFE CYCLE

Structured systems analysis and design is keyed to the *system life cycle* (Fig. 1.4). As a system moves from concept to implementation, it must pass through each of these steps. When a structured approach is used, the systems analyst must progress from step to step in a careful, methodical fashion, completing a number of well defined exit criteria for each step. Let's briefly consider this process. In the descriptions that follow, the exit criteria for each step will be mentioned; these exit criteria will be defined in subsequent chapters and in the tutorials of Part IV.

Fig. 1.4: *The steps in the system life cycle.*

Problem definition

Feasibility study

Analysis

System design

Detailed design

Implementation

Maintenance

Problem Definition

What is the problem? This is the key question that must be answered during *problem definition*; it makes little sense to try to solve a problem if you don't know what the problem is. Although the need for problem definition may seem obvious, this is perhaps the most frequently bypassed step in the entire systems analysis and design process.

What is the source of the problem definition? Obviously, someone must recognize that a problem exists. Often the user will encounter difficulties and ask for help. Perhaps management will identify an area of poor performance within the user's function; frequently the systems analyst will spot the problem. Initial discussions concerning the problem are often quite informal. Eventually, however, these discussions reach the point where the user, management, and the systems analyst agree: "Yes, we really do have a problem."

If the problem is judged significant, management and the user may want the analyst to look into it. Once an analyst is assigned, the process undergoes a subtle change. Assigning an analyst implies a commitment of funds; the informal discussions suddenly become a defined project.

The systems analyst's first responsibility is to prepare a written statement of the objectives and the scope of the problem (Fig. 1.5). Based on interviews with management and the user, the analyst writes a brief description of his or her understanding of the problem, and reviews it with both groups, ideally in a joint user/management meeting. People respond to written statements. They ask for clarification; they correct obvious errors or misunderstandings. This is why a clear statement of objectives is so important.

Management (and perhaps the user) is about to commit funds to support the analyst's work. How much will it cost? This is a reasonable question that all too often is not asked until quite late in the system development process. The analyst should provide a rough estimate of the scope of this financial commitment. Clearly, at this early stage, no one can provide an accurate estimate of the final cost of a project, but "ballpark" estimates or "order of magnitude" estimates are certainly possible. The analyst should also be able to provide an accurate assessment of both the cost and the schedule for the next phase in the system life cycle: the feasibility study. These estimates will give management and the user a sense of the scope of the project.

Problem definition can be quite brief, lasting a single day or even less. The intent is to define the objectives and the scope of the proposed system. Communication breakdowns do occur, and it is essential that the user, management, and the systems analyst agree on a general direction very early in the probject. A misunderstood problem definition virtually guarantees that the system will fail to solve the problem.

Fig. 1.5: *Problem definition.*

Step	Key Question	Exit Criteria
Problem definition	What is the problem?	Statement of scope and objectives

The Feasibility Study

What exactly is a *feasibility study*? Basically, it is a high level, capsule version of the entire process, intended to answer a number of questions (Fig. 1.6). What is the problem? *Is there a feasible solution to the problem*? Is the problem even worth solving? The feasibility study should be relatively brief; the task is not to solve the problem, but to gain a sense of its scope. The user will be asked to react to the feasibility study; how else could we be sure that we are attacking the right problem? Management is vitally concerned with the results, as they will be asked to commit funds and personnel based on the feasibility study.

The statement of scope and objectives prepared during problem definition is generally rather vague. Essentially, the analyst promises to investigate (not to solve) a broadly defined problem. During the feasibility study, the problem definition is brought into sharper focus. Specific system objectives are set, and aspects of the problem that will be excluded from the system are clearly identified. As a result, the analyst should be able to estimate the costs and benefits of the system with greater accuracy. A cost/benefit analysis of the proposed system is an important part of the study.

The feasibility study ends with a formal presentation to the user and to management. This presentation marks a crucial decision point in the life of the project. Many projects will die right here; only those promising a significant return on investment should be pursued. Assuming that management approval is granted, the feasibility study represents an excellent model of the systems analyst's understanding of the problem (comparable to the engineer's design sketches), and provides a clear sense of direction for the subsequent development of the system.

Fig. 1.6: *The feasibility study.*

Step	Key Question	Exit Criteria
Feasibility study	Is there a feasible solution?	Rough cost/benefit analysis System scope and objectives

Analysis

Analysis is a logical process. The objective of this phase is not to actually solve the problem, but to determine exactly *what must be done to solve the problem*. The user knows what must be done, but does not know how to do it. During analysis, the systems analyst works with the user to develop a logical model of the system (Fig. 1.7).

Many systems analysts have a technical background. The temptation of many technically trained people is to move too quickly to program design, to become *prematurely physical*. This temptation must be avoided. The objective, don't forget, is to solve the user's problem. The user knows the problem, and is the analyst's primary source of information at this stage. If the analyst begins talking about programming details, the user may become lost, and unable to contribute. As a result, the analyst may develop a system that fails to solve the problem.

How can an analyst avoid becoming prematurely physical? A structured methodology can help. With a structured approach, specific exit criteria must be completed for each step in the process. The basic objective of the analysis stage is to develop a logical model of the system, using such tools as data flow diagrams, an elementary data dictionary, and rough descriptions of the relevant algorithms. This logical model is subject to review by both the user and management, who must agree that the model does in fact reflect what should be done to solve the problem.

Fig. 1.7: *Analysis.*

Step	Key Question	Exit Criteria
Analysis	What must be done to solve the problem?	Logical model of system Data flow diagram Data dictionary Algorithms

System Design or High-Level Design

Once the analysis stage is completed, the systems analyst knows what must be done. The next step is to determine, in broad outline form, how the problem might be solved. During *system design*, we are beginning to move from the logical to the physical (Fig. 1.8).

Several alternative solutions might be considered. For example, a given system might be implemented by computer or manual means. If a computer is used, the system might be either batch or interactive. We might use traditional data files; a data base is another option. It is important that the analyst avoid simply picking a system implementation and moving on to program design. Once again, the use of a structured methodology can help.

The question to be answered during system design is: *How, in general, should the problem be solved?* The answer to this question is crucial to both the user and the programmer; thus the exit criteria for this phase must be aimed at both groups.

Management is particularly concerned about the future direction of the system. Up to this point, the project has involved the time of a few systems analysts (Fig. 1.9), and the cost has been limited. Detailed design may well involve these analysts plus a few more, and the detailed design stage will take longer; thus the costs begin to accelerate. As we move to implementation, programmers, operators, technical writers, and computer time must be committed, and the accumulated cost increases dramatically. To make matters worse, following implementation is a long-range, potentially expensive maintenance commitment. Before management can support the system, they must have an idea of what it will cost. Some sense of the likely cost/benefit ratio is essential.

The cost of a system accelerates as we move through the development process; this is another argument for using a careful, structured approach. An error detected during analysis means that a month or two of a systems analyst's time has been wasted. What if the error is not detected until the implementation stage? More time will have elapsed, and many more people will be involved; the cost of the wasted effort will be much higher. Even worse is the danger that a bad system will be installed in spite of crucial errors, simply because "We've already spent all this money, and we can't quit now." The methodical approach of structured systems analysis and design increases the likelihood that significant errors will be detected early.

As exit criteria from the system design stage, the analyst is often asked to outline several alternative solutions to the problem including, as a minimum, the following:

1. A low cost solution that does the job and nothing more.
2. An intermediate cost solution that does the job well, and is convenient for the user. This system may include several features that the user did not specifically request, but that the analyst, based on experience and knowledge, knows will prove valuable.
3. A high cost, "Cadillac" system, with everything the user could possibly want.

Supporting each alternative should be a system flow diagram or other system description, and an estimate of costs and benefits.

The system design phase ends with a choice. Perhaps, management or the user will decide that the benefits to be gained from this system are not worth the cost, or programming may be unable to support the system. Thus the project ends. If the system is worth supporting, one alternative will be selected. The systems analyst's description of this alternative will be used as a high level model for developing the physical system.

Fig. 1.8: *System design.*

Step	Key Question	Exit Criteria
System design	How, in general, should the problem be solved?	Alternative solutions System flow diagrams Cost/benefit analysis

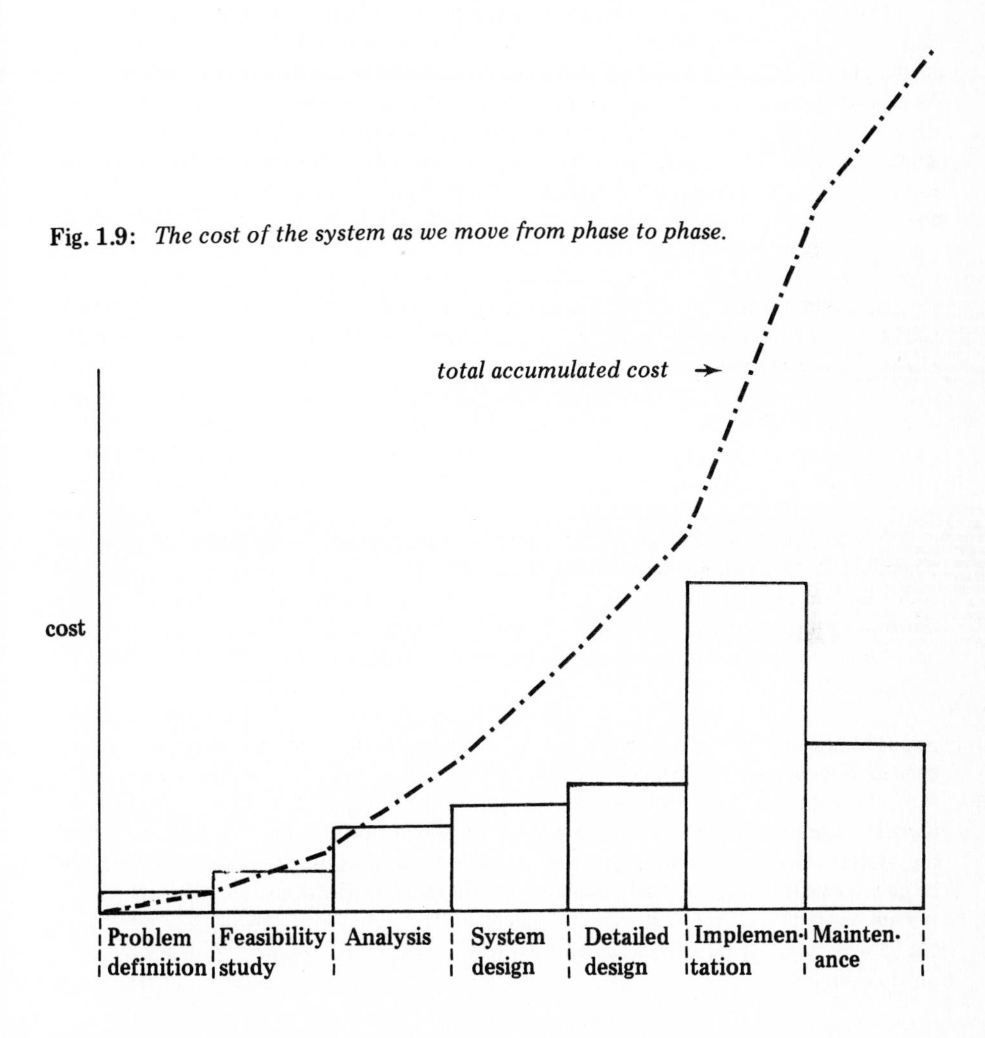

Fig. 1.9: *The cost of the system as we move from phase to phase.*

Detailed Design

As we move into *detailed design*, management, the user, and programming have agreed on a general strategy for solving the problem. We know, for example, what programs will be needed by this system. We have not begun to write the programs, nor have we given serious consideration to how they might be written, but we do know that a program will be needed to perform certain functions. We also know, again in general terms, what hardware will be required. The basic question that must be answered during detailed design is: *How, specifically, should the system be implemented* (Fig. 1.10)?

Consider, for example, the programs. During detailed design, the systems analyst must develop, (ideally with the help of the programming department) a set of specifications for each program. These specifications should contain enough detail to support writing the actual code. If the system calls for new hardware, hardware specifications must be written in a form acceptable to the purchasing department. Generally, during detailed design, the systems analyst must define each component of the system to whatever level of detail is required for the implementation step.

Until very recently, writing a computer program was considered a highly personal task. The programmer was given a rather broad, functional level overview of the program, and turned loose. Some time later, the program was finished. Sometimes it performed the necessary functions. Occasionally, a creative programmer produced a true "gem," exceeding everyone's expectations. All too often, however, the finished product was late, over budget, barely acceptable, and impossible to maintain. No one was satisfield: not the programmers, not the users, and certainly not management. Structured programming was, in part, a reaction to such inconsistent programmer performance.

When a structured approach to systems analysis and design is used, one of our objectives is to develop program specifications that make it easy to write structured code. In this text, we'll prepare a hierarchy chart and a set of input/process/output charts for each program in the system. Algorithms will be defined using pseudo code or structured English. An alternative, Warnier/Orr diagrams is explained in Part IV (Module K).

The specifications developed during detailed design are analogous to the engineer's blueprints. Given these detailed specifications, it is possible to generate highly accurate cost estimates and implementation schedules, two factors extremely important to management. The programmers (more generally, those who will implement the system) will use the specifications to guide their actions. Finally, the user should be able to see an image of the finished system in the specifications.

Fig. 1.10: *Detailed design.*

Step	Key Question	Exit Criteria
Detailed design	How, specifically, should the system be implemented?	Implementation specifications HIPO pseudo code Warnier/Orr diagrams Hardware specifications Cost estimates Preliminary test plan Implementation schedule

Implementation

During the *implementation* stage, the system is physically created (Fig. 1.11). Necessary programs are coded, debugged, and documented. New hardware is selected, ordered, and installed. In many organizations, the systems analyst is also the programmer; in this text, however, we will assume that analysis and programming are done by different groups, and we will not consider the details of implementing the system.

There are, however, certain functions that the systems analyst must perform as the programs are being written and the hardware is being ordered and installed. Operating procedures must be developed. Security and auditing procedures will probably be required. The test plan must be established. It is easy to overlook the procedures and the test plan, but without these elements, there is no system. The implementation stage normally ends with a formal system test involving all components and procedures.

Maintenance

Following implementation, the system enters a *maintenance* stage. The objective of maintenance is to keep the system functioning at an acceptable level. Occasionally, program bugs will slip through the system test undetected; correcting such errors is a maintenance function. More often, the parameters and algorithms used to develop the original programs will change, meaning that the programs must be updated. Hardware maintenance is an obvious requirement. Even procedures change.

What does the systems analyst do during the maintenance stage? Typically, very little. In fact, it is likely that the analyst who designs a system will be working on another as the maintenance stage begins. The analyst does, however, have a significant impact on maintenance. Change may well be the only constant in the computer field; thus a well-designed system must anticipate and allow for change. Designing for change is the analyst's responsibility, and the structured approach to systems analysis and design tends to support this objective.

Fig. 1.11: *Implementation and maintenance.*

Step	Key Question	Exit Criteria
Implementation	Do it!	Programs Code Documentation Hardware Operating procedures Security procedures Auditing procedures Test plan Formal system test
Maintenance	Modify the system as necessary.	Continuing support

THE PROCESS

We have just covered the key steps in the structured systems analysis and design process (Fig. 1.12). Note the steady progression from the general to the specific. Note how we started with a purely logical view of the system, and gradually added details, layer by layer, until the physical system was implemented. This is an example of a top-down approach.

From our discussion, it might seem that the development of a system progresses smoothly, from step to step. Problem definition is followed by analysis; analysis by system design; in general, the end of one stage marks the beginning of the next. Unfortunately, it rarely happens that way. What happens, for example, if an error in problem definition is discovered during analysis? What if implementation difficulties force a change in design? A more realistic view of the process would map a less smooth path, with frequent returns to earlier steps. The important thing is that a sense of progression from step to step be maintained.

Monitoring the Process

Structured systems analysis and design defines a series of steps that must be followed in designing a system. Specific documentation standards exist for each step. How can we be sure that the analyst actually follows the process? It is essential that specific, measurable milestones or objectives be set for each step in the process. To be effective, such milestones must be enforceable. In the text, we will use a formal inspection process, coupled with management reviews and the system test, to enforce the structured methodology. The inspections, management reviews, and the system test are summarized in Fig. 1.13. Formal inspections are described in Module A.

Fig. 1.12: *A summary of the structured systems analysis and design process.*

Step	Key Question	Exit Criteria
Problem definition	What is the problem?	Statement of scope and objectives
Feasibility study	Is there a feasible solution?	Rough cost/benefit analysis System scope and objectives
Analysis	What must be done to solve the problem?	Logical model of system Data flow diagram Data dictionary Algorithms
System design	How, in general, should the problem be solved?	Alternative solutions System flow diagrams Cost/benefit analysis
Detailed design	How, specifically, should the system be implemented?	Implementation specifications HIPO pseudo code Warnier/Orr diagrams Hardware specifications Cost estimates Preliminary test plan Implementation schedule
Implementation	Do it!	Programs Code Documentation Hardware Operating procedures Security procedures Auditing procedures Test plan Formal system test
Maintenance	Modify the system as necessary.	Continuing support

Fig. 1.13: *Key milestones in the systems analysis and design process.*

Step	Milestone
Problem definition	Management/user review
Feasibility study	Management review
Analysis	Inspection Management review
System design	Inspection Management review
Detailed design	Inspection Management review
Implementation	Inspection or walkthrough Formal test Management review

The Process and the Tools

Each step in the process calls for the completion of one or more specific documentation standards. In discussing this documentation, we will be referring to a variety of systems analysis and design tools and techniques. It is easy to become bogged down in the details of a given tool. You may, for example, find yourself concentrating so heavily on completing the HIPO specifications for a single program, that you will lose sight of the overall process. Don't let that happen. Understand the process; that's the key. The individual documentation standards are details. You can learn how to use data flow diagrams, a data dictionary, a hierarchy chart, pseudo code, or any of the other tools described in this text; they are not difficult. Given enough practice, you will learn how to use the tools.

The process is different. It is almost a philosophy, a point of view. You can't really memorize it. Of course you can learn the names of the steps, but that's not the point. The idea is the smooth, methodical progression from the logical to the physical, from the top, down. This ability to sense the flow of the process, and to complete the documentation within the context of this process, is the essential skill of the systems analyst.

When you eventually begin working as a systems analyst, you may discover that your employer uses structured systems analysis and design, but requires several exit criteria that were not covered in this text. The tools may change, but the philosophy will not. If you learn only the tools, your knowledge may quickly become obsolete. If you understand why, rather than simply how, you will have learned a lifetime skill.

AN OVERVIEW OF THE TEXT

The text is organized to help you keep the process and the tools separate. Parts I, II, and III are case studies of three typical systems analysis problems. As you read the case studies, concentrate on the process. The specific tools needed to support each step in the process will be referenced and used, but will not be explained in detail within a case study. Instead, the details have been grouped in Part IV of the text, where you will find a number of brief tutorials, each explaining how to use one tool or technique. By moving the details out of the mainline, we avoid interrupting the flow of the systems analysis and design process. Later in the term, the process will seem almost second nature to you. As you are working on a class project, you will want information on a specific tool or technique, and it will be the more philosophical discussion of the process that will seem to get in the way. When you reach this point, Part IV should be a valuable reference.

The case studies cover a variety of typical systems analysis and design problems. In the remainder of Part I, we will discuss a project intended to improve an existing system: payroll. Part II considers the design and implementation of a brand new, small business computer system for a doctor. Finally, in Part III we turn our attention to the development of an on-line games and recreation system. These three cases have been selected because they illustrate a variety of situations. Both existing and new systems will be considered, as will batch and on-line systems. Traditional files and data bases will be discussed. Our analysts will be asked to deal with both highly sophisticated and unsophisticated users. Hardware as well as software will be selected.

One final comment before we begin: the best way to learn how to analyze and design systems is to analyze and design systems. Do the exercises in the text. Your instructor (assuming you have one) will make additional assignments. Do them. Do them yourself or, if you are working as part of a team, know what each member of your team is doing. This text will help you understand systems analysis and design, but the text alone will not make you a systems analyst. There is simply no substitute for experience.

SUMMARY

The chapter began with a definition of "system." We saw that a system contains many components, including hardware, software, people, and procedures. The person who plans and designs systems is called a systems analyst. An example from engineering was developed to illustrate, through analogy, the kind of work done by a systems analyst. The structured approach to systems analysis and design provides the analyst with a methodology, which is tied to the system life cycle.

The steps in the system life cycle include problem definition, the feasibility study, analysis, system design, detailed design, implementation, and maintenance. We briefly discussed the functions performed during each step, and the exit criteria for each step; an excellent summary can be found in Fig. 1.12. We then summarized the process, discussed how the process might be monitored, and pointed out the importance of distinguishing between the process and the tools. The chapter ended with a brief overview of the balance of the text.

EXERCISES

1. What is a system? Consider a system about which you know something—for example your school's registration system or the system for compiling and distributing end-of-term grades. See if you can identify the hardware, software, people, and procedures involved in this system.

2. Why is a systems analyst needed?

3. Why is it so important that a methodical approach such as structured systems analysis and design be used in developing a large, complex system?

4. List the steps in the system life cycle. What is the objective of each step? What are the exit criteria for each step? Why are the steps performed in the specified order?

5. What is a feasibility study? Why is something like a feasibility study needed early in the systems analysis and design process?

6. What does it mean when a systems analyst is prematurely physical? Why is this a problem?

7. Explain how the cost of a system development project increases as we progress through the system life cycle. Why is it so important that errors be detected early in the process?

8. The systems analysis and design process rarely progresses smoothly, from step to step. It is often necessary to backtrack to a previously completed step. Why?

9. Why is it important that specific, enforceable milestones be defined to monitor the structured systems analysis and design process? What milestones will be used in this text?

10. Why is it so important that the reader distinguish between the process and the tools?

2

Case A Problem Definition

OVERVIEW

This chapter introduces our first case study. We begin with a brief overview of the company, and then introduce the problem: payroll. We'll consider how a systems analyst might define the problem and estimate its scope. The output of this first stage is a statement of scope and objectives.

Payroll is a traditional data processing application. The intent of this first case study is to introduce the student to the systems analysis and design process. Concern yourself with the process. Concentrate on how the analyst attacks this problem, and not on the details of the solution.

THE PRINT SHOP

Carl Jones established THE PRINT SHOP twenty years ago. Until recently, it was a typical small town printing shop specializing in wedding invitations, graduation announcements, business cards, menus, and similar items. Carl did good work, and developed a reputation for quality and dependability, but THE PRINT SHOP was almost unknown outside the town.

Five years ago, everything changed. Carl's daughter, Carla, returned from college with a degree in computer science, and began selling her father on the potential for computer graphics. He purchased a small computer primarily (he admits) to indulge his daughter. She was serious, however. Combining skill in programming with a good eye for graphic art, she began producing eye-catching graphic effects to accompany her father's already high quality printing. The real story, however, was the cost; these designs could be planned and prepared at a fraction of the cost of manual techniques. Quality and effect at a competitive price is tough to beat.

The rest is history. In just five years, the demand for THE PRINT SHOP's product has increased dramatically. Employment has gone from 10 to 50 people, and the firm has just moved into a new plant. Carl now has four functional manager's working for him (Fig. 2.1). Carla is the Vice President in charge of design; working for her are four other designers, a systems analyst, and a programmer/operator. Initially, she worked on a microcomputer system, and an artist copied her designs by hand from a CRT screen. Today, however THE PRINT SHOP owns five microcomputers, one for each designer, and a minicomputer system with a plotter. The designers work on their own personal micros. When a design is complete, a diskette is loaded onto the minicomputer system, and a precise copy is drawn with no human intervention. The process is fast and inexpensive.

THE PRINT SHOP is no longer a small town business; its reputation has spread throughout its corner of the state. In fact, the first out-of-state order arrived just a few days ago, an annual report for a major company. The future looks bright—added expansion is a definite possibility.

PAYROLL AT THE PRINT SHOP

For the past twenty years, a local bank has prepared the payroll for THE PRINT SHOP. Carl feels a certain sense of loyalty to the bank (they gave him his first business loan), but lately, the cost of their payroll service has increased dramatically. These increased bank charges, coupled with the firm's higher level of employment, result in a substantial weekly bill. Future growth is expected. Since the bank charges a fixed rate for each check, the payroll charge will certainly increase, which concerns accounting.

THE PRINT SHOP has its own minicomputer system, so why not do the payroll internally? We know how much the bank charges, and we can estimate how much it will charge in the future. Perhaps THE PRINT SHOP can save some money. Thus the systems analyst is asked to investigate the possibility of moving payroll onto the minicomputer system.

Fig. 2.1: *An organization chart of THE PRINT SHOP's management.*

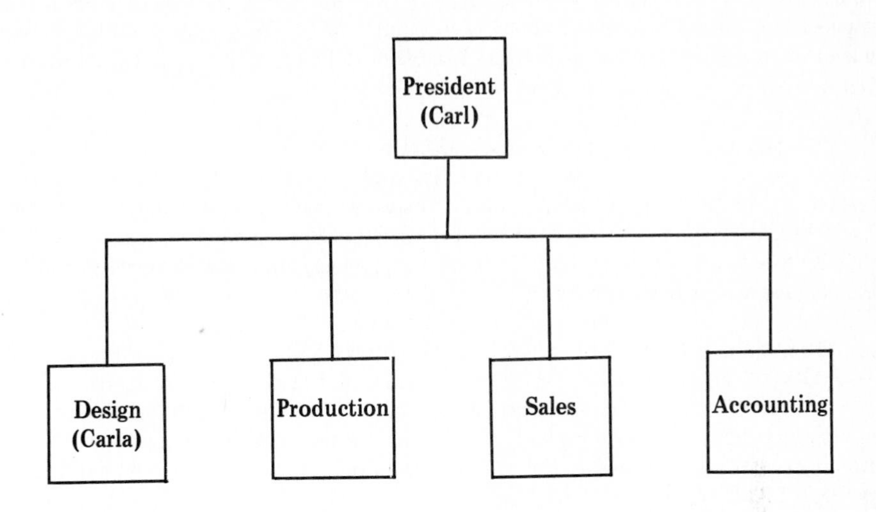

PROBLEM DEFINITION

Where should the analyst start? It is tempting to begin considering the details of implementing the payroll system immediately. The technical questions are both interesting and challenging. Do we need any new hardware? If so, what kind? Should we consider a commercial computing service? What about software? Should the program be written in BASIC, Pascal, or COBOL? Should we buy an already-written payroll program? How should the data files be defined?

Such questions, at least at this early stage, tend to miss the point. Accounting (the user) has not asked us to implement a payroll system on our own computer, but to investigate the *possiblity* of implementing such a system on our computer. That's a very different problem. It implies a very important question. Are the benefits we might expect from implementing payroll on our own computer greater than the cost of developing the system? In other words, *is the job worth doing* at all?

The good analyst should go a step further. What is the problem? Payroll costs are too high. The real objective is to find a less expensive way to generate paychecks, not necessarily to implement payroll on our computer. When accounting asked the analyst to "investigate the possibility of moving payroll onto the minicomputer system," they were not stating a problem, they were suggesting a solution. The solution may well be a good one, and the analyst should certainly consider it, but other possible solutions should be considered as well. A good problem definition identifies the real problem, not an implied solution to that problem.

Another key concern is the project's expected scope. How much is an improved payroll system worth? There are no explicit targets, but there certainly are some implied limits. Payroll is currently done by the bank. If the proposed new system is more expensive than the bank, forget it. More to the point, three different cost figures should be considered: the present bank charges, the projected operating cost of the new system, and the development cost of the new system. The new system's operating cost must be less than the bank's charges. How much less? Developing a new system is an investment. The cost savings must allow THE PRINT SHOP to recover its investment in a reasonable period of time.

Although the present bank charges are known, the analyst, at this early stage, can only guess at the new system's operating and development costs. It is possible, however, to "scope" the proposed system. Assume, for example, that the bank currently charges $2.00 per check. THE PRINT SHOP has 50 employees; thus the cost is $100 per week or (roughly) $5000 per year. Obviously, no new system can reduce this cost below zero. Thus the maximum possible benefit is $5000 per year.

What could we afford to spend in order to save $5000 per year? $50,000? No. It would take ten years to recover the investment, and most firms like to recover investments within two or three years. Thus a reasonable upper limit on the payroll project might be a development cost of $15,000. Clearly, this is a rough figure, but it does give the analyst a sense of the scope of the project. If the job can't be done for less than $15,000, it is probably not worth doing.

THE STATEMENT OF SCOPE AND OBJECTIVES

The analyst now has a sense of the problem definition and the scope of the proposed system. Is this what accounting has in mind? Isn't it possible that the analyst has misunderstood something? Certainly. Have you ever taken an examination and confidently filled a bluebook, only to discover that you misread the question? What is the value of an excellent answer to the wrong question? If the analyst's sense of the problem does not agree with the user's (or management's), there is no way the analyst can design a system to solve the problem. A system, even a "good" system, that fails to solve the problem is worth nothing; it is a waste of the money and resources expended in developing it. It is essential that the analyst clearly communicate his or her sense of the problem at this early stage in the system life cycle. Typically, this is done with a simple written memorandum called a *statement of scope and objectives* (Fig. 2.2).

Read the statement of scope and objectives carefully. After the project is given a name, the problem is broadly defined, and the objectives of the project are stated. Next comes a reasonable upper limit on the project's development cost. Accounting earlier hinted at a possible solution to the problem; this potential solution is included, but is clearly defined as a preliminary idea, rather than as the problem. Finally, a feasibility study is proposed, and its cost estimated.

What is the value of such a statement? The function of the statement of scope and objectives is communication. It is always possible that the analyst misunderstood the problem. The statement of scope and objectives is a formal way of saying "Here is what I think you want." Ideally, management, the user, and the analyst should read the statement together. Now is the time to clear up any misunderstandings. If communication is accurate, the analyst can begin the feasibility study with confidence.

Fig. 2.2: *A statement of scope and objectives.*

STATEMENT OF SCOPE AND OBJECTIVES: *April 25, 1983*

THE PROJECT: *PAYROLL*

THE PROBLEM: *The present cost of preparing payroll is too high.*

PROJECT OBJECTIVES: *To investigate the potential for a new, lower cost payroll system.*

PROJECT SCOPE: *The development cost of this project should not exceed $15,000 (± 50 percent).*

PRELIMINARY IDEAS: *One possible solution would be to transfer payroll to our own, internal computer system.*

THE FEASIBILITY STUDY: *In order to more fully investigate the potential of the PAYROLL project, a feasibility study lasting approximately two weeks is suggested. The cost of this study should not exceed $1000.*

There is no standard format for a statement of scope and objectives; every project is different. The analyst should state his or her understanding as clearly and concisely as possible, ideally on a single page. The project should be given a name, the problem should be defined, and the projected scope of both the entire project and the feasibility study (if any) should be clearly stated. Often, a simple memo will suffice; the point is communicating ideas, and not following a standard form.

How accurate are the cost estimates included in the statement of scope and objectives? The estimated project scope is merely a ballpark figure; this is why the ±50 percent factor has been included (Fig. 2.2). The estimated cost of the feasibility study, the next phase in the system life cycle, should, however, be quite accurate. Essentially, the analyst is asking management to commit $1000 to the PAYROLL project so that the costs and benefits can be more accurately estimated. At the end of the feasibility study, we should be able to decide if the benefits justify the investment.

Let's assume that management and accounting agree with the statement of scope and objectives, and commit funds to pay for a two-week feasibility study. We can now begin to investigate the PAYROLL project seriously.

SUMMARY

In this chapter, we developed a problem definition for our first case study, a payroll application. The chapter opened with a brief background discussion on the firm,

THE PRINT SHOP. Rapid growth had resulted in a significant increase in the cost of preparing payroll, and the systems analyst was asked to look into this problem. The analyst carefully identified the real problem, estimated the scope of the project, and prepared a statement of scope and objectives. In Chapter 3, we will consider a feasibility study of the proposed payroll project.

EXERCISES

1. Why is it important that the analyst avoid beginning to work with implementation details before developing an adequate problem definition? Why do you suppose that so many analysts do, in fact, move prematurely to the details?

2. It is important that the analyst distinguish between the real problem and an implied solution to that problem. Why?

3. Why is it so important that the analyst have a sense of the scope of a project early in the system life cycle?

4. What is the purpose of a statement of scope and objectives?

5. In the statement of scope and objectives of Fig. 2.2, a preliminary idea for solving the problem was identified. The analyst was not obligated to include this potential solution, but chose to do so. Why? This question was not explicitely answered in the text. Think about it. Why would the analyst choose to include the user's suggestion?

6. The rapid growth of a firm causes problems in areas other than payroll. Listed below are several potential problem areas. Prepare a statement of scope and objectives for each. Note that you may not have complete information in every case; if you feel that you are missing something, note the information you would like to have, and indicate why you feel you need it.

   a. *The order backlog.* When THE PRINT SHOP was a small, ten-employee company, orders were simply entered in a log book by the office secretary, and crossed off as they were completed. Today, with the increasing volume of orders, the old system has simply broken down. On one occasion, an order was lost; if this problem continues, it could begin to cost the firm some business, or even damage its reputation. Management would like the systems analyst to look into the possibility of a computerized order entry system.

   b. *Billing.* With the increase in sales, the volume of customer billings threatens to overwhelm the two billing clerks. If customers don't get their bills on time, payment will be delayed, and interest that might be earned on that money will be lost. Would computerized billing help?

c. *Inventory.* As sales volume has increased, it has become necessary to keep more supplies in inventory, and keeping track of that inventory is a problem. For example, just last week the purchasing department ordered a very expensive red ink; two days later a stock clerk discovered that the ink was already in stock, but had been overlooked. A firm can't afford to continue spending money on things it doesn't need. Can the computer be used to help control inventory?

d. *The ledger.* The accounting department currently maintains a set of ledgers in which they record all the incomes and expenditures of THE PRINT SHOP. The two clerks are having trouble keeping up with the paperwork. Before a third clerk is hired, should moving the ledger to the computer be considered? A related problem involves using the information recorded in the ledger; in its present form, it is simply not accessible to management.

7. Perhaps your instructor can arrange for you to conduct a class project with a local small business concern. If so, prepare a statement of scope and objectives for a proposed new payroll system, or for any of the application areas mentioned in exercise 6.

8. Another source of possible class projects is your school. Potential problem areas include scheduling, preparing grade reports, tracking library circulation, billing student charges, and many others. With your instructor's permission, prepare a statement of scope and objectives for a school problem.

3

Case A
The Feasibility Study

OVERVIEW

This chapter describes a feasibility study for the proposed payroll system. First, the system's scope and objectives are clarified. The existing system is an important reference for the new system, and thus is studied next. After gaining an understanding of the existing system, the analyst uses this knowledge to model the proposed payroll system. This model is then taken back to the user and to management, where the scope and objectives are further clarified.

Eventually, the analyst reaches the point where his or her understanding of the system objectives is acceptable; at this time, a number of alternative physical solutions to the problem are considered and analyzed for feasibility. Based on this analysis, a course of action is recommended, and a rough development plan established. Finally, the finished feasibility study is presented to management and to the user.

An in-depth discussion of feasibility studies can be found in Module C.

THE FEASIBILITY STUDY

A systems analyst often deals with problems that do not have simple or obvious solutions. In fact, many cannot be solved within the system's established scope. The purpose of a feasibility study is to determine if there is a feasible solution; if not, any time, effort, or money spent on the project is wasted.

A feasibility study is a high-level capsule version of the entire systems analysis and design process. The objective is to determine quickly, and at minimum expense, if the problem can be solved. It might be reasonable to view the feasibility study as an insurance policy. Any investment carries a certain risk, and it makes sense to investigate the likelihood of success before committing funds or effort.

Remember that the purpose of the feasibility study is not to solve the problem, but to determine if the problem is worth solving. It is easy to get carried away with technical details, but the analyst must remember that time and money are limited. The expected time and cost of the feasibility study were clearly stated as part of the statement of scope and objectives. It is important that the analyst stay within these limits. The analyst who cannot accurately estimate or control costs on a feasibility study will enjoy little credibility on a large system project.

Clarify the Scope and Objectives

During problem definition, a brief, general statement of scope and objectives was prepared. How accurate is this document? What do management and the user really want? The analyst's first responsibility is to clarify the problem definition.

The study begins with a series of interviews (see Module B). The problem involves payroll at THE PRINT SHOP. The user is accounting; thus the head accountant is the analyst's initial contact. The first question is an open one: "What did you have in mind when you suggested that I look into payroll?" The chief accountant's response begins to give the analyst a sense of the real problems: increasing costs, concern with the bank's recent price increase, a personality clash with the bank's newly hired data processing manager, and a sense that THE PRINT SHOP does not have adequate control over its payroll. Follow-up questions deal with constraints that the analyst must consider, such as union agreements, legal regulations, and management committments.

Few systems exist in a vacuum; thus the analyst asks if any other systems use the payroll data. The chief accountant indicates that payroll totals are entered into the ledger journals; clearly, the propoosed payroll system cannot ignore the ledger system. As the interview ends, the analyst asks, "Who works with payroll every day?" The chief accountant names two payroll clerks, thus identifying two future interview subjects.

Carl Jones, the owner, is interviewed next. He is going to be asked to invest in a payroll system. What would he like to see? The scope of the project was estimated at roughly $15,000; how firm is that figure? Does Carl consider it a ballpark estimate, or an absolute upper limit? Rumor suggests that the owner feels a sense of responsibility to the bank. Does he? How strong is it? How much better than the bank does the proposed system have to be before the boss will buy it? It is important that the analyst

Fig. 3.1: *As the feasibility study begins, the payroll system is seen as a large black box.*

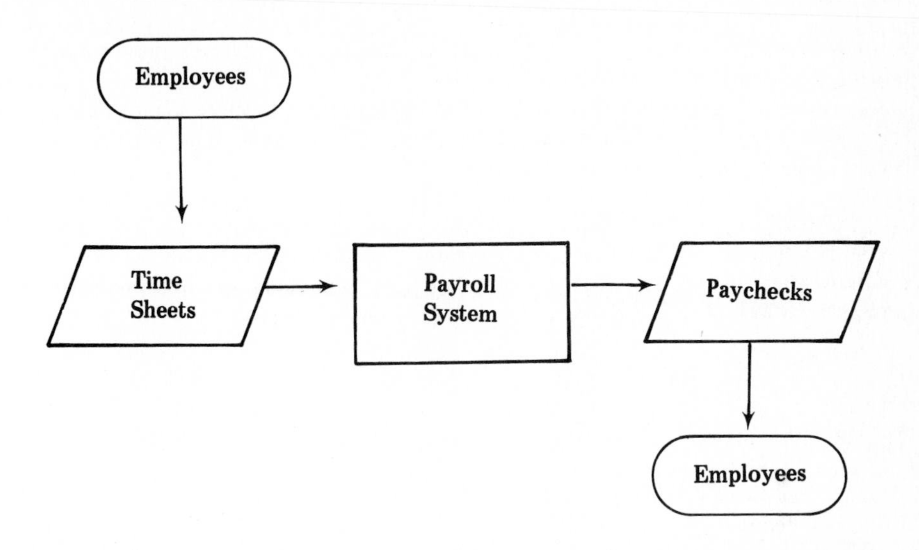

clearly understand what the scope and objectives really mean, before digging into the system.

Finally, the analyst should do some homework. The problem to be investigated is payroll. Payroll has its own terminology, and body of assumed knowledge. The analyst must learn certain basics about the application before attempting to analyze it. Key sources might be internal documentation or, perhaps, general material about payroll found in a textbook or a professional magazine. The objective is not to become an expert in payroll, but to gain a sense of the application.

Study the Existing System

Perhaps the quickest way to understand any application is to study the existing system. There is a danger, however. Most analysts are technically trained, and the technical details of a system can be fascinating. The analyst must remember that the objective is not to document the existing system, but to understand it.

The first step is to interview key people, but how can these people be identified? The earlier interview with the chief accountant yielded the names of two payroll clerks, so the analyst might start with them. Who else might be interviewed? At this stage, the analyst probably views THE PRINT SHOP's payroll system as a black box (Fig. 3.1). Employees record their hours on time sheets which enter the payroll system; later, the system returns and distributes checks to the employees. The objective is to define the contents of that black box. Let's begin at the edges. Who or what accepts those time sheets? Who distributes checks? The answers to these questions may identify people who are inside the black box, and thus clearly know more about the payroll system than the analyst. The idea is simple. Start with what you know. Talk to the people and study the functions at the edge of what you know, and let them suggest the next step. Gradually, you should find the unknown portion of the system shrinking.

Eventually, you may want to summarize your understanding of the system by drawing a system flowchart (Fig. 3.2). (Note: see Module F for an introduction to system flowcharts.) The analyst now has a good idea of what happens to payroll inside THE PRINT SHOP. The only remaining unknown is the bank system: what happens inside the bank? The only way to find out is to ask.

The bank represents a different problem. It is a different organization. Certain functions performed within its payroll system may well be viewed as confidential. The knowledge that the analyst represents a customer who could, as a result of learning too much, be lost, complicates discussions with the bank personnel.

Fig. 3.2: *The logical flow through the payroll system inside THE PRINT SHOP can be defined, leaving the bank system as a smaller black box.*

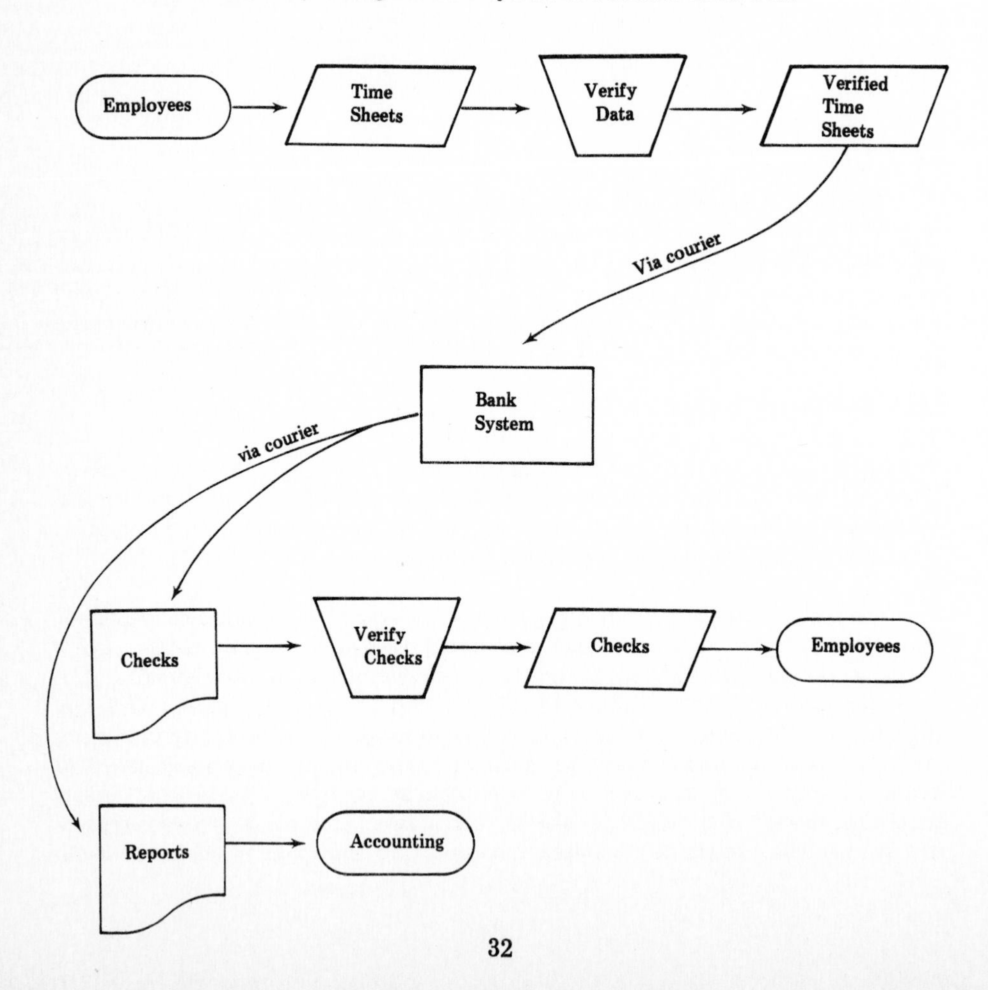

Fig. 3.3: *The payroll system inside the bank.*

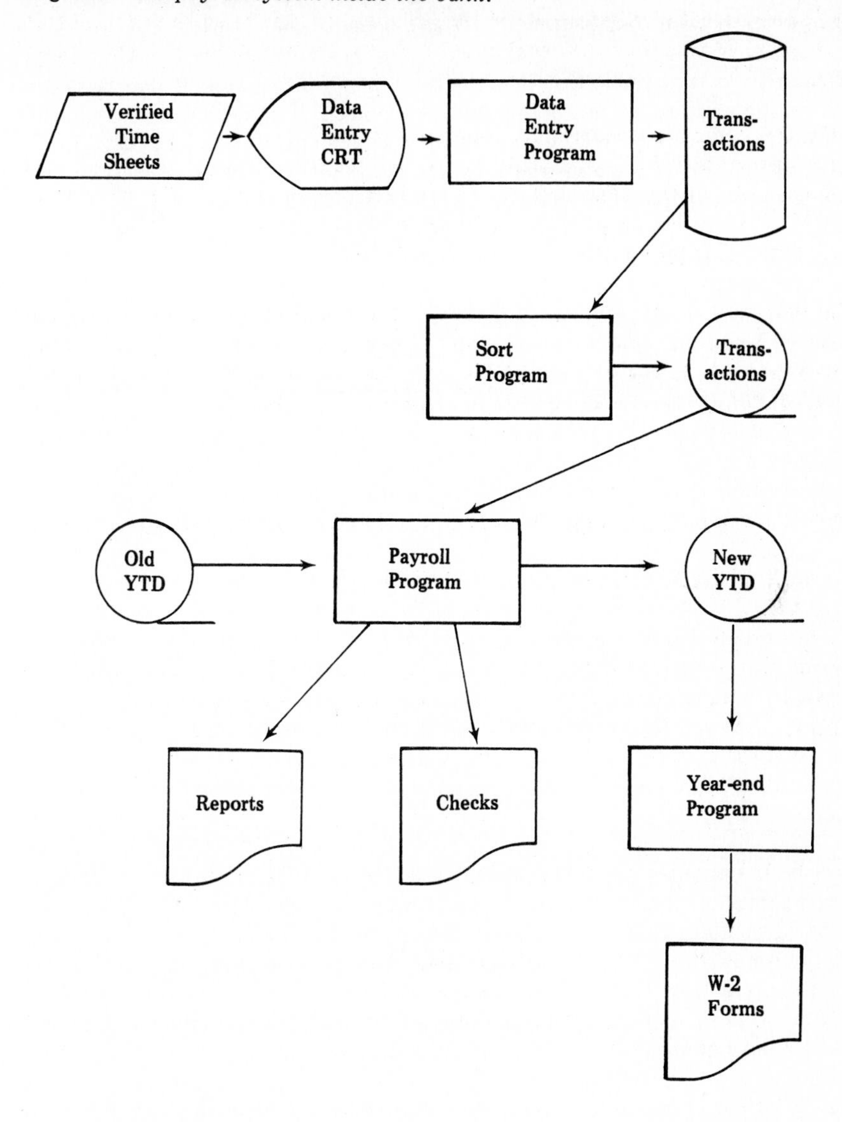

Let's assume, through a combination of interviews, a tour of the bank's facilities, and a careful reading of advertising and technical literature, that the analyst is able to sketch the system flow diagram of Fig. 3.3 When the bank receives the time slips, they go immediately to a data entry clerk, where they are entered via CRT terminal and stored on disk. Later, shortly before the actual payroll is to be run, these data are sorted by social security number and stored on magnetic tape. One hour before the checks are due to be delivered to THE PRINT SHOP, this transactions tape and an old year-to-date earnings tape are mounted on tape drives, and the payroll program is run. The checks, along with several accounting reports, are then given to a bonded courier, and delivered to the chief accountant for distribution. Another program generates W-2 forms from the year-to-date earnings file at the end of each year.

We now have two system flowcharts (Figs. 3.2 and 3.3); together they represent a picture of the existing system. Is this picture accurate? Has the analyst really identified the key functions of payroll? One problem with an existing physical system is that it is often difficult to distinguish what happens from how it happens. Thus, the next step is to develop a high-level logical model of the payroll system.

Develop a High-Level Model

A system flowchart is a good way to describe a physical system. The symbols, however, convey very precise physical meanings. For example, in Fig. 3.3, a circle with a line under it represents magnetic tape. There is no tape drive on THE PRINT SHOP's computer, and thus there is simply no way the bank's system could be reproduced exactly. The objective is not to reproduce the existing physical system, but to create a new system that performs the same functions; thus the analyst would like to summarize the existing system in a way that stresses function rather than physical implementation. Data flow diagrams (see Module D) are excellent for this purpose.

A data flow diagram shows data sources and destinations, data flows, data stores, and processes, but says nothing about how these elements are implemented. For example, a process can be a program, a subroutine, a mechanical sort, a manual process—in short any step that changes or moves data. Contrast the general nature of data flow diagram symbols with the precise, physical meanings of system flowchart symbols. Figure 3.3 clearly shows transaction data stored on magnetic tape. On a data flow diagram, transaction data will be placed in a data store, which might represent tape, disk, drum, diskette, a hand-written ledger book, or any other medium you might care to imagine.

How does the analyst develop a data flow diagram? Begin with the four basic elements: sources or destinations, processes, data stores, and data flows. What data sources or destinations can be identified in the system flowcharts of Figs. 3.2 and 3.3? Clearly, the employees provide the basic labor data through their time sheets, and receive a paycheck from the system; thus the employees represent both a data source and a destination (Fig. 3.4). Are there any others? Accounting gets a number of reports generated by the system, so accounting must be a destination.

Let's consider processes next (Fig. 3.4). What functions does the payroll system perform? First, data are collected. Next, these data are verified. The payroll is then processed, and checks distributed to the employees. At the end of the year, tax statements (W-2 forms) are prepared and distributed. Finally, as part of a related system, the accounting ledgers are updated.

What about data stores? Transactions are collected from the employees. Year-to-date earnings data must be maintained. The checks and various reports are also data stores (Fig. 3.4, again). Why haven't we distinguished between the old year-to-date data, and the new year-to-date data? The two-file, sequential update is a very common solution to such problems as payroll. It is not, however, the only solution. We want our data flow diagram to reflect the functions that must be performed, and not the tech-

Fig. 3.4: *Elements for a data flow diagram of the payroll system.*

Sources and Destinations

- Employee
- Accounting

Processes

- Collect Data
- Verify Data
- Process Payroll
- Process Year-end
- Distribute Checks
- Update Ledger

Data Stores

- Transaction
- Year-to-date Data
- Checks
- Reports
- W-2 Forms

Data Flows

- See Data Stores

nique that the bank has selected to perform them. We may choose a very different solution. A data store is simply a place where data are held for a time; it can represent any physical device or devices, and any logical data organization.

What is a data flow, and how does it differ from a data store? A data flow is data in motion; a data store is data at rest. A data store and a data flow are two different forms of the same thing—data. Data stores are filled by data flows; the source of a data flow is often a data store. At a more detailed level, we may want to study individual data elements, and consider how several data flows can combine to form one data store, or how a single store can provide different elements to each of several flows. During the feasibility study, however, we are still working at a very high level, and thus can view the data flows and data stores as merely different forms of the same data.

We now have a list of the necessary components, and can draw a data flow diagram (Fig. 3.5). Follow the data flows carefully. Note how data are collected from the employees and stored, and how they are subsequently verified. Note how these same data, along with the year-to-date data, flow into the *process payroll* function. Checks and reports flow from *process payroll*; where do they go? How are W-2 forms generated at the end of the year? You should be able to answer these questions simply by glancing at the data flow diagram.

Fig. 3.5: *A data flow diagram of the payroll system.*

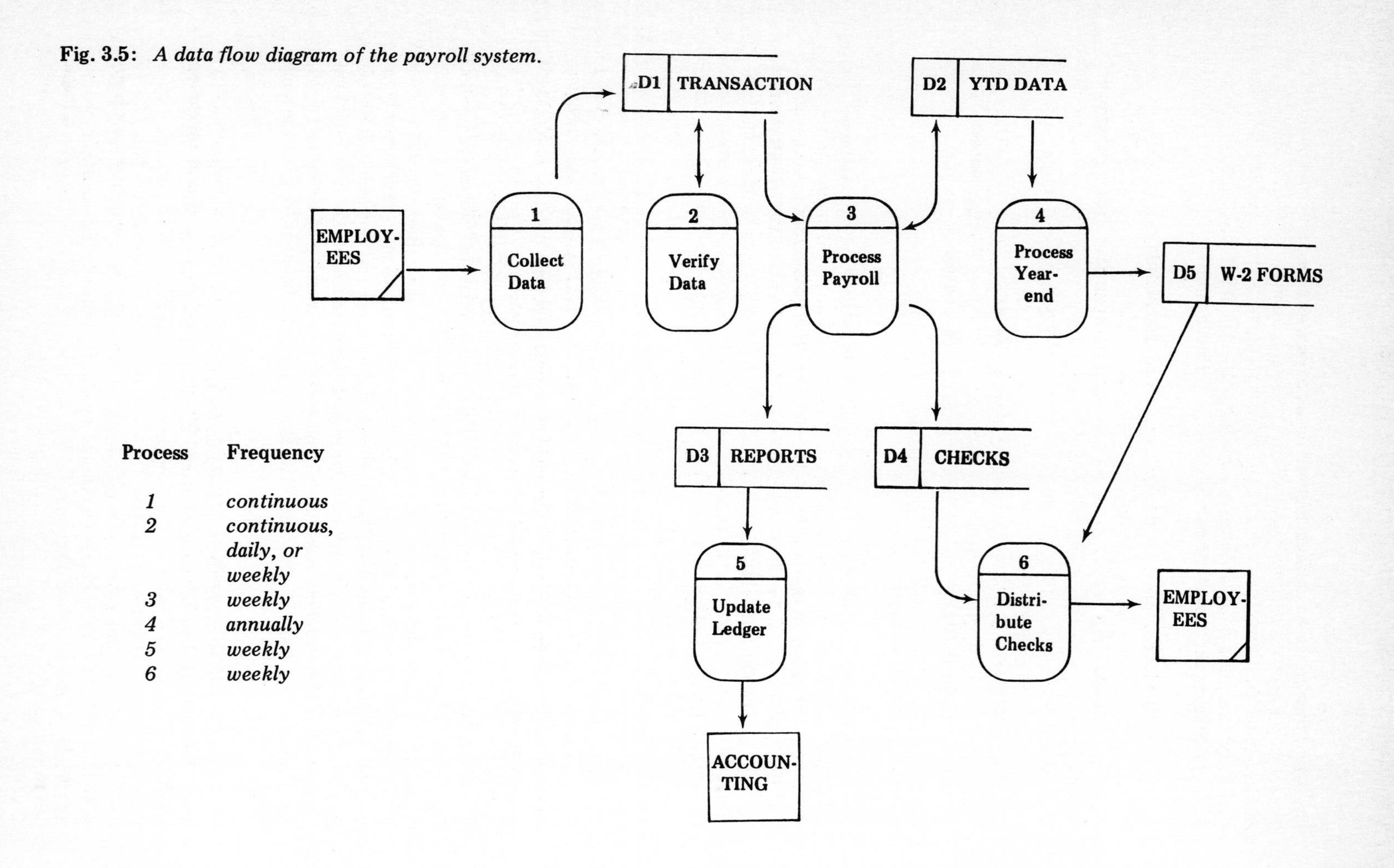

Process	Frequency
1	*continuous*
2	*continuous, daily, or weekly*
3	*weekly*
4	*annually*
5	*weekly*
6	*weekly*

Note that the reports flowing from *process payroll* eventually enter another process called *update ledger*; accounting is shown as the ultimate destination (Fig. 3.5). Responsibility for the ledger lies outside the scope of the payroll system. Still, the payroll system must communicate with at least this one other system, and it is important to identify key points of interface.

As a final point, the analyst should note key timing assumptions directly on the data flow diagram; for example, data collection is continuous, while payroll is processed on Friday. These assumptions will play an important role in the subsequent system design. By clearly identifying such assumptions, the analyst increases the chances that misunderstandings will be corrected.

The data flow diagram represents a logical model of the system. With it, the analyst can approach the user and management, and request added feedback. This logical model will also prove useful in developing the new or proposed system.

Redefine the Scope and Objectives

The time has come for followup interviews with both the user and management. The data flow diagram is the focus of these interviews; it represents the analyst's understanding of the system. Has anything been overlooked? Are the assumptions valid? A logical model such as a data flow diagram can summarize a great deal of information in an easy-to-digest form. Mr. Jones should be able to quickly spot errors involving broad investment options. The chief accountant might identify a related system that the analyst has ignored. The two payroll clerks should spot any missing operational details. Perhaps even the bank will comment; THE PRINT SHOP, after all, purchases other services than payroll.

Note that, at this stage, the analyst's understanding of the system is much deeper than it was during problem definition. Is the initial statement of scope and objectives still valid? If the analyst has already learned enough to believe that there is no way a payroll system can be developed for less than $15,000, management (Carl Jones) should be told. Perhaps the limit will be increased; perhaps the payroll project will be scrapped; that is management's decision. If the system's scope and/or objectives change, however, the analyst must be prepared to start over.

View the first four steps in the feasibility study as a loop. The analyst studies the problem, develops a logical model of the problem, identifies any missing elements, and repeats the process until the logical model is accurate. When this point is reached, the analyst can finally begin to consider *how* the system might be designed.

Develop Alternative Solutions

By now, the analyst understands the problem. Can it be solved? Is there a feasible solution? The only way to answer these questions is to develop a number of alternative solutions and analyze them.

How can an analyst develop these alternative solutions? One approach is to begin with the data flow diagram (Fig. 3.5); assume several sets of automation boundaries, and try to imagine a system to fit each pattern (see Module D). For example, it might be reasonable to group the collection and verification processes within the same automation boundary, yielding an on-line data collection program (Fig. 3.6); the payroll process would fall within a different boundary, and would thus represent a different program. Or, consider enclosing data verification and the payroll process within a single automation boundary. This might suggest a batch program that first verifies the data and then processes the payroll; data might be collected off-line, and submitted to the payroll program without verification. Each time the analyst selects a different set of boundaries, a different alternative is suggested.

Another option is a checklist (Fig. 3.7). Across the top of this matrix are a number of hardware options; to the left are several general types of systems that might be implemented on that hardware. Each block within the matrix defines one general system type; for example, the first column suggests a batch system on a microcomputer, an interactive system on a micro, a real-time system on a micro, and so on. The basic idea is to try to imagine a system to fit each block. Is a batch payroll system implemented on a microcomputer technically feasible? Yes, although a batch system on our minicomputer might be better. How about an interactive system on a minicomputer? Although the once-a-week nature of the payroll process would argue against an interactive solution, it would certainly be possible to use interactive techniques for the data collection and/or data verification processes; there is no reason why the same type of design must be used for every element of a system. The checklist suggests possible implementations; it is an excellent way to generate a variety of technically feasible alternatives quickly.

As the analyst considers these possible solutions, the concern is with *technical feasibility*. Options are rejected because they can't be done on existing hardware, or

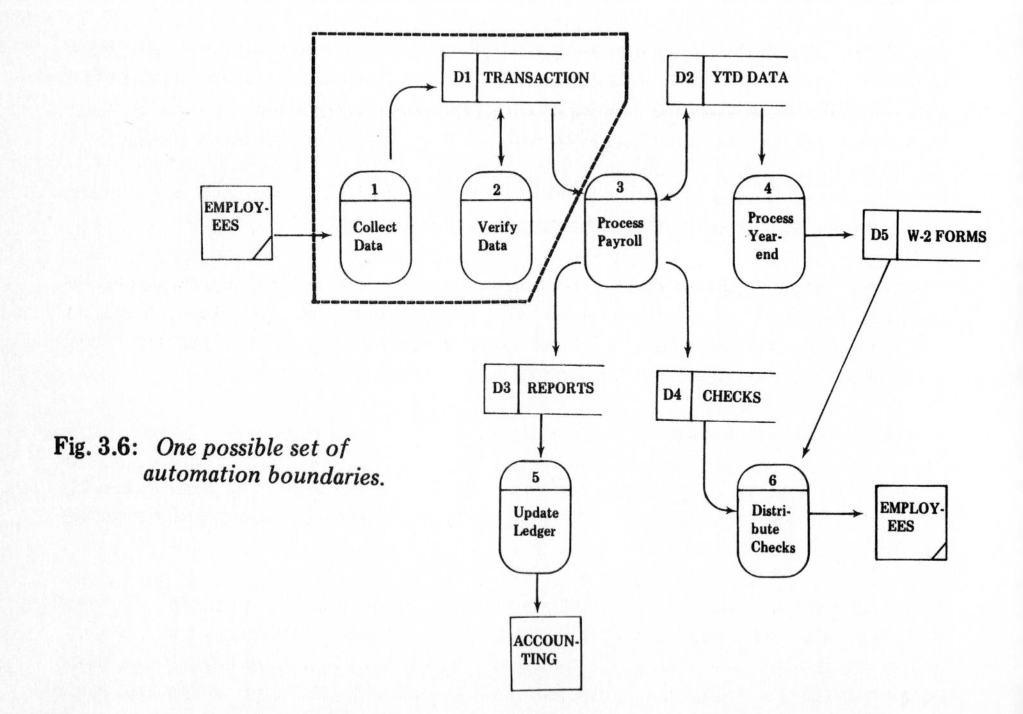

Fig. 3.6: *One possible set of automation boundaries.*

Fig. 3.7: *A checklist of system types.*

		COMPUTER				
SOURCE	TYPE	MICRO-COMPUTER	MINI-COMPUTER	MAIN-FRAME	SERVICE BUREAU	TIME-SHARING SERVICE
INTERNAL	BATCH					
	INTERACTIVE					
	REAL-TIME					
EXTERNAL	BATCH					
	INTERACTIVE					
	REAL-TIME					
TURNKEY	BATCH					
	INTERACTIVE					
	REAL-TIME					

EXISTING SYSTEM____________

MANUAL SYSTEM____________

because of timing conflicts with the application. Clearly, there is no sense considering a technically impossible option, but technical feasibility is only a first screen. Before an alternative can be accepted as "feasible", other tests must be passed as well.

One key concern is with *operational feasibility*. A solution might make a great deal of technical sense, but if it conflicts with the way THE PRINT SHOP does business, that solution is not feasible. Consider, for example, on-line data collection. In many organizations, a minicomputer system is used for precisely this purpose. In THE PRINT SHOP, however, the primary application of the minicomputer is graphics. When the plotter is running, the computer is not available to support on-line data collection, and the company is not about to disregard its primary application in order to support payroll. Technical feasibility asks, "Can we do it?" Operational feasibility asks, "Can we do it here?"

After an alternative has passed the technical and operational tests, it still has one more hurdle to clear: *economic feasibility*. The question is simple: do benefits outweigh cost? There is an existing payroll system. If the new system is more expensive than the present one, then why bother developing the new system? Thus the analyst must perform a cost/benefit analysis on those alternative solutions that have passed the technical and operational tests (see Module G).

It is always a good idea to give management and the user a range of alternatives: a low cost system, an intermediate system that does the job well, and a high cost system that contains everything anyone could possibly want. The analyst must also consider the present system. It is in place, and it works. There is very little risk involved, and no new money must be invested in this system. The operating costs seem a bit too high, but if nothing else, these costs represent a target against which alternatives can be measured.

Is there any way to reduce the cost with little or no new investment? Certainly. THE PRINT SHOP could shift to a biweekly or monthly payroll. Paying people every other week instead of every week would cut the cost of producing payroll in half; if a weekly payroll costs $5000 per year, a biweekly payroll should save the firm about $2500 annually. Except for the already authorized feasibility study, no new investment would be needed. This is an excellent low cost alternative.

There are, of course, a few disadvantages to consider. Employees might object to the change, and THE PRINT SHOP certainly doesn't need labor trouble. A longer pay cycle makes it more difficult to keep track of the workers' performance; a certain amount of control will be lost. Finally, this solution really doesn't solve the basic problem; the bank will still control THE PRINT SHOP's payroll costs. In spite of these problems, however, the idea of substantially cutting payroll costs with little or no investment is a good one that is certainly worth considering.

As an intermediate cost alternative, the analyst suggests essentially duplicating the existing system (Fig. 3.8). The data collection and data verification processes will continue without change, but the time sheets will be delivered to THE PRINT SHOP's computer room, rather than to the bank. A payroll clerk will enter the data through a CRT screen, and a data collection program will verify the transactions and store them on disk. After all transactions have been entered, the payroll program will be loaded and executed; it will read the transactions, compute pay, update the year-to-date file,

Fig. 3.8: *A system flowchart of the proposed intermediate cost solution.*

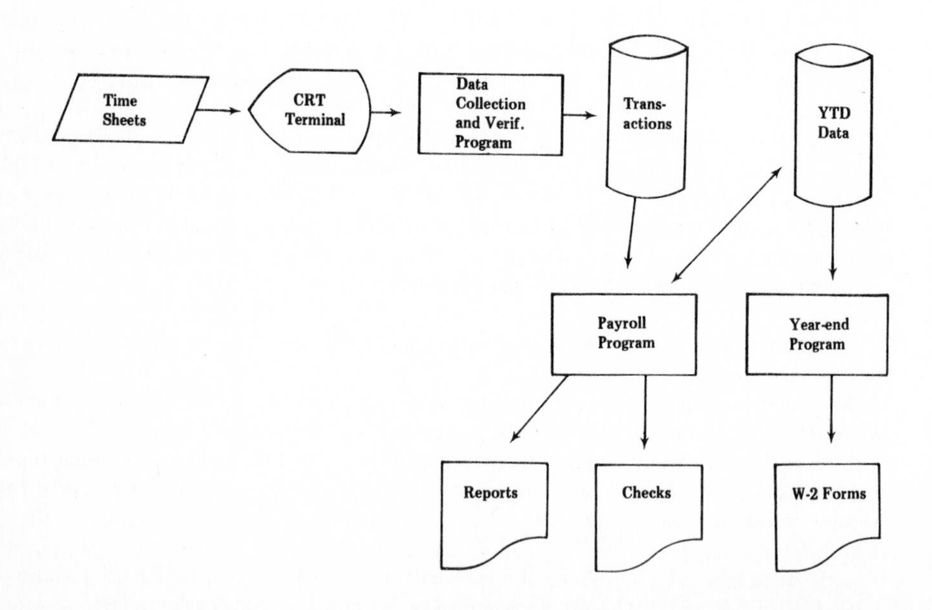

and write the checks and accounting reports. Another program will generate W-2 forms from the year-to-date file at the end of the year.

This alternative seems realistic, so the analyst performs a complete cost/benefit analysis (Fig. 3.9). (Note: a similar cost/benefit analysis should be performed for each alternative seriously considered.) A reasonable estimate of the personnel time needed to investigate, select, customize, install, and test a commercial payroll system is about four months; at $1000 per employee month, that's $8000. A typical commercial payroll system sells for about $2000 (source: *The Datapro Reports*); thus the total development cost is $10,000. It is certainly possible that THE PRINT SHOP will decide to write its own payroll system, but the make or buy decision will not have to be made until the analysis or system design phase. For now, it is easier to estimate accurately the cost of a commercial system.

To recover this investment, THE PRINT SHOP will have to save money on the new system. The analyst estimates the operating cost of the new system at $25 per week, or about $1250 per year (Fig. 3.9). Maintenance is estimated at $500 per year, yielding an annual operating cost of $1750. The present system costs $5000 per year; thus the annual savings are $3250. Figure 3.9 summarizes a number of financial calculations based on these figures; see Module G for a detailed explanation. The important statistic is an estimated internal rate of return of between 18 and 19 percent. With a prime rate of 16 percent, the intermediate alternative seems a very reasonable investment. It is economically feasible.

There are, of course, disadvantages. Responsibility for the system shifts from the bank to THE PRINT SHOP, and, although the new workload is not enough to require hiring another person, someone's job will be affected. The cost and benefit estimates could be wrong; there is a risk. Finally, THE PRINT SHOP could find it slightly more difficult to obtain a loan from the bank in the future. However, the analyst feels that these disadvantages are relatively minor.

Finally, why not consider a more costly option? THE PRINT SHOP is just beginning to move into electronic data processing. Why not do things right? Why not lay the foundation for a central data base, and treat payroll as the first application in an integrated accounting system? The development cost would probably triple to about $30,000, and the benefits derived from the payroll application would not change; thus, based only on the payroll application, the development cost could not be justified. Future applications, however, could be implemented at a much lower cost, and in the long run, the payoff could be substantial. If management is interested, the analyst can perform a more thorough feasibility study for about $4000.

Recommend a Course of Action

The low cost option, while intriguing, seems to call for too much change in THE PRINT SHOP's established operating procedures. The high cost system, with its committment to a data base and its integrated approach to a variety of applications, represents a long-term ideal, and the analyst is prepared to argue long and hard for this option. Given the established scope and objectives of the payroll system, however, the intermediate cost solution is clearly the best choice, and this will be the analyst's recommendation.

Fig. 3.9: *A cost/benefit analysis of the proposed intermediate cost solution.*

DEVELOPMENT COSTS:

Labor (4 months @ $2000/month)		$8000
Payroll package (purchase)		2000
	TOTAL	$10,000

OPERATING COSTS—NEW SYSTEM:

Labor, supplies ($25/week)		$1250
Maintenance		500
	TOTAL	$1750/year

OPERATING COSTS—EXISTING SYSTEM:

Accounting payments $5000/year

ANNUAL COSTS SAVINGS: $3250/year

Year	Savings	Present Value (at 12%)	Cumulative Present Value
1	*3250*	*2901.79*	*2901.79*
2	*3250*	*2590.88*	*5492.67*
3	*3250*	*2313.29*	*7805.96*
4	*3250*	*2065.43*	*9871.39*
5	*3250*	*1844.14*	*11,715.53*

RETURN ON INVESTMENT: 18 to 19%

NET PRESENT VALUE: $1715.53

PAYBACK PERIOD: 4.07 years

Rough Out a Development Plan

The analyst's next step is to rough out a development plan based on the recommended course of action. At this very early stage, it is difficult to estimate accurately the time, effort, and expense required to plan, design, and implement a new payroll system, but reasonable estimates based on the system life cycle are certainly possible. The analyst's best guesses are summarized in Fig. 3.10; let's consider each step individually. Note that the two-week feasibility study is included.

During the analysis stage, the analyst will be faced with the task of defining the functional requirements of the system. Among other things, the analyst must learn a great deal about payroll. A month seems reasonable. During system design, a set of broad, functional specifications for the new system will be developed. These specifications will prove essential in selecting a commercial payroll package; once again, a month of full-time work seems appropriate.

Detailed design comes next. Using the functional specifications, the analyst plans to investigate a number of payroll packages, and select the one that best meets THE PRINT SHOP's needs. There should be about two weeks of actual work, but, since much of this effort will involve studying material provided by other companies, some time delays are expected; thus the elapsed time is set at a full month. Possibly, no commercial package will meet the needs of THE PRINT SHOP. Following the proposed two week study of the available software products, a make or buy decision will be made; it should be noted that a decision to write an original system will change the schedule.

The final stage is implementation (maintenance is included in the continuing operating costs). Once a payroll system is selected, it must be ordered; some paperwork

Fig. 3.10: *A rough implementation plan for the proposed intermediate cost solution.*

	Personnel Time (months)	Elapsed Time (months)
Feasibility study	0.5	0.5
Analysis	1.0	1.0
System Design	1.0	1.0
Detailed Design	0.5	1.0
Implementation	1.0	2.0
TOTALS:	**4.0**	**6.0**

and a time delay of perhaps a week can be expected. It is unlikely that any commercial program will meet THE PRINT SHOP's needs perfectly, so anticipate at least three weeks of programming effort (customizing). Before the new system can be accepted, it must, of course, be tested, so a parallel run with the existing system is planned for the first two weeks of operation. There will, of course, be unanticipated delays, so the implementation step is estimated at one month of actual work spread over two months of elapsed time (Fig. 3.10).

This simple implementation plan provides management with the answers to two very important questions: how long will the project take (six months), and how many people will be involved (probably just one). Given the cost/benefit analysis and this rough schedule, an intelligent go/no go decision can be made.

Write and Present the Feasibility Study

The feasibility study is now essentially over. Thus the analyst collects the information and writes a formal report (see Module C for an outline). The report is presented to management and the user in a joint meeting. They like what they see, and agree with the analyst's recommendation. Thus it's on to the analysis step, with the feasibility study providing a clear sense of the project's technical direction.

THE SKILLS OF THE ANALYST

Note the variety of skills the analyst used in preparing this feasibility study. Some, of course, were technical: system flowcharts and data flow diagrams were drawn, a cost/benefit analysis developed, and a number of computer-related ideas considered. Other problems, however, were much less technical. As the feasibility study began, the analyst had to quickly learn the basics of an unfamiliar function: payroll. Numerous interviews were conducted. Dealing with the bank personnel required a certain political sense. A formal report was written, and at least one oral presentation made. The non-technical aspects of the feasibility study probably consumed more of the analyst's time than the technical factors. A good systems analyst must have a solid technical base, but must be able to work beyond that base. A key non-technical skill is an ability to communicate; if you hope to become a systems analyst, you should work very hard on your written and oral communication skills.

SUMMARY

This chapter concerned a feasibility study for a proposed payroll system; the purpose was to determine if there was a feasible solution. The analyst began by clarifying the statement of scope and objectives, primarily by conducting interviews and reading available documentation. Next came a study of the existing system. Initially, the system was viewed as a black box; gradually, the contents of this box were defined and summarized in system flowcharts. Based on an understanding of the existing system, the analyst then developed a logical model using a data flow diagram, and this logical model became the basis of a redefinition of the system's scope and objectives. These first four steps formed a loop, with the analyst redefining the problem, analyzing it, redefining it in the light of new knowledge, and so on, until an acceptable logical model emerged.

The logical model was used to develop a series of alternative solutions to the problem; low intermediate, and high cost options were considered. A cost/benefit analysis was performed for each alternative. (Note: the text covered the intermediate solution). The analyst then recommended a course of action, and roughed out a development plan. Finally, the feasibility study was written and presented to management and the user.

To be effective, a systems analyst must have a strong technical base, but must possess a number of nontechnical skills as well. This chapter highlighted interviewing, creativity, writing, and presenting material orally. If you plan to be a systems analyst, work on your communication skills.

For additional material on feasibility studies, see Module C.

EXERCISES

1. What is the purpose of a feasibility study? Why is the feasibility study so important?

2. The first four steps in a feasibility study were presented as a loop, with the analyst defining, studying, redefining, and restudying the problem again and again, until an acceptable logical model was developed. Why? Why does this make sense?

3. Distinguish between technical, operational, and economic feasibility.

4. What is the value of a cost/benefit analysis?

5. Why is it so important that the analyst give management and the user a variety of reasonable options? This question was not explicitly answered in the book, so think about your response.

6. Why should the analyst recommend a course of action? After all, the decision to commit funds to a project should be made by management. Again, the text does not directly answer this question.

7. The systems analyst who possesses only technical skills is not likely to be very successful. Why?

8. In Chapter 2, exercise 6, you were asked to prepare a statement of scope and objectives for another function (or other functions) in THE PRINT SHOP. Explain how you would conduct a feasibility study for this function (or these functions). Where would you begin? What information would you need?

9. In Chapter 2, exercise 7, you were asked to prepare a statement of scope and objectives for a small business problem. Now, conduct a feasibility study.

10. In Chapter 2, exercise 8, you were asked to prepare a statement of scope and and objectives for a problem at your school. Conduct a feasibility study.

4

Case A Analysis

OVERVIEW

In this chapter, we consider the analysis step in THE PRINT SHOP's payroll project. The intent is to determine what the system must do. The starting point is the documentation developed during the feasibility study, the data flow diagram in particular. During this step, the analyst will develop a more precise data flow diagram, a data dictionary, and a series of brief algorithm descriptions. Following management and user approval, the results will be used to help design the physical system.

Analysis begins with a careful consideration of the data. Data elements that must be computed identify needed algorithms; the algorithms, in turn, identify additional data elements. Data descriptions are recorded in a data dictionary, and algorithm descriptions are documented on a set of preliminary IPO charts. Additional insight is gained by exploding the data flow diagram. Step by step, the logical system is more fully defined. Analysis ends with a technical inspection and a management review.

ANALYSIS

Based on the results of the feasibility study, management (Carl Jones) has authorized the analyst to proceed with the payroll project. Everyone seems to agree on the scope and objectives, and there are a number of feasible solutions. What happens next? If the analyst follows a structured methodology, the next step should be analysis.

The objective of analysis is to answer the question: *Exactly what must the system do?* Wasn't that question answered during the feasibility study? Not really. Remember the purpose of the study: to determine, at relatively little cost and in a relatively brief time, if there is a feasible solution. Many details were disregarded in the feasibility study, details that cannot be overlooked in the final system. During analysis, the systems analyst attempts to develop a complete functional understanding of the proposed system. The intent is not to determine how the system will work, but what it must do.

Analysis begins with the output from the feasibility study, in particular the data flow diagram (Fig. 4.1). This document identifies a number of functions or processes that must be performed by the payroll system. During analysis, the analyst will study these functions in greater detail. The structured methodology used by THE PRINT SHOP specifies that the analyst must prepare a data flow diagram (Module D), a data

Fig. 4.1: *A data flow diagram of the payroll system.*

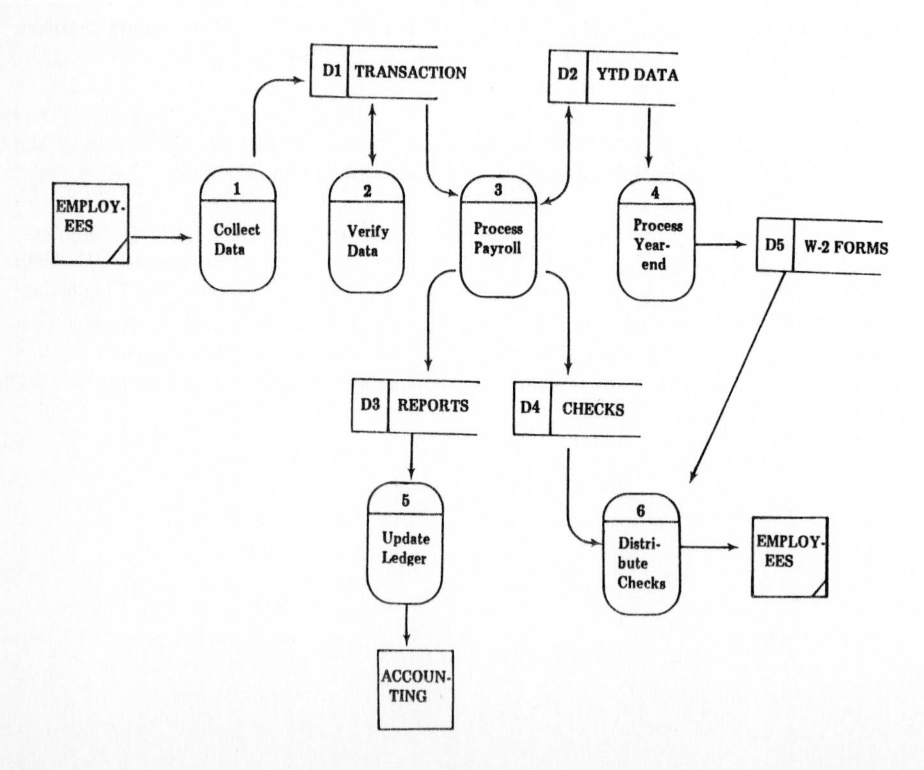

dictionary (Module E), and a set of brief algorithm descriptions as exit criteria from the analysis step; these exit criteria will be used to document the analyst's understanding of the system.

At the end of the analysis stage, THE PRINT SHOP will conduct an inspection, a formal technical review, of the exit criteria (see Module A). The objective of analysis is to determine what must be done to solve the problem. Clearly, the user is in the best position to judge if the analyst has met this objective; thus, user representatives will conduct this first inspection. A management review will follow.

THE ANALYSIS PROCESS

What does the analyst do during analysis? The objective is to define each of the logical functions more fully, but what exactly does this mean? What are these functions? How can the analyst find out? Where should he or she begin?

The basic function of any data processing application is to convert the input data into the desired output information. The data define the processes and algorithms; thus the data seem the obvious starting point. During the feasibility study, the actual data elements were largely ignored; the analyst simply did not need those details yet. The time has come to define the data elements.

Figure 4.1 shows the data flow diagram from the feasibility study. One objective of analysis is to define the data flows and stores down to the element level. Where should the analyst begin? Depending on the nature of a given application, the answer might be anywhere. Generally, however, it makes sense to start at the end, with the output. Why? The purpose of the system is to generate that output. The data elements that appear in the output define the minimum content of the system—they *must* be present.

Thus the analyst begins with the data destination EMPLOYEES. Flowing to this destination are the checks. What data elements (or fields) compose a check? A study of the check and stub used by THE PRINT SHOP reveals several (Fig. 4.2). Where do these data elements come from? Obviously, if they are output by the payroll system, they must be input to or computed by the payroll system. By backtracking through the data flow diagram, the analyst should be able to define the source of each data

Fig. 4.2: *The data elements found on a check.*

Employee name	Social security number
Current gross pay	Year-to-date gross pay
Current federal tax	Year-to-date federal tax
Current state tax	Year-to-date state tax
Current local tax	Year-to-date local tax
Current FICA tax	Year-to-date FICA tax
Current net pay	Year-to-date net pay

element. What happens if the analyst cannot identify the source of a given element? The payroll experts should know; additional interviews will be needed. Gradually, the functional details of the system will be defined.

As an example, let's select one data element from Fig. 4.2: gross pay. What is the source of gross pay? The checks come from process 6, *Distribute checks* (Fig. 4.1). Does *Distribute checks* change the data? No; this process merely distributes already prepared checks. Move more deeply into the data flow diagram. Data store D4, CHECKS, changes nothing; it's just a data store, and must contain the same elements as its input and output data flows. Move on to *Process payroll*. Gross pay must be computed within this process, calling for an algorithm. At this stage, let's assume the analyst knows that gross pay is, basically, the product of hours worked and an hourly pay rate. What are the sources of these two data elements? Hours worked is part of the transaction data flow, and comes ultimately from the employee. What about the hourly pay rate, however? At first, the analyst might assume that it too is part of the transaction, but the pay rate is not on the employee time sheets. It must come from somewhere. The analyst temporarily admits ignorance, notes that the source of the pay rate must be defined, and moves on to another data element, federal tax, for example (Fig. 4.2 again). The process of tracing data elements back to their source is repeated for each element. A list of data elements identified for the payroll system is shown in Fig. 4.3.

Formal Documentation

Up to this point, the analyst has been working largely with scratch paper, lists, and sketches. The analyst's ideas at this stage are preliminary; they will change. How do you feel when you compile a list, and then discover that it is inaccurate? Do you enjoy recopying it? Lists and sketches will be maintained to a point; then they will be discarded or ignored. Even worse is the possibility that an analyst, reluctant to modify the documentation once again, will change the system requirements to fit the existing model. As the analyst collects information, it should be recorded in a form that is easy to maintain.

Fig. 4.3: *A preliminary list of the data elements found in other data stores and data flows.*

YTD DATA
- Employee name
- Social security number
- Year-to-date gross pay
- Year-to-date federal tax
- Year-to-date state tax
- Year-to-date local tax
- Year-to-date FICA tax
- Year-to-date net pay

TRANSACTION
- Employee name
- Social security number
- Hours worked

The analyst is not the only one who must understand the system documentation, however; following analysis, the user and management must review the analyst's work, and others may need it for future reference. Systems typically involve many people. If the system is to do the job, it is essential that these individuals understand each other. Thus the analyst's work must be documented in a consistent and easy to understand form. Our structured methodology calls for the completion of three formal documents: a data flow diagram, a data dictionary, and a set of black box algorithm descriptions.

A data dictionary (Module E) is a collection of data about data. For each data element, such information as the element name, description, format, source, and use is recorded. During analysis, the data dictionary helps the analyst organize information about the data, and is an excellent tool for communicating with the user. Additionally, the data dictionary serves as a memory aid. Certain key information must be recorded for each data element; missing information clearly alerts the analyst of a need to know.

The data dictionary is still quite valuable after analysis is complete. As the source of data about data, it can be used to generate record, file, and data base formats during system design or detailed design. During implementation, the data dictionary serves as a common base against which all the programmers working on a system can compare their data descriptions. Following implementation, the dictionary gives the maintenance programmer a clear idea of how a particular data element is used, invaluable information when a program must be changed.

A number of computerized data dictionary systems are commercially available. Many allow the analyst (or a data entry clerk) to enter the data about data through a CRT screen, with one record stored for each data element. Since not all readers have access to data dictionary software, we will use a simulated data dictionary in this text. One 3x5 card will be filled out for each data element; Fig. 4.4 shows the data to be collected, while Fig. 4.5 shows completed data dictionary entries for several

Fig. 4.4: *The information to be collected for each data element on the simulated data dictionary.*

Name:
Aliases:
Description:

Format:

Location:

Fig. 4.5: *Several elements in THE PRINT SHOP's payroll system data dictionary.*

Name: Employee name
Aliases: Name
Description: The name of an employee in the form: Last, First Initial.
Format: Character
Location: Check
Transaction
YTD Data

Name: Current Gross Pay
Aliases: Gross pay, Gross
Description: An employee's computed pay before taxes and other deductions.
Format: Numeric; maximum value = 9999.99.
Location: Check

Name: Current federal tax
Aliases: Income tax
Description: Computed amount of income tax to be withheld for the federal government.
Format: Numeric; maximum value = 9999.99.
Location: Check

Fig. 4.6: *The structure of a black box.*

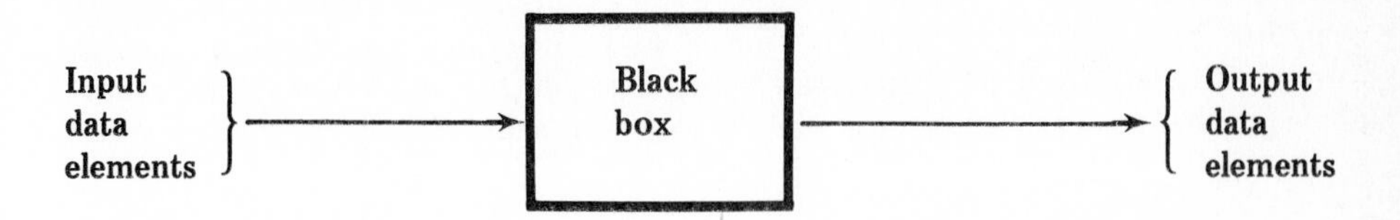

payroll system data elements. Why use 3x5 cards instead of lists or tables? Cards can be processed, changed, added, or deleted individually; a list cannot. The cards can be resequenced or regrouped without recopying the data; a list cannot. Simply, cards are easier to maintain. (Note: the simulated data dictionary used in this text contains only minimal information present in almost any computerized data dictionary system. If you have data dictionary software available, use it.)

The analyst should also record, at a black box level, information about the algorithms. What is a *black box*? Consider the square root function available in many high-level languages. Few programmers are concerned with the instructions that compose this subroutine. Instead, most simply call the subroutine by coding the data elements passed to and returned by it (Fig. 4.6). The function is seen as a black box which, when given certain inputs produces certain outputs. This is how the analyst should view the algorithms at this early stage; the details can wait until later. If an algorithm is obvious, or if, in the course of an interview a user explains a particular algorithm, the analyst should, of course, make a note, but describing the contents of the black box is not a priority yet.

How should the analyst document the algorithms? One possibility is to use 3x5 cards or sheets of paper (one per algorithm), and sketch the inputs and outputs flowing through a black box (Fig. 4.7); as details concerning the algorithm become known,

Fig. 4.7: *The gross pay algorithm illustrated as a black box.*

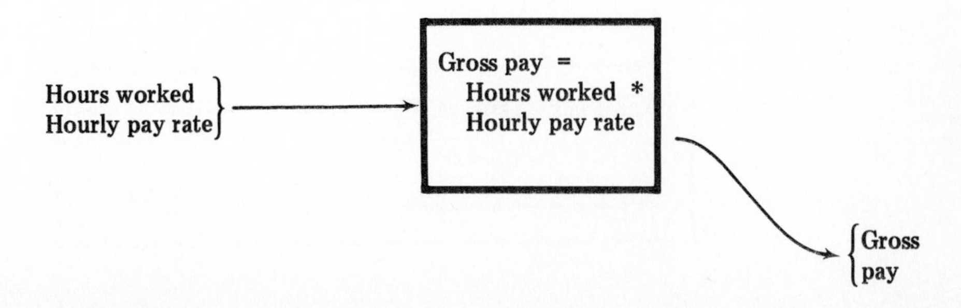

they can be added, later. One problem with this approach, however, is that the black box sketches have no future use in the system development process. They will eventually be discarded, and no one likes to create documentation only to throw it away. As an alternative, the preliminary algorithm descriptions can be recorded on an IPO (Input/Process/Output) chart (Fig. 4.8), which can be refined and later used during detailed design as part of the HIPO package (Module H). Some organizations use the computer to prepare HIPO documentation; if so, by describing the algorithms in IPO format, the analyst can get a head start on detailed design. In a good structured methodology, the documentation produced in one step should be directly relevant to future steps.

Fig. 4.8: *The gross pay algorithm documented on a preliminary IPO (Input/Process/Output) chart.*

IPO Chart

SYSTEM: Payroll
PREPARED BY: Davis
MODULE: Gross Pay Algorithm
DATE: 4/14

CALLED OR INVOKED BY:

CALLS OR INVOKES:

INPUTS:
Hours Worked
Hourly Pay Rate

OUTPUTS:
Gross Pay

PROCESS:
Gross Pay = Hours Worked * Hourly Pay Rate

LOCAL DATA ELEMENTS:

NOTES:

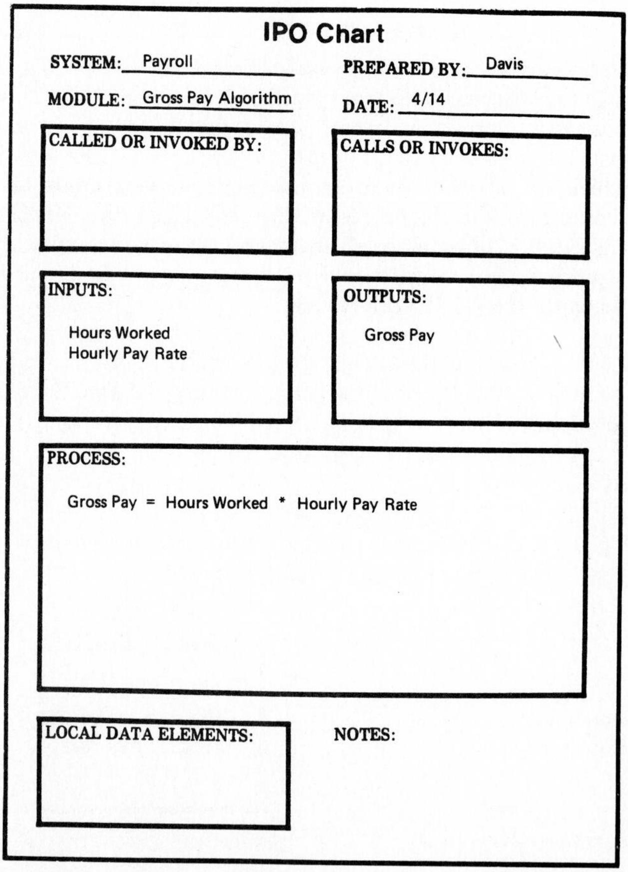

Defining the Logical System

What next? The analyst has identified many of the data elements that must flow through the payroll system, and has recorded them in a preliminary data dictionary. Several algorithms have been documented at a black box level. Almost certainly, however, questions remain. Is the data dictionary accurate? Is it complete? Are the algorithms accurate and complete? Are any missing? What is the source of certain data elements? The analyst must obtain answers to these questions.

As before, the source of detailed information about payroll is the people who work directly with the system—the users. Thus another cycle of interviews with the chief accountant and the two payroll clerks is scheduled. The data flow diagram is an excellent tool to focus these interviews, with the discussion following the data flow, from beginning to end. Transaction data flow from the employee into the *Collect data* process. The analyst has identified, in the data dictionary, the elements that make up a transaction. Are they correct? Is anything missing? Look back at the data flow diagram. What happens in the *Collect data* process? In *Verify data*, certain algorithms seem to be required. Are they correct? Is anything missing? For the analyst, the data flow diagram, data dictionary, and algorithm descriptions serve as check lists or memory aids. Known elements must be verified. Unknown factors, blank spaces, must be filled in.

Consider, for example, the gross pay algorithm. Basically, the rule says to multiply hours worked and an hourly pay rate. Assume that the analyst and a clerk have gone through the first two processes on the data flow diagram, and are discussing *Process payroll* (Fig. 4.1). They have already verified that hours worked enters the system as part of a transaction, but where does the pay rate come from? The user answers: from the personnel file. Look back at Fig. 4.1. Do you see a data store to hold personnel data? It's not there. The step-by-step approach of analyzing the data flows identified an unknown data element. This unknown element led to an obvious interview question. The answer highlighted a previously unknown component of the payroll system: a PERSONNEL data store.

This revelation, in turn, leads to still more questions. Where does the personnel data store come from? Obviously, if these data exist, they must enter the system somewhere. Let's assume that the ultimate source is accounting, and that a new process, *Maintain personnel*, will be required; the source, process, and data store are thus added to the data flow diagram (Fig. 4.9).

Try to sense the spiral nature of this structured methodology. Early analysis generates questions. The answers provide a more detailed understanding of the system, and may generate more questions. Subsequent answers lead to an even better understanding, and perhaps even more questions. Each cycle yields more and more detail about the logical system. What is the role of the data flow diagram, data dictionary, and algorithm descriptions? To complete the documentation, certain information must be collected. Missing pieces are obvious. A system can be quite complex, making it easy to overlook something. By highlighting missing pieces, formal documentation helps the analyst minimize this risk. The documentation also serves as an excellent communication tool, and will become the basis for the next step, system design.

What else might the analyst learn by discussing the gross pay algorithm with the user? Is it complete? No. The payroll clerk points out a need for computing overtime pay—hours over 40 in any given week are paid at 1.5 times the regular rate. Bonuses are paid too; the boss occasionally rewards outstanding performances with extra cash. The fact that a bonus must be paid is indicated on the transaction. The bonus is a new data dictionary entry; the need for overtime and bonus pay computations must be noted on the gross pay algorithm description.

What next? The other algorithms might be discussed. Additional algorithms might be identified; for example, there are two for computing federal income tax, one for single, and another for married taxpayers. Note how carefully and methodically the analyst works. Such effort is essential if a good system is to be designed.

Exploding the Data Flow Diagram

To this point, the analyst has been treating the algorithms and data elements as independent entities, but anyone who has written a computer program knows that they are not. Functions must be performed and data supplied in a certain sequence. The time has come for the analyst to expand his or her view of the logical flow through the system by exploding the data flow diagram (see Module D).

The data flow diagram is exploded through *functional decomposition*. Essentially, one process on the diagram is selected and broken down into its subfunctions. These lower-level functions then become processes on a new data flow diagram, complete with their own data stores and data flows.

For example, consider *Process payroll* (Fig. 4.9). What steps are involved in processing payroll? The analyst should have a good sense of the basic algorithms involved in this process, and the inputs and outputs of these algorithms may suggest a sequence of functions. Assume that the analyst is able to define the following key functional groups:

1. Get data.
2. Compute current pay.
3. Compute year-to-date pay.
4. Update year-to-date pay.
5. Write checks.
6. Write reports.

Each function contains one or more of the algorithms. Their relationship can be graphically represented by a partial data flow diagram (Fig. 4.10); note that only the decomposition of *Process payroll* is shown. The decomposed steps can now be added to the system data flow diagram to generate a new, more detailed version (Fig. 4.11). Note that the steps exploded from process 3 are identified as 3.1 through 3.6.

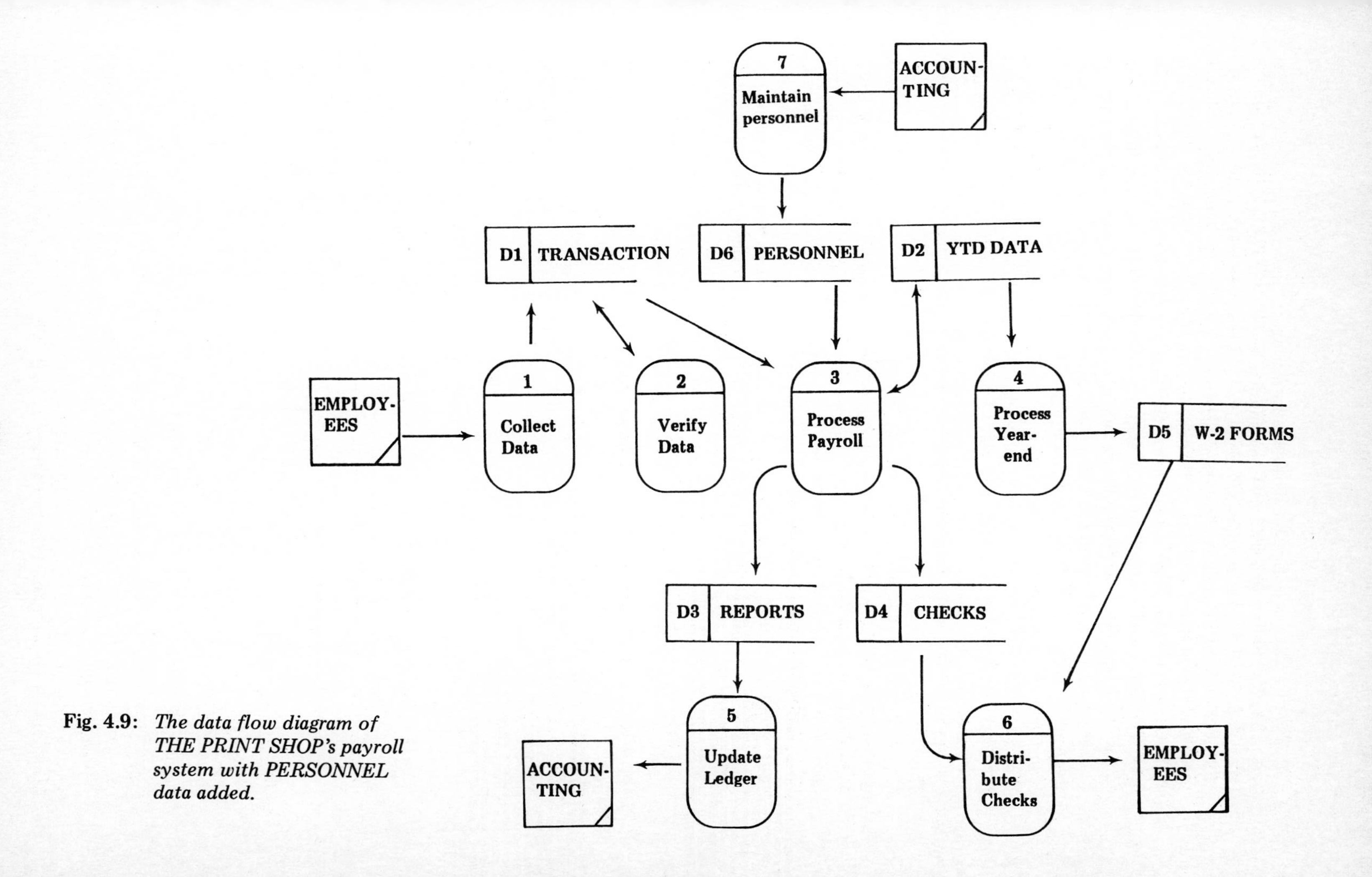

Fig. 4.9: *The data flow diagram of THE PRINT SHOP's payroll system with PERSONNEL data added.*

The new data flow diagram shows the logical payroll system in greater depth; in particular, it illustrates the sequential relationships among the various major functions. Other functions can be decomposed similarly; however, given the relative simplicity of this payroll application, we'll assume that this is not necessary. Should any of the processes exploded from process 3 be further decomposed; for example, should the analyst break *Get data* into independent steps that get the transaction and personnel data? Perhaps, but this may be going a step too far. What do you envision when you encounter a process that does nothing but get data from a particular data store? You probably think of a read instruction, and perhaps wonder about error processing. Remember the intent of the analysis stage; the concern is with *what* happens, and not with *how* it happens. When further decomposition causes you to begin thinking about the code that might be written to implement a function, that decomposition is unnecessary. Code is physical. It is still too early to worry about the physical details. During analysis, the analyst should work at a *functional* level.

The Exit Criteria

As the data flow diagram is decomposed, the relationships between the various system components become more clear. Each change suggests new questions, and the answers may in turn suggest new data dictionary entries and new or refined algorithm descriptions, but as analysis progresses, the questions and answers tend to define more and more detail. Eventually, the analyst will be satisfied with his or her understanding of the functional requirements of the system. The data flow diagram (exploded to an

Fig. 4.10: *An explosion of the Process Payroll function.*

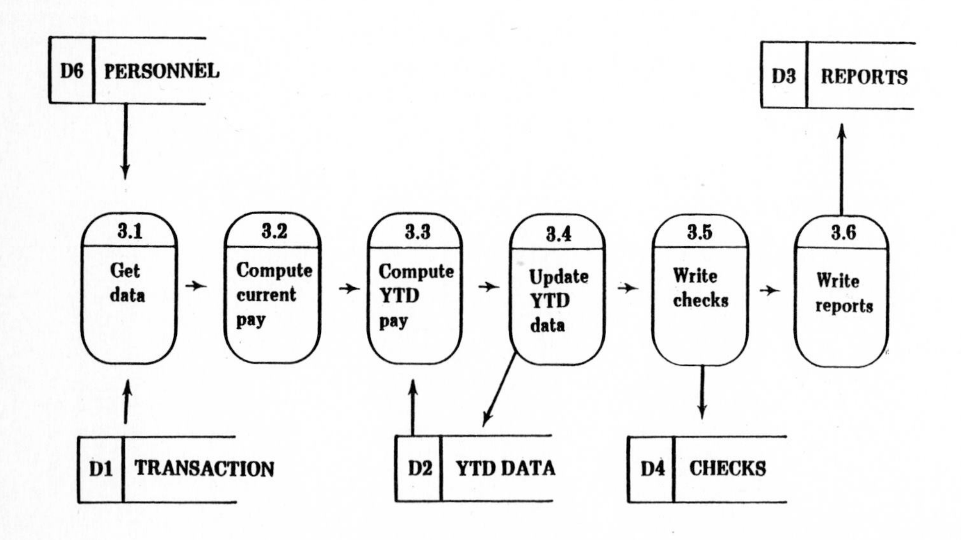

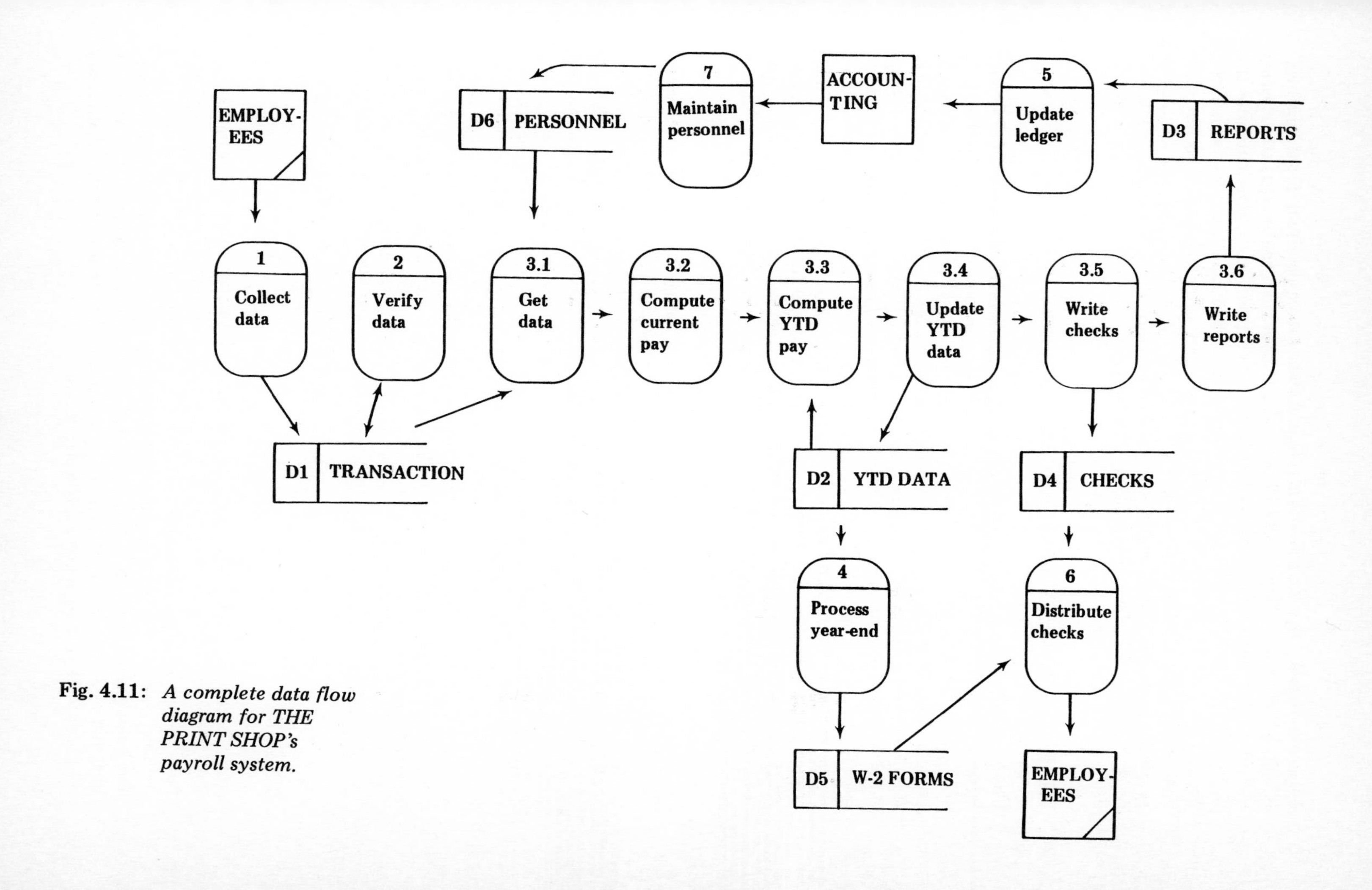

Fig. 4.11: *A complete data flow diagram for THE PRINT SHOP's payroll system.*

appropriate level), the data dictionary, and the black box algorithm descriptions will document this understanding, and serve as exit criteria from the analysis step. During analysis, these documents were used as memory aids, and as a convenient recording medium for facts concerning the system. During the transition from analysis to system design, these same documents will become communication tools, allowing the users and management to understand clearly the analyst's view of the system. Later, as the analyst moves into system design and subsequent steps, these same documents will form a foundation for the development of the system.

THE INSPECTION AND THE MANAGEMENT REVIEW

The final milestone in the analysis step is a formal technical inspection of the exit criteria (Module A). A team composed of a moderator (perhaps another analyst who was not involved in the payroll project), the two payroll clerks, and the analyst is assembled, and one of the clerks is selected as the reader. Step by step, using the data flow diagram as a primary document and the data dictionary and algorithm descriptions as support, the team studies the documentation. The analyst does not lead the team, but instead answers technical questions; the objective is to find errors or oversights, and not to defend or attack the work. A few minor errors are noted and subsequently corrected by the analyst. Now, all the inspectors can agree: the analyst really understands the problem and seems to know what must be done to solve it. The team members' signatures on the inspection form indicate that the work to date is technically sound.

Before moving on to system design, however, a management review is necessary. The technical inspection has indicated that an acceptable payroll system can be developed for THE PRINT SHOP, but management has other concerns as well. What about the schedule? Can the job still be done on time, or must plans be changed? What about the cost? During the feasibility study, a development cost of roughly $10,000 was estimated. The analyst now has a better understanding of the system; is $10,000 still a reasonable estimate? Let's assume that the analyst has encountered no real surprises during analysis, and thus the initial cost estimate and schedule (from Figs. 3.9 and 3.10, respectively) still seem reasonable. The next step, system design, should take about a month. The inspection still makes economic and operational sense. Thus management approves funding for the next step in the system life cycle: system design.

SUMMARY

This chapter considered the analysis step in THE PRINT SHOP's payroll system project. The starting point was the data flow diagram developed during the feasibility study. Beginning with the output data flow, the analyst defined the individual data elements, and traced them through the data flow diagram, back to their sources. As a consequence, algorithms were highlighted, and questions generated. Information concerning the data was recorded on a data dictionary; black boxes and preliminary IPO charts were used to document the algorithms.

The process of preparing the formal documentation identified unknown data and algorithm parameters. Combined with the questions generated while tracing the data flows, these unknowns formed the basis of a series of interviews with the users. Gradually, the functions performed by the payroll system were defined in more and more detail. In order to trace these more detailed data flows, the analyst exploded the data flow diagram to a lower level. The cycle of tracing data flows, identifying data elements and algorithms, generating new questions, seeking answers to these questions, exploding the data flow diagram, and retracing the data flows continued until the analyst obtained an acceptable functional understanding of the payroll system. A formal inspection verified the accuracy of this understanding, and a management review authorized the analyst to continue with the project.

EXERCISES

1. What is the objective of analysis?

2. What are the exit criteria from the analysis step? Briefly describe each exit criterion. How do the criteria help the analyst during analysis? How are these criteria used after analysis?

3. In our example, the systems analyst started the analysis step by examining the output data flow. Why does it make sense to begin with the output?

4. Describe the cyclic nature of analysis.

5. Why is it so important that information collected by the analyst be formally documented?

6. What is a black box? Why, do you suppose, a black box is such a convenient tool for describing algorithms during analysis?

7. What is functional decomposition? What does it mean when an analyst "explodes" a data flow diagram?

8. What is an inspection? What is its purpose?

9. Figures 4.2 and 4.3 list a number of basic data elements for the payroll system. Figure 4.5 shows the data dictionary entries for several of these data elements. Complete the data dictionary.

10. In discussing the analysis of the payroll system, we intentionally ignored process 4, *Process year-end* (Fig. 4.9) and data store D5, W-2 FORMS. Assume that a W-2 form contains all the year-to-date data as it exists on the last pay date of a year. Trace this data flow. What data elements must D5 contain? What algorithms, if any, must be in *Process year-end*? Would adding this information change your view of the functions performed in *Process payroll*?

11. In the text, we followed the analyst as he or she traced the flow of the data element gross pay, and defined, at a black box level, the algorithm. Do the same thing for:

    a. federal tax.
    b. state tax (use your own state).
    c. local tax, if any.
    d. social security or FICA tax.
    e. any other deductions you might care to imagine.
    f. net pay.
    g. each of the year-to-date data elements (see Fig. 4.2).

12. The chapter also ignored another branch of the initial data flow diagram (Fig. 4.1), the reports and the *Update ledger* process. Assume that the reports call for the following information, and explain how each new data element might change the analyst's view of the logical payroll system. In other words, what new data elements, algorithms, processes, and/or data stores would each of the following require? Remember, don't worry about how the reports might be generated; concentrate instead on what must be done to generate the reports.

    a. Gross pay totals summarized by department.

    b. Totals for each of the current pay statistics.

    c. Counts, including how many people were paid, how many received overtime, and how many received a bonus.

    d. Hours worked for each employee, summarized by department.

    e. An exception report listing, by department, all employees who worked less than 20 or more than 45 hours.

13. Continue with Chapter 3, exercise 8. Explain how you would conduct the analysis step.

14. Continue with Chapter 3, exercise 9. Conduct an analysis.

15. Continue with Chapter 3, exercise 10. Conduct an analysis.

5

Case A System Design

OVERVIEW

Continuing with THE PRINT SHOP's payroll example, this chapter describes system design. The primary input document is the data flow diagram prepared during analysis. The objective of system design is to determine, in general, how the system will be implemented. During this phase, the analyst develops a number of alternative strategies, and documents each of these options with a system flowchart, a list of physical components, a cost/benefit analysis, and a schedule. A project network is used to prepare a detailed schedule for the recommended alternative. System design ends with an inspection and a management review.

SYSTEM DESIGN

Analysis has just been completed. The functional requirements of THE PRINT SHOP's payroll system have been documented with a data flow diagram, a data dictionary, and a set of algorithm descriptions. What must be done is now clear. The time has come to consider how to do it, and thus system design begins. The objective is to decide, in general, *how* the system should be implemented. Individual physical components—programs, files, manual procedures, documents—will be identified, but only at a black box level; the contents of these black boxes will be planned in the next step: detailed design. In effect, during system design the analyst prepares a blueprint for the system.

System design begins with a search for alternative solutions. Using the data flow diagram as a primary reference, the analyst imagines a number of automation boundaries, and develops preliminary implementation strategies. Checklists, brainstorming, and creativity are some of the tools brought to bear on this problem. Eventually, one or more reasonable strategies are selected from these ideas, and a system flowchart, cost/benefit analysis, and implementation schedule are prepared for each. These exit criteria, accompanied by the analyst's recommendations, are subjected to an inspection, and to management review. Assuming they are approved, the exit criteria form the basis for subsequent steps.

DESIGNING THE PRINT SHOP'S PAYROLL SYSTEM

Generating Alternatives

How might the payroll system be implemented? During system design, the analyst must devise a set of alternatives. The data flow diagram is an excellent starting point. Certain processes might be logically grouped within a single automation boundary, while others might form other automation boundaries (see Module D). These boundaries, in turn, suggest implementation strategies.

Working methodically, the analyst begins with process 1 (Fig. 5.1, ignore the dashed lines for now). Can processes 1 and 2 be grouped to form a single program or manual procedure? Certainly; whoever (or whatever) collects the data could certainly verify them. Could process 3.1, *Get data*, be included within this same automation boundary? No. *Get data* is the first process in the weekly preparation of payroll; *Collect data* takes place throughout the week. Since the timings of processes 1 and 3.1 are incompatible, they cannot be grouped. Processes 1 and 2 could be treated as independent modules, or could be grouped and implemented together.

Next, the analyst moves on to process 2. If the *Verify data* process were independent of *Collect data*, could it be grouped with any other processes? Yes. Figure 5.1 shows a single automation boundary surrounding *Verify data* and the six processes (3.1 through 3.6) that prepare the weekly payroll; imagine a program that first verifies the data, and then processes them. Could process 4, *Process year-end*, be included within this same boundary? No. Process 4 is needed annually; the automation boundary defines a module needed weekly. What about process 6, *Distribute checks* (Fig. 5.1)? The processes included within the automation boundary could all be performed by a single program, but the distribution of checks requires manual interven-

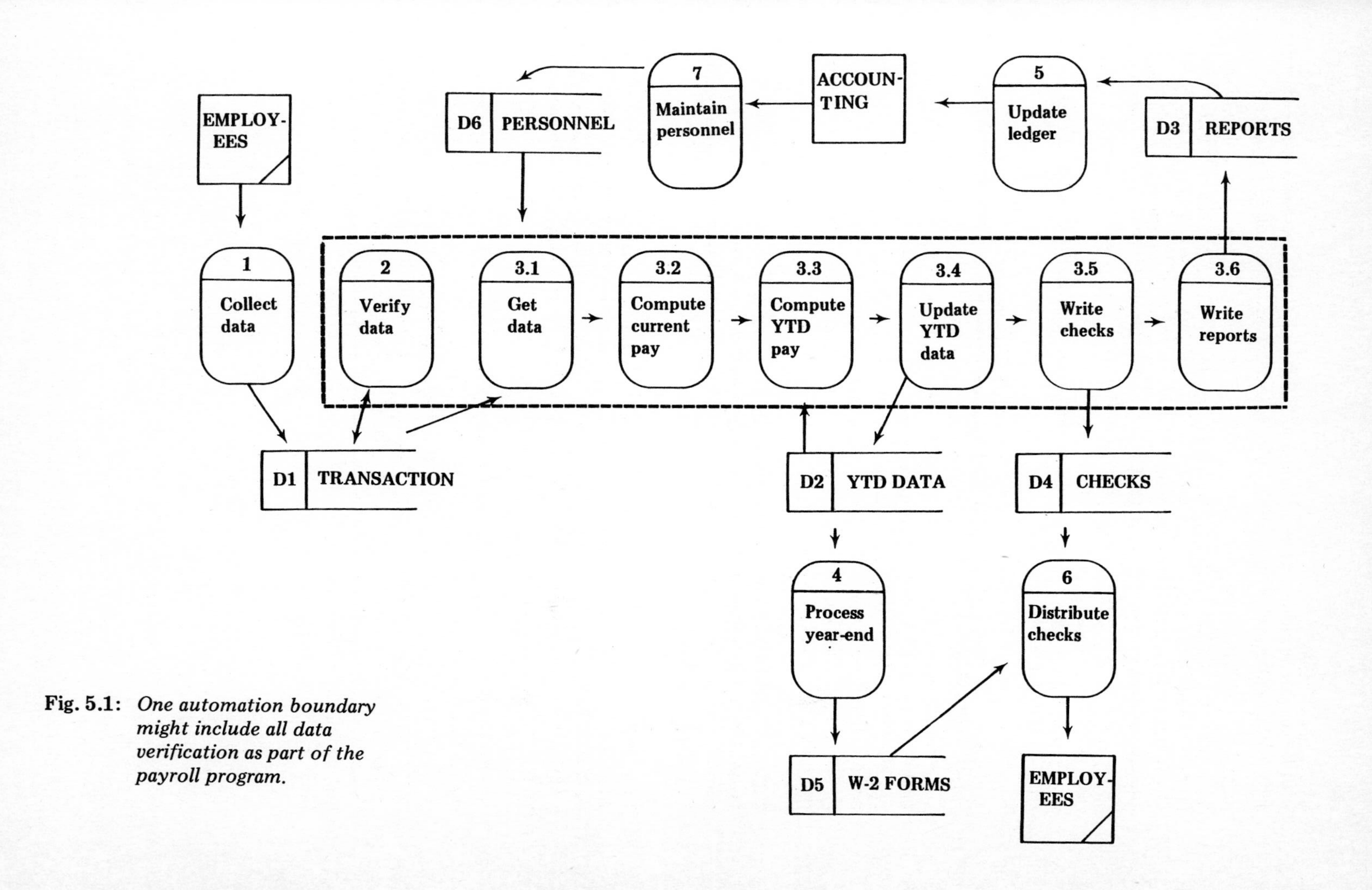

Fig. 5.1: *One automation boundary might include all data verification as part of the payroll program.*

tion, if only for security reasons. Think about the other two processes, *Update ledger* and *Maintain personnel*. Why are they outside the automation boundary?

The idea is simple. Using the data flow diagram as a reference, imagine every possible grouping of processes. Discard those that are technically infeasible; for example, eliminate any containing processes that have timing conflicts. Those sets of automation boundaries that remain represent possible implementation strategies, and can be used to suggest alternative physical system designs.

For example, Fig. 5.2 shows a set of automation boundaries in which six primary groupings can be identified:

1. Data collection/verification.
2. Payroll processing.
3. Year-end processing.
4. Check distribution.
5. Personnel maintenance.
6. Ledger update.

Begin with the first boundary enclosing processes 1 and 2. How might these functions be implemented? A manual data collection and verification subsystem (such as the one currently in use at THE PRINT SHOP) is one option. Partial automation, with weekly

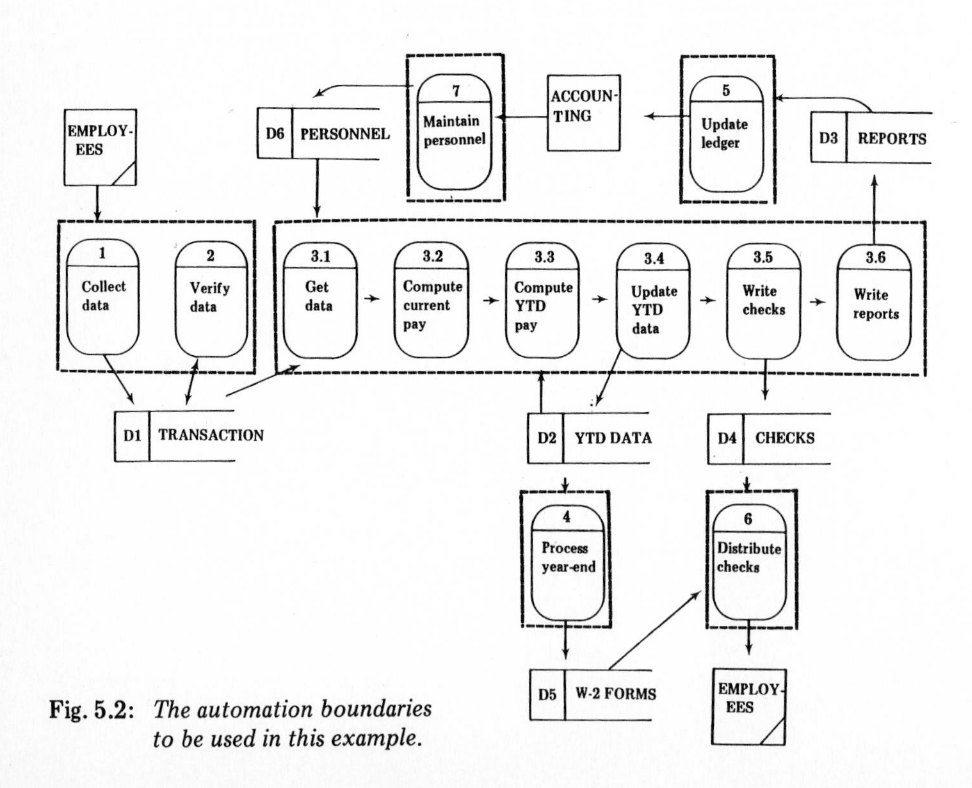

Fig. 5.2: *The automation boundaries to be used in this example.*

data entered through a CRT terminal controlled by a verification program, is another. A third option is complete automation, with a data entry terminal (or microcomputer) collecting and verifying data submitted directly by the employees. Other alternatives might be imagined as well.

Once a reasonable number of options have been listed for one automation boundary, the analyst moves along to another. At this stage, the idea is to imagine and list options, boundary by boundary, without stopping to evaluate them. A checklist such as the one used during the feasibility study (Fig. 3.7) might help. Brainstorming (see Module C) is another useful tool. However, there is simply no substitute for the experience and creativity of a good systems analyst.

Selecting Reasonable Alternatives

The next step is to select one or more reasonable alternative strategies from the resulting list of feasible options. The initial statement of scope and objectives and the results of prior steps should certainly be considered; for example, using an on-line data collection terminal might be rejected because of management's earlier reaction to this option. Identifying reasonable alternatives may require additional interviews with the users. Since plans are being developed to implement the system, the programmers and technical management should be interviewed as well.

Let's assume that the analyst has identified three options for THE PRINT SHOP's payroll system, all based on the automation boundaries of Fig. 5.2. The basic system will eventually reproduce the present system. As an alternative, a microcomputer system can be purchased to collect and verify the data. Finally, the process of writing the checks can be broken into two separate steps for security purposes. Management will also be asked to decide if the software to support these options should be purchased or written internally.

The Exit Criteria

For each alternative strategy, the analyst prepares:

1. a system flowchart (Module F).
2. a list of system components.
3. a cost/benefit analysis (Module G).
4. an implementation schedule (Module L).

These documents represent the system design step's exit criteria. Note that the documentation is prepared for each alternative.

The basic system is pictured in Fig. 5.3. The existing (manual) data collection system will be used, although a new program to enter these data to a transactions file will be needed. The payroll program will access transaction, year-to-date, and personnel

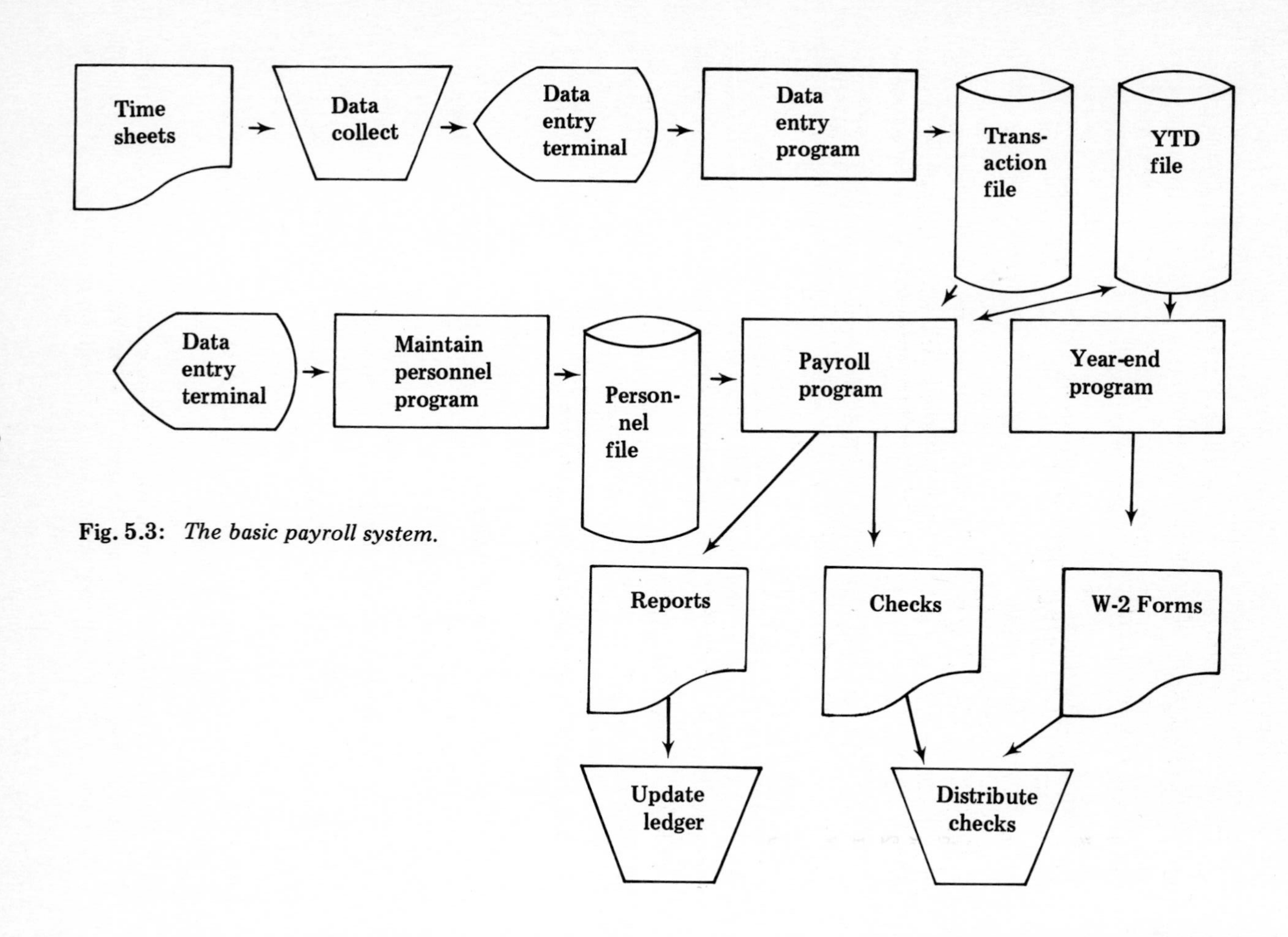

Fig. 5.3: *The basic payroll system.*

files from disk, and will write the checks and a variety of reports. A year-end program will generate W-2 forms from the data stored on the year-to-date file. Another new program will be needed to maintain the personnel file; a payroll clerk will enter the data via CRT. Figure 5.4 lists the physical elements—programs, manual procedures, files, and forms—that will compose this system; we'll assume that the hardware already exists.

What advantages are gained by outlining the physical components of the system? Several. These components will actually be implemented, and only by defining them can the analyst begin to plan their implementation. During the early stages of the design process, the analyst was working with a logical image of the system—pure ideas. Now, time and cost estimates can be based on concrete physical components, rather than on imaginary functions. The analyst can envision a payroll program instead of loosely defined processes. Instead of data stores, specific files or a data base will be accessed. Each element of the system can be studied, one at a time; the system cost and the system schedule are compiled by summing the component costs and schedules.

The cost/benefit analysis for the basic payroll system (Fig. 5.5), presents management with a major decision: make or buy. Option A assumes that all software will be purchased at a total system cost of $9000 ($1000 below the initial estimate). If original software is written (Option B of Fig. 5.5), the development cost will increase to $10,250, but that's not the whole story. The analyst believes that a well-structured, internally developed program will be easier and cheaper to maintain than purchased software; thus, operating costs should drop from an estimated $1750 to $1500 per year. The computed return on investment (see Module G) for Option A is 23 percent; for Option B, it's 20 percent; given the inaccuracies inherent in estimating costs, those

Fig. 5.4: *The physical elements of THE PRINT SHOP's payroll system.*

Programs

- Data entry
- Payroll
- Year-end
- Maintain personnel

Manual Procedures and Training

- Data collection
- Update ledger
- Distribute checks
- Use of data entry terminal

Files

- Transaction
- Year-to-date
- Personnel

Forms

- Time sheets
- Reports
- Checks
- W-2 forms

Fig. 5.5: *A cost/benefit analysis for the basic payroll system.*

Development Costs

Prior estimate (from feasibility study)			**$10,000**
New estimate (system design phase)			
Option A: Purchase all software			
Already spent (labor)		$4000	
Detailed design		1000	
Implementation and training		2000	
Purchase price—software		2000	**$ 9,000**
Option B: Write original software			
Already spent (labor)		$4000	
Detailed design			
Data entry	$ 250		
Payroll	1000		
Year-end	250		
Personnel	500		
File design	250		
Forms design	250	$2500	
Inplementation			
Data entry	$ 250		
Payroll	2000		
Year-end	250		
Personnel	500		
Training	500		
File creation	250	$3750	**$10,250**

Operating Costs (Annual)

	Option A	*Option B*
Labor	1250	1250
Maintenance	500	250
TOTAL	**$1750/year**	**$1500/year**

Cost Savings and Return on Investment

	Existing	*Option A*	*Option B*
Operating cost	5000	1750	1500
Cost savings	---	3250	3500
Investment	---	9000	10,250
Return on investment	---	23%+	20%+

numbers are very close. (Note: the $4000 identified as already spent is for the problem definition, feasibility study, analysis, and system design steps.)

Other non-economic arguments might be advanced for writing original programs, too. THE PRINT SHOP currently employs one programmer/operator, a young man who was promoted from the stock room two years ago. Writing a system like payroll could be an excellent training exercise for him. The schedule (Fig. 5.6) shows that, while only six weeks of labor will be billed under Option A, twelve weeks of elapsed time is expected—there is simply no way to control the schedules of people who work for other organizations. Option B will involve the efforts of two employees, the analyst and the programmer/operator. Although a total of 12.5 weeks will be billed, careful scheduling could make the payroll system available sooner. We'll return to this idea later in the chapter.

As an alternative, a microcomputer might be used for data collection and verification (Fig. 5.7). Earlier, management rejected the use of an on-line terminal because the main business of THE PRINT SHOP is printing and design, not payroll. This microcomputer is different, however. It is an independent computer that need not be linked to the central minicomputer. As employees enter labor data through the specially designed terminal unit, these data are verified and stored on a disk file. At

Fig. 5.6: *Implementation schedule for the basic system. All time estimates are shown in weeks.*

	Option A		Option B
	Labor	elapsed	
Detailed design	2.0	4.0	
Data entry			0.5
Payroll			2.0
Year-end			0.5
Personnel			1.0
File design			0.5
Forms design			0.5
Implementation	4.0	8.0	
Data entry			0.5
Payroll			4.0
Year-end			0.5
Personnel			1.0
Training			1.0
File creation			0.5
Totals	6.0	12.0	12.0

Fig. 5.7: *An alternative is to use an on-line data collection terminal driven by its own microcomputer.*

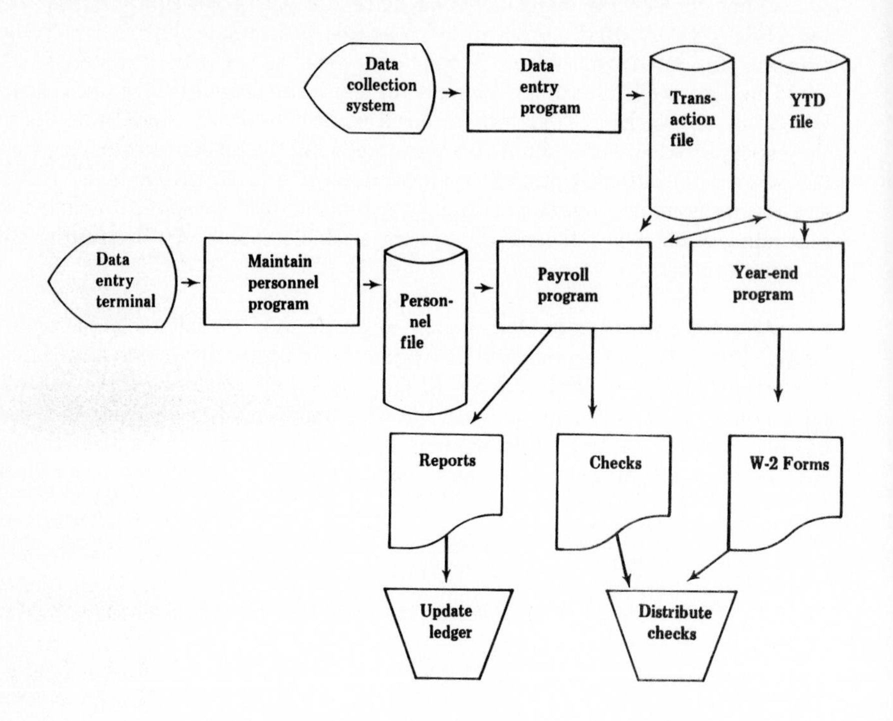

the end of the week, this disk file can be carried to the central computer, and used as input to the payroll program.

Since only the data entry portion of the basic payroll system is changed, the analyst can support this option with an incremental cost/benefit analysis and schedule (Fig. 5.8). The microcomputer system will cost $8500, and the data entry program should cost $2000 more than the simple version envisioned in the basic system; thus the incremental development cost is $10,500. Savings will come from an unusual source. In discussing the system with the payroll clerks and the chief accountant, the analyst sensed that paperwork is increasing throughout THE PRINT SHOP. The demands of the new payroll system will add to that workload, but installing a data collection system could relieve the clerks of some of their burden. A third clerk will have to be hired soon; the data collection terminal might allow this action to be postponed for at least six months, saving THE PRINT SHOP $10,000 in salary that will not have to be paid. Coupled with an estimated $500 in annual labor savings, the return on investment for the data entry system lies between 13 and 14 percent. While the analyst does not intend to recommend this option, management should be aware of it. If the basic system is designed with future expansion in mind, microcomputer-controlled data entry can perhaps be integrated later.

Fig. 5.8: *An incremental cost/benefit analysis for adding on-line data collection to the payroll system.*

Incremental Development Costs (Investment)		
System purchase	$8500	
Data entry program	2000	**$10,500**
Incremental Cost Savings		
1-time personnel savings	**$10,000**	
Annual labor savings	**500/year**	

Return on investment lies between 13% and 14%.

Schedule Changes	
System delivery	8 weeks
Data entry program	
Planning	2 weeks
Coding	2 weeks
Add total of	12 weeks to schedule

As a third option, the analyst suggests that the check writing process be broken into two steps (Fig. 5.9). The information for a check will first be written to disk. After the payroll is complete, the contents of this disk file will then be dumped to produce the paychecks. The primary reason is security. Printing checks is a weak spot, and a dishonest person could conceivably slip an extra check or two through the system. If computing and printing are separated, however, the computer room can be cleared of all personnel except the operator, the chief accountant, and an auditor as soon as the payroll program is complete. The number of checks written to disk can then be compared with the expected check count based on accounting records, and the correct number of prenumbered blank checks can be loaded on the printer. With such controls, printing bogus checks is very difficult.

Fig. 5.9: *Another alternative is to write an intermediate check data file, and write the actual checks through an independent program. This is a security feature.*

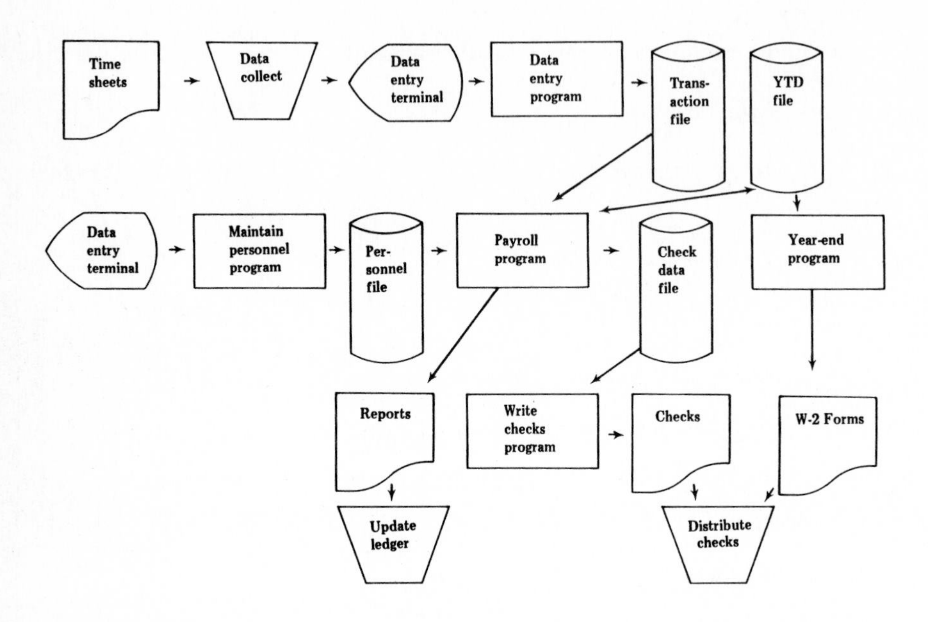

What would this option cost? Very little. The check writer program is a simple module that reads, reformats, and writes records; a few counters and accumulators add little to the complexity. The analyst would have to develop the check formats, counters, and accumulators whether the separate print module were used or not, so almost no additional development time will be needed. Programming might require an extra half week, but the incremental cost is almost not worth considering. This is a good addition to the system, and the analyst is prepared to recommend it.

The Analyst's Recommendation

The analyst feels that the basic system should be implemented, along with the separate check writing program. The data collection microcomputer is interesting, but the extra $10,500 investment violates the system's intended scope. Perhaps it can be added later. Additionally, the analyst believes that THE PRINT SHOP should write rather than purchase the payroll programs. The cost/benefit analyses for both options are comparable, but the cost and time estimates associated with Option B seem more dependable. Furthermore, the need to train the programmer/operator is an important argument; without a realistic career path, he may leave for a better job.

As the final step in the system design process, a detailed implementation plan for the recommended option is prepared. The key to this plan is a *project network* (Fig. 5.10: see Module L). The circles represent events or milestones in the design

Fig. 5.10: *A project network for THE PRINT SHOP's payroll system. Times are expressed in weeks. The critical path is highlighted with a dark line.*

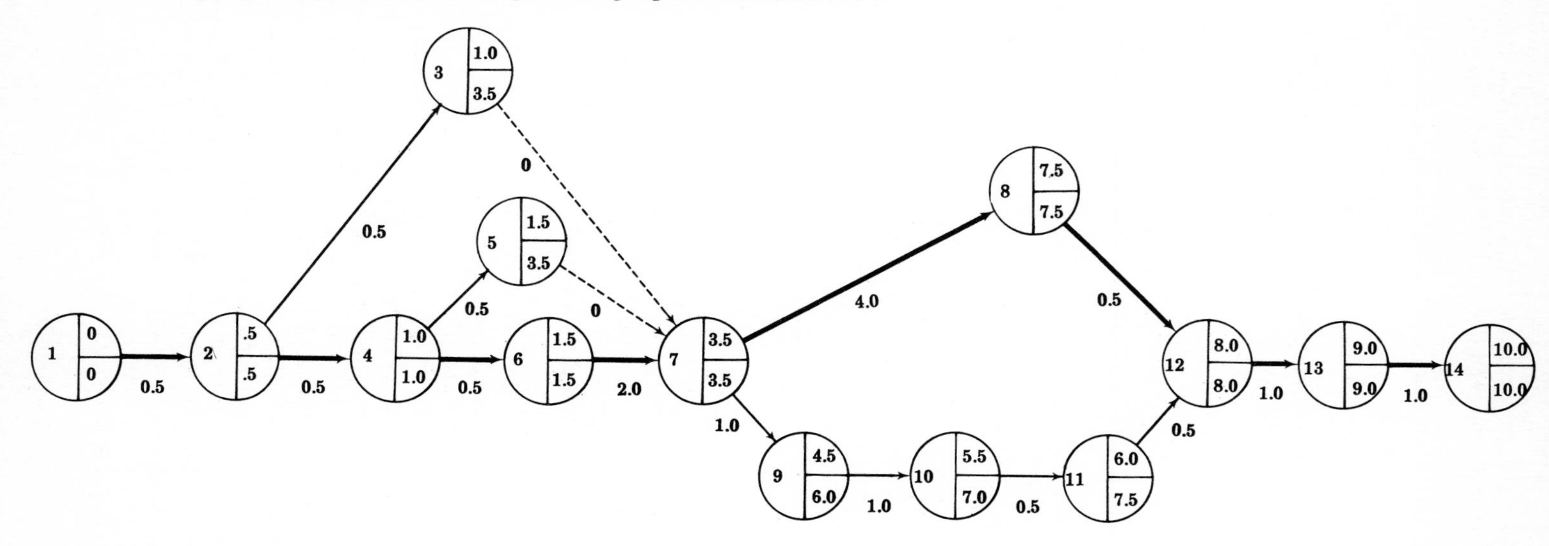

1-2	*File design (analyst)*	*7-8*	*Write payroll program (programmer)*
2-3	*File creation (programmer)*	*7-9*	*Design personnel program (analyst)*
2-4	*Design data entry program (analyst)*	*8-12*	*Write check writer program (programmer)*
3-7	*Dummy*	*9-10*	*Write personnel program (programmer)*
4-5	*Write data entry program (analyst)*	*10-11*	*Design year-end program (analyst)*
4-6	*Design payroll reports (analyst)*	*11-12*	*Write year-end program (analyst)*
5-7	*Dummy*	*12-13*	*System test (both)*
6-7	*Design payroll program (analyst)*	*13-14*	*Training (analyst)*

and implementation of the system; for example, the event numbered 12 marks the completion of the check writer and year-end programs and the beginning of the system test. The lines linking the events represent activities; for example, between events 7 and 8 (activity 7-8) is the process of writing the payroll program. Note the numbers under the lines; they define the duration (in weeks) of the activity. A list of activities is shown at the bottom of Fig. 5.10.

The project network graphically illustrates the order in which events must occur, and also shows how those events are related. Follow the flow through Fig. 5.10. Begin with event 1. The first activity is to design the files (activity 1-2). Once his step has been completed, the analyst can begin to design the data entry program (activity 2-4) while, at the same time, the programmer/operator runs the utility programs that create the files (activity 2-3). Event 4 occurs when the analyst completes the design of the data entry program. At this point, the programmer can begin to write the data entry program (activity 4-5), while the analyst works on designing the payroll reports (activity 4-6). Using the event descriptions at the bottom of Fig. 5.10 as a guide, follow the rest of the project network yourself. Why must events occur in a certain order? When do the analyst and the programmer work in parallel? Once you learn how to read a project network, it can convey a wealth of information.

Note the two numbers recorded at the right of each event. On top is the earliest possible event time; on the bottom is the latest time an event can occur without impacting the expected project completion time (see Module L for an explanation of how to compute these numbers). Event 1 occurs at time zero, while the earliest possible completion time for the last event (number 14) is ten weeks. If this schedule is followed, the payroll system can be delivered in ten weeks. During the design and implementation stages, this project network will prove a valuable control tool. The dark path through Fig. 5.10 represents the critical path; if an event along the critical path is late, the system will be late. With such a document serving as an early warning device, management should have little trouble monitoring progress.

THE INSPECTION AND THE MANAGEMENT REVIEW

The analyst has prepared a set of formal exit criteria, including a system flow diagram, list of physical components, cost/benefit analysis, and rough schedule for each of several alternatives. A project network has also been prepared for the recommended option. To this point, the analyst has worked alone. The next two steps (detailed design and implementation) will involve the programmer/operator and a considerable amount of computer time as well; the cost will begin to accelerate. Before committing additional funds, management wants to be sure that the proposed system really meets the user's needs and can actually be implemented. Thus, a formal inspection is scheduled. Both user and technical representatives are on the inspection team. Carefully, they read through the exit criteria. Will the proposed system do the job? Yes, says the user, it makes sense. Can it be implemented? Certainly, says the programmer. Are the cost estimates reasonable? Does the projected time frame make sense? Have obvious alternatives been ignored; in other words, is the analyst's recommendation based on bias, rather than on careful consideration of the facts? Following the inspection, it is clear that the analyst has done a good technical job on the payroll system.

A management review comes next. The owner, Carl Jones, his daughter Carla (the manager of both the analyst and the programmer/operator), and the chief accountant are all present. They are impressed with the analyst's work, and inclined to accept the recommendation. Can the programmer/operator be spared for the time demanded by the schedule? Would it be better to limit him to perhaps three-quarters time so that his other responsibilities don't suffer? This would, of course, change the estimated completion time; does accounting really want the system within ten weeks? Note how the documentation provided by the analyst allows management to weigh the consequences of such decisions.

Let's assume that management's decision is to support the analyst's recommendations without change. Funds have now been authorized for detailed design, and, unless unanticipated factors are uncovered, will almost certainly be authorized for implementation. The system flow diagram is a blueprint of the proposed physical system. The data flow diagram, data dictionary, and algorithm descriptions (from an earlier step) provide additional guidance. Thus it's on to detailed design.

Note how the step by step approach of structured systems analysis and design allows for clear management control of a project. Each step ends with an inspection that assures technical accuracy. Given the results of the inspection, management can ignore the technical details and concentrate on funds and schedule, the true responsibilities of management.

SUMMARY

In this chapter, we studied the system design phase of THE PRINT SHOP's payroll project. Using the data flow diagram (from analysis), the analyst outlined a series of possible automation boundaries, and used them to suggest physical implementation strategies. Several reasonable alternatives were then identified, and a system flow diagram, list of physical components, cost/benefit analysis, and rough schedule were prepared for each. A project network was prepared for the recommended strategy, and exit criteria were subjected to an inspection and to management review. In the next stage, detailed design, these exit criteria will be used as a blueprint of the system.

EXERCISES

1. What is the objective of system design?

2. In the chapter, the documentation prepared during system design was compared to a blueprint of the system. What parallels can you see between system design documentation and a blueprint?

3. What is an automation boundary? Explain how automation boundaries can be used to suggest alternative physical implementation strategies.

4. Figure 5.2 illustrated a set of automation boundaries for the payroll system. The chapter then described the process of generating ideas for implementing the logic enclosed by one boundary containing the data collection and verification steps. Suggest alternative strategies for the other automation boundaries of Fig. 5.2. After preparing a list, decide if any of your strategies are reasonable, and thus worthy of further consideration.

5. Describe each of the system design exit criteria. What is the purpose of each document? Why is each important?

6. Develop a set of exit criteria for the alternatives you generated in exercise 4.

7. Imagine that THE PRINT SHOP'S computer system used tape rather than disk as secondary storage. How would this change the system flowchart?

8. Draw a system flowchart showing the equipment in your school's computer center.

9. Why is it so much easier to accurately estimate a project's cost and schedule at the end of system design?

10. Imagine that a second programmer were available to work with the analyst and the programmer/operator on the payroll project. How might this change the project network of Fig 5.10? Develop a new project network to reflect the new schedule.

11. The structured approach to systems analysis and design allows management to control the system development process. How? Why is this important?

12. Throughout the system design step, the analyst made it a point to communicate with technical people as well as the user, and both user and technical personnel were on the inspection team. Why?

13. Continue with Chapter 4, exercise 13.

14. Continue with Chapter 4, exercise 14.

15. Continue with Chapter 4, exercise 15.

6

Case A Detailed Design

OVERVIEW

Continuing with the payroll system example, this chapter describes detailed design. The objective is to determine how, specifically, the system should be implemented. After a brief discussion in which the system design documentation is compared to a blueprint, the chapter turns to file design and the development of test data. Next, the payroll program is designed using the HIPO technique. A high-level hierarchy chart is drawn and, gradually, through functional decomposition, a detailed hierarchy chart is developed. IPO (Input/Process/Output) charts are then prepared for each module on the hierarchy chart. The resulting documentation is checked against the data flow diagram, the data dictionary, and the algorithm descriptions. A structure chart is prepared to help the analyst evaluate coupling and module independence. The detailed design of the payroll program ends with a formal technical inspection.

DETAILED DESIGN

System design has just been completed. The physical components making up the system—programs, files, forms, manual procedures, and (in many cases) hardware—have been identified. How, specifically, should the system be implemented? The time has come to plan the details.

The system design exit criteria represent the starting point for detailed design. Earlier, we compared them to a blueprint. To extend the analogy, consider the process of designing and constructing a house. A blueprint provides a general plan for the location of the walls, plumbing, heating ducts, electrical wires, and other components. Using it as a guide, the contractor divides the work among numerous carpenters, plumbers, electricians, and other subcontractors. The blueprint allows the contractor to plan and coordinate their activties.

The system flowchart (Fig. 6.1) serves a similar purpose. Specific physical components are identified, and their interrelationships defined; thus the work can be divided among several analysts or programmers. With several people working on the project, the schedule can be advanced, and the user can have the system sooner. The project network of Fig. 5.10 shows how the analyst plans to divide the work.

Fig. 6.1: *A system flowchart of the payroll system.*

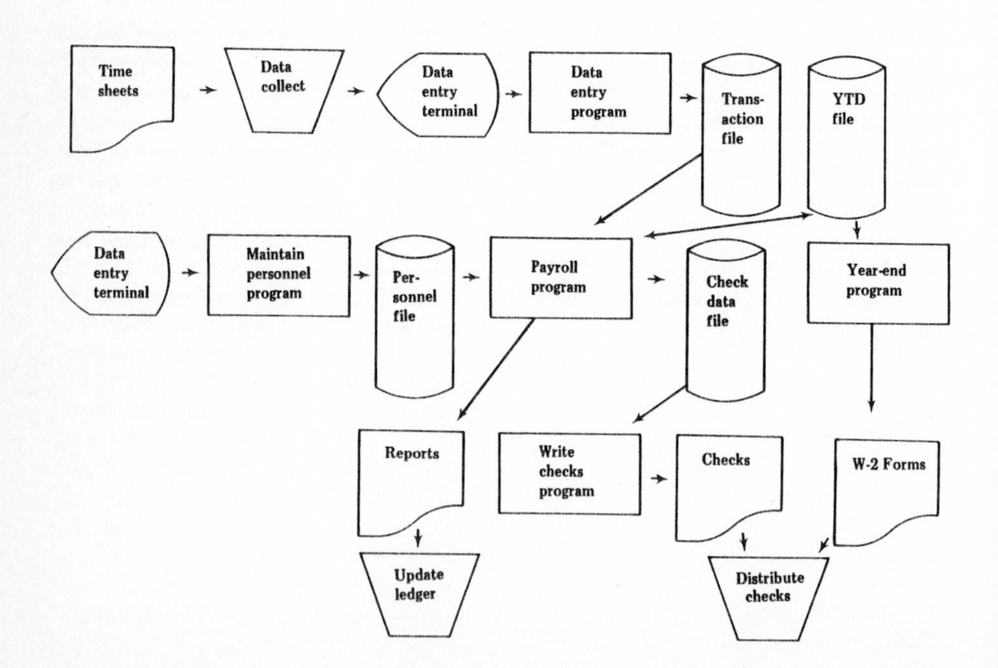

The analyst and the programmer/operator will work together in designing and implementing the payroll system. The analyst will begin by designing a program using the *HIPO* (for *Hierarchy plus Input/Process/Output*) technique (see Module H). Following a technical inspection, the programmer will write, debug, and document the code. Meanwhile, the analyst will design the next program. The project network and the system flow diagram will help to coordinate their efforts. After all the programs have been designed and written, a formal system test will be conducted.

DESIGNING THE FILES AND THE TEST DATA

When building a home, certain tasks (such as installing the plumbing and the wiring) can be done in parallel. Others, however, must be performed in a fixed sequence; for example, the foundation must be constructed first, and the drywall must be installed before the painters can begin. Creating a system presents a similar problem. Certain functions represent the foundation on which the rest of the system is constructed, so they must be completed first. Where should the analyst begin?

Consider the system flowchart of Fig. 6.1. It identifies many components including programs, procedures, files, and forms. What is the framework that links all these components? The *data*. The data entry program produces the transaction file, which is input to the payroll program. The year-to-date file is common to both the payroll and year-end programs. The programs cannot be written until the data are defined. The files (or the data base) must be designed first.

The starting point is the data dictionary. The records, data structures, and elements to be stored on each file are identified and extracted. Next, the analyst codes source statements for the record descriptions and other data structures, and places them on a source statement library. Eventually, the programmer will incorporate this source code into the various programs, thus assuring consistency and simplifying the coding process.

The transaction, year-to-date, personnel, and check data files will all be stored on disk. Magnetic disk is a limited resource; only so much room is available. How much space will each file require? Will the files all fit? Given the record formats and an estimate of the number of records on each file, the analyst can easily find out (see Module M for details). We'll assume that space is not a problem, and move on.

Many systems require command language or job control language statements to formally describe each program and each input or output device a program accesses. The system flowchart (Fig. 6.1, again) is a graphic representation of these physical system components, and can thus be used as a checklist for coding the job control language or for naming the programs and files. (Note: the rules vary considerably from system to system.) The analyst is responsible for describing the system formally; the programmer should not have to worry about job control statements or file naming conventions. The reports, checks, and time sheets should also be defined before the programs and procedures are planned (see Module N for suggestions).

Finally, before the analyst turns to the programs, test data must be generated. Why? Without good test data, there is simply no way to know if a program or a system is correct until after it is released. Imagine the reaction of management and the user if the fact that the payroll system overpaid everyone by one thosand dollars were not discovered until after the first checks were printed and cashed! Imagine, to use a less dramatic example, if the system's inability to handle bonus payments were not discovered until the third week of use. Good, realistic test data can help to minimize such errors. Unfortunately, the test data are often overlooked or treated as an afterthought. Don't make this mistake.

Good test data must be realistic; for example, the data used in processing THE PRINT SHOP's most recent payroll might serve as an excellent base. There is one problem with historical data, however; they rarely cover everything that could possibly go wrong. In order to test the entire program, it is essential that the test data reflect other than normal conditions. Thus, additional data representing extreme conditions are added. Some sets of test data should contain high values for all data elements. Others should contain low values, while still others should mix high and low values. The user might suggest some extreme conditions, too. Include bad data; for example, bury an occasional comma in the middle of a numeric field. A good system should be able to handle (or at least flag) anything it might get.

It is important that the user be involved in generating and defining the test data. Systems are designed for users, not for programmers and analysts. The user is the customer. When it comes to the application, the user is the expert. The analyst has no more right to dictate the test data than the user has to define the code. The argument that "the user doesn't understand" is little more than technical chauvinism.

Once the files have been designed and the test data specified, the files can be created and initialized with the test data. Meanwhile, the analyst can turn to designing the program.

DESIGNING THE PAYROLL PROGRAM

Constraints of time and space do not permit us to investigate the design of each program in the payroll system. Instead, we will concentrate on one, the payroll program; the others follow a similar pattern. The HIPO (Hierarchy plus Input/Process/Output) technique will be used to document the analyst's plan; see Module H for details.

The High-level Hierarchy Chart

In designing the system, the analyst used a set of automation boundaries drawn on the data flow diagram to suggest implementation strategies. The payroll program represents one of those automation boundaries (Fig. 6.2). Note that six primary functions can be identified (processes 3.1 through 3.6); clearly, the payroll program must perform these functions. The first step in designing the program is to draw a high-level *hierarchy chart* showing the primary functions as defined on the data flow diagram (Fig. 6.3).

What exactly does the hierarchy chart of Fig. 6.3 mean? It describes a control structure. On top, is the main control module, *Process payroll*. Below it are six functional modules. The main control module determines the order in which the lower level modules are executed; each lower level module performs a single function and then gives control back to the main control module. The payroll program begins with *Process payroll*. It tells *Get data* to execute. In response, this module performs its

Fig. 6.2: *The payroll program is a physical representation of one of the analyst's automation boundaries. The processes define key functions of the program, while the data stores and data flows define inputs and outputs.*

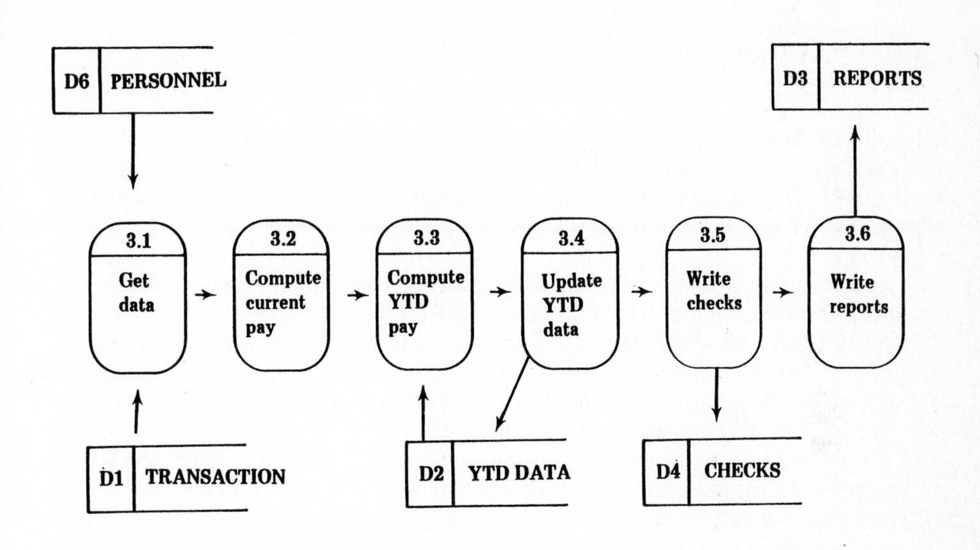

Fig. 6.3: *A high-level hierarchy chart of the payroll program.*

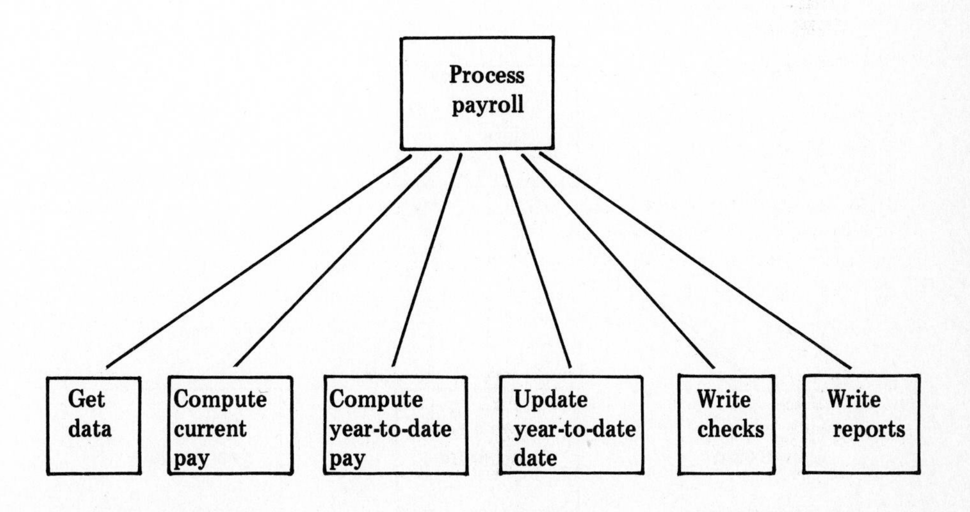

function and sends control back to *Process payroll*. Next, the main control module transfers control to *Compute current pay* which, as you may have guessed, computes the current pay. Then it's back to *Process payroll*, which invokes the next of the lower level modules. To use a military analogy, the main control module is a general, and the six other modules carry out the general's orders. Control and coordination are exercised at the top; tasks are performed at the bottom.

Functional Decomposition

By itself, the initial hierarchy chart is not very useful; it is just a starting point. Consider the six level II functions of Fig. 6.3. They are very general. What, exactly does *Get data* mean? What does *Compute current pay* mean? If a program is to be written, these questions must be answered; thus the analyst must break down each function to a lower level of detail. The process is known as *functional decomposition*.

For example, consider *Get data*. What functions must be performed (directly or indirectly) by this module? The data flow diagram (Fig. 6.2) shows data flowing from TRANSACTION and PERSONNEL. These data stores become files in the system design, and *Get data* should read a record from each of the files. Let's assume that during system design (after the data flow diagram was completed), the analyst learned of a payroll operating procedure: an employee's pay is not to be computed unless all data are available. Using this rule, transaction, personnel, *and* year-to-date data would all be needed to compute pay. Although the data flow diagram shows the year-to-date data entering the system at process 3.3, new knowledge always supersedes old; thus *Get data* should read a record from the year-to-date file as well. The functions performed under control of *Get data* are summarized in Fig. 6.4.

Fig. 6.4: *A functional decomposition of the get data module.*

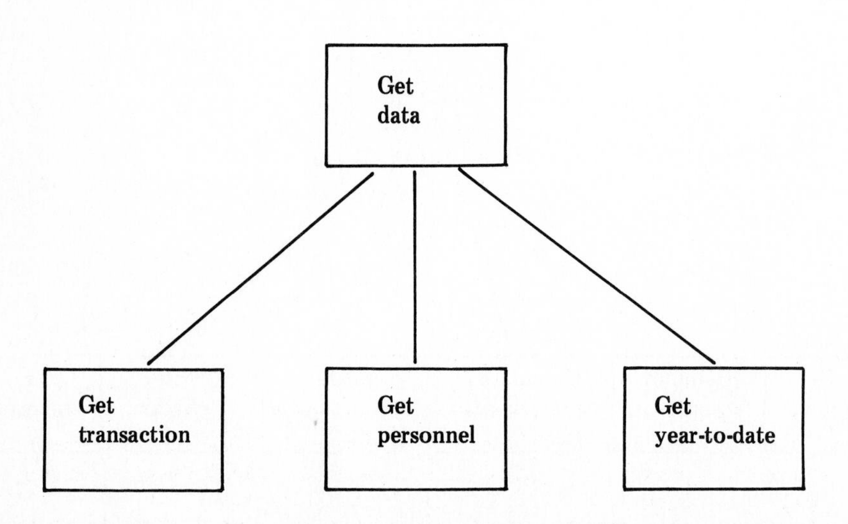

Should any of these functions be further decomposed? How complex is the logic needed to "get a transaction"? Not very. Basically, each of the lower-level functions of Fig. 6.4 is obvious to most programmers. No further decomposition is necessary.

Next, the analyst moves to *Compute current pay*. What must be done in this module? The algorithm descriptions prepared during analysis provide a clue. Each algorithm represents one function, and each function should occupy a separate module. The key steps are summarized in Fig. 6.5; other deductions might, of course, be added, but we'll assume that THE PRINT SHOP runs a very basic payroll operation.

Should any of these functions be further decomposed? Is the logic obvious? Consider, for example, *Compute local tax*. Often, local tax is a fixed percentage of gross pay; if this is the case, the logic is obvious and further decomposition is unnecessary. What about *Compute gross pay*, however? Basically, gross pay is the product of hours worked and an hourly pay rate, but what about overtime? How are bonuses computed? Are Sunday and holiday hours paid at a special rate? What happens with salaried employees? How is vacation pay handled? Given so many unanswered questions, the logic of *Compute gross pay* is clearly not obvious, and additional decomposition is necessary. Figure 6.6 shows a functional decomposition of *Compute gross pay*.

IPO Charts

Functional decomposition generates numerous modules, each one representing a set of detailed logic. To keep track of these details, an *IPO (Input/Process/Output) chart* is prepared for each module on the hierarchy chart. (Note: at THE PRINT SHOP,

Fig. 6.5: *A decomposition of the compute current pay module.*

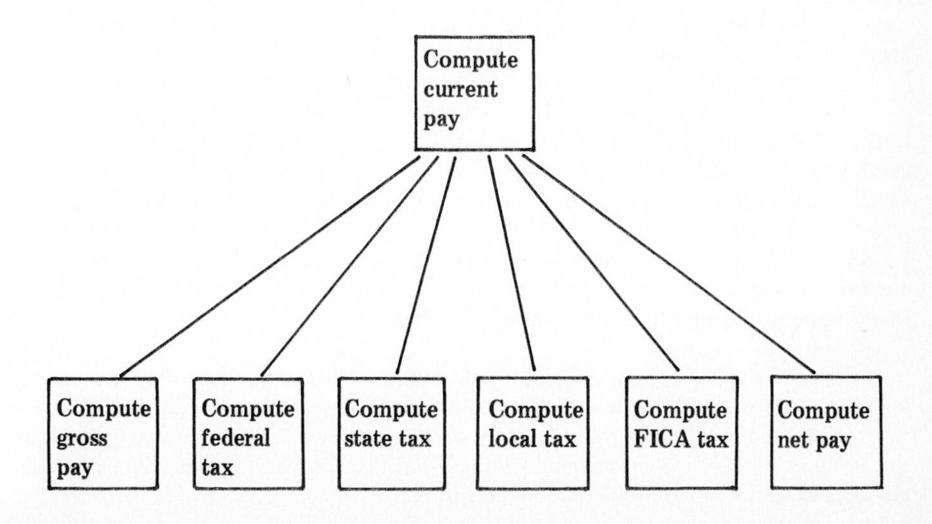

Fig. 6.6: *A further decomposition of the compute gross pay module.*

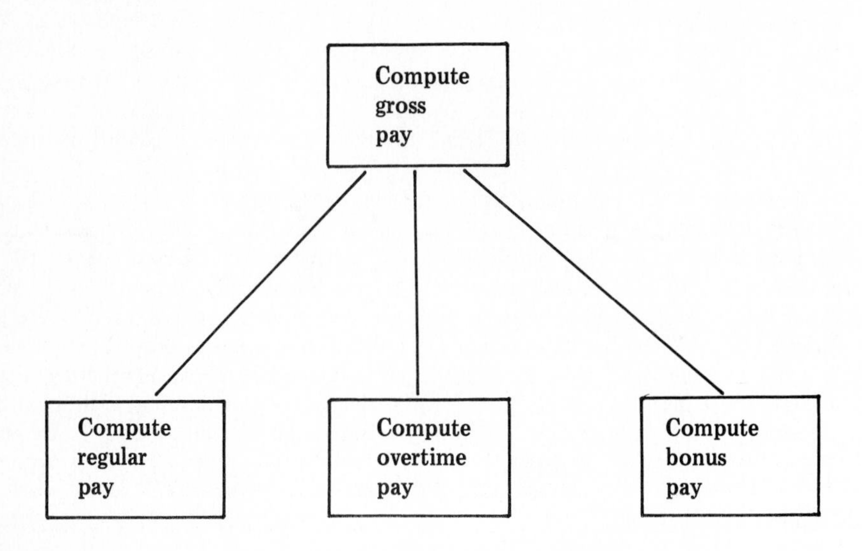

algorithm descriptions were compiled on partially completed IPO charts. We'll be using them shortly.)

An IPO chart shows the inputs to, outputs from, and process performed by a routine. For example, consider *Compute gross pay* (Fig. 6.7). The top blocks show how the module fits into the program hierarchy. *Compute gross pay* is called or invoked by a higher level module, *Compute current pay*. It calls or invokes three lower level modules. Next come the inputs and outputs. *Compute gross pay* accepts values for hours worked, hourly pay rate, and bonus from *Compute current pay*, and sends back a value for gross pay.

The large block just below the middle of the IPO chart describes, in structured English (see Module H), the process performed by this module. First, regular pay is computed. If hours worked is greater than forty, overtime pay is then computed. Finally, if a bonus is shown on the transaction, the bonus pay is computed. As alternatives to structured English, many analysts prefer pseudo code (Module I) or logic flowcharts (Module J), and some use a combination. The key idea is to clearly and concisely describe the process performed by a module, not to force every process to fit the structure of a particular documentation form.

Figure 6.8 shows the IPO chart for *Compute regular pay*. Note that it is called by *Compute gross pay*, and that it calls no lower level modules. Inputs are hours worked and the hourly pay rate; the only output is the computed gross pay, which is the product of the two input parameters. Given the IPO chart, here should be little question as to the function the module performs.

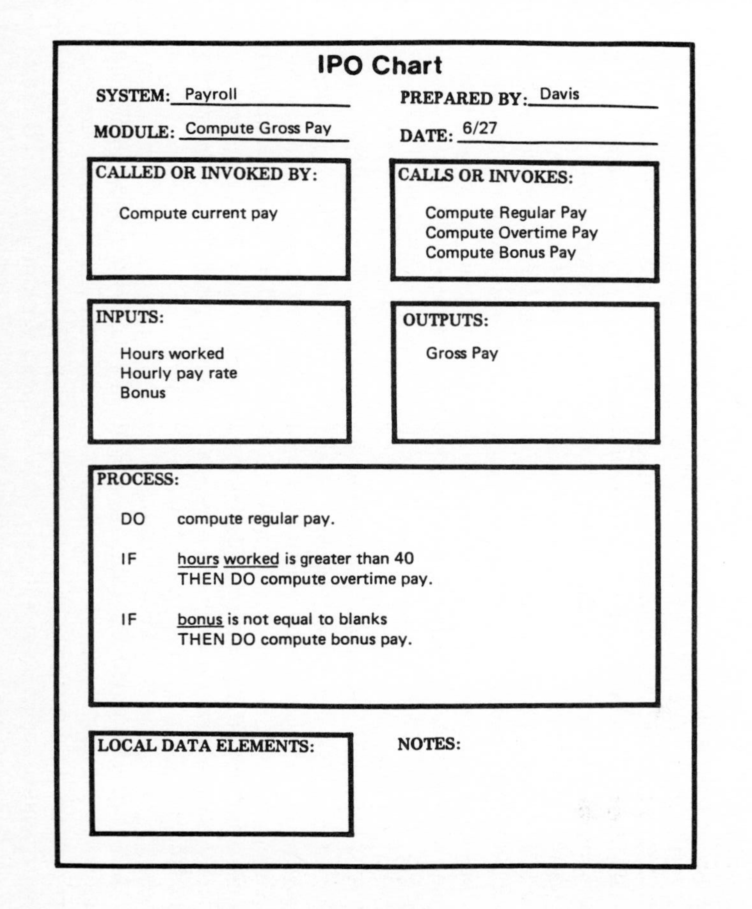

IPO Chart

SYSTEM: Payroll
PREPARED BY: Davis
MODULE: Compute Gross Pay
DATE: 6/27

CALLED OR INVOKED BY:
Compute current pay

CALLS OR INVOKES:
Compute Regular Pay
Compute Overtime Pay
Compute Bonus Pay

INPUTS:
Hours worked
Hourly pay rate
Bonus

OUTPUTS:
Gross Pay

PROCESS:
DO compute regular pay.
IF hours worked is greater than 40
THEN DO compute overtime pay.
IF bonus is not equal to blanks
THEN DO compute bonus pay.

LOCAL DATA ELEMENTS:

NOTES:

Fig. 6.7: *An IPO chart for the compute gross pay module.*

Fig. 6.8: *The compute regular pay module.*

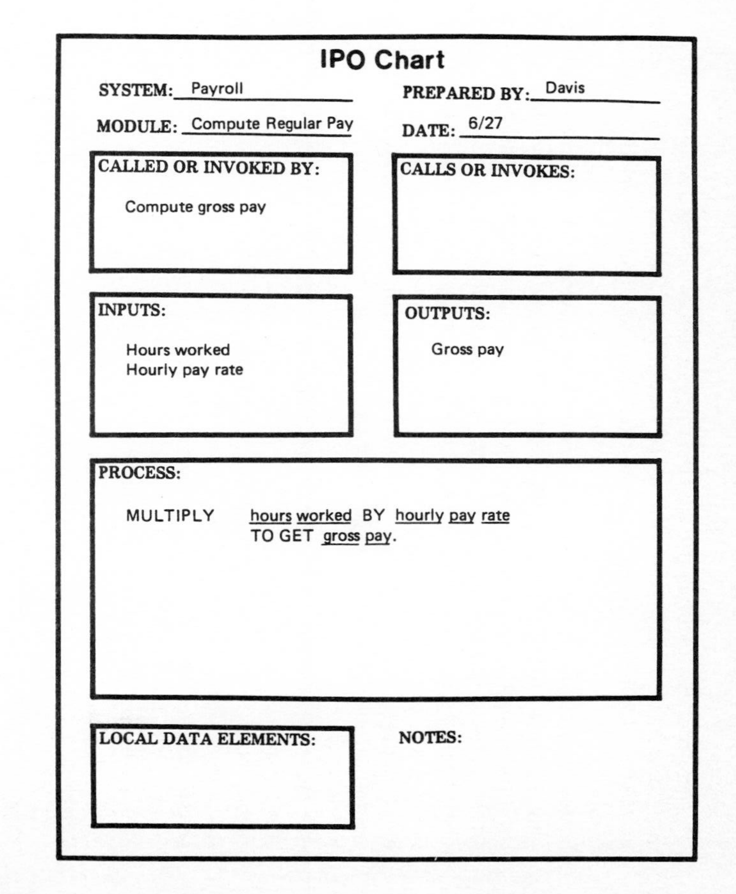

IPO Chart

SYSTEM: Payroll
PREPARED BY: Davis
MODULE: Compute Regular Pay
DATE: 6/27

CALLED OR INVOKED BY:
Compute gross pay

CALLS OR INVOKES:

INPUTS:
Hours worked
Hourly pay rate

OUTPUTS:
Gross pay

PROCESS:
MULTIPLY hours worked BY hourly pay rate
TO GET gross pay.

LOCAL DATA ELEMENTS:

NOTES:

The process for *Compute overtime pay* (Fig. 6.9) is a bit more complex. Hours over forty have already been paid at the regular rate. The rule is that such hours must be paid at "time and a half," so the extra half time bonus must be computed. This involves the use of several *local* data elements, which are listed in the block near the bottom left of Fig. 6.9. What exactly is a local data element? It is a data element used only within a single module. View *Compute overtime pay* as a black box. To use it, the programmer must know about the inputs (hours worked, hourly pay rate, and gross pay) and the outputs (gross pay), but the local data elements can be ignored. They hold the results of intermediate computations, and are irrelevant to the rest of the program. In contrast, the inputs to and outputs from this module are used by many other modules; they are *global*. Global data are shared by several modules. Local data are defined, stored, and used exclusively within a single module.

At the bottom right of the IPO chart is space for notes (see Fig. 6.9). The algorithm used for overtime pay is a bit unusual, in that hours over forty have already been paid at the regular rate and only the overtime bonus must be computed; the note clarified the logic by describing the analyst's approach. Notes are not always needed, but can be helpful.

The IPO chart for *Compute bonus pay* is shown as Fig. 6.10.

Cohesion

Figure 6.6 shows the decomposition of *Compute gross pay* into three computational modules. Should any of these modules be further decomposed? Probably not. Earlier, we discussed a rough criterion of complexity; a programmer looking at Fig. 6.6 and at the IPO charts of Figs. 6.7 through 6.10 would certainly consider the logic obvious, and would probably regard additional decomposition as overkill. Let's state this principle more precisely.

In designing a program, an analyst should strive for *cohesion*. Each module in the program should perform a *single*, *complete* logical function. Both adjectives are important. *Compute gross pay* was subdivided because three separate algorithms are called for; that's three functions, not one. *Compute overtime pay* was not decomposed because it performs a single function, computing overtime pay. The overtime module could be decomposed into separate routines to compute the overtime premium, overtime hours, and overtime pay, but each of these low level modules would be performing only part of a complete logical function. Thus decomposition is not a good idea.

Why is cohesion important? The objective is to produce programs that are easy to document, debug, and maintain. All three objectives can be met if the program logic is easy to follow. Lengthy, multiple-function modules are confusing. Excessive modularization, with its multiple, nested calls and returns, produces equally incomprehensible code. The principle of cohesion represents a compromise. Each module must perform a single, complete function. A program composed of such modules will be easy to follow.

Several rules of thumb can guide the analyst. In a finished program, no module should exceed a single page of code. Each structured English statement represents (depending on the programming language used) four or five lines of actual code; thus, if the structured English logic will not fit in the process block of a single IPO chart,

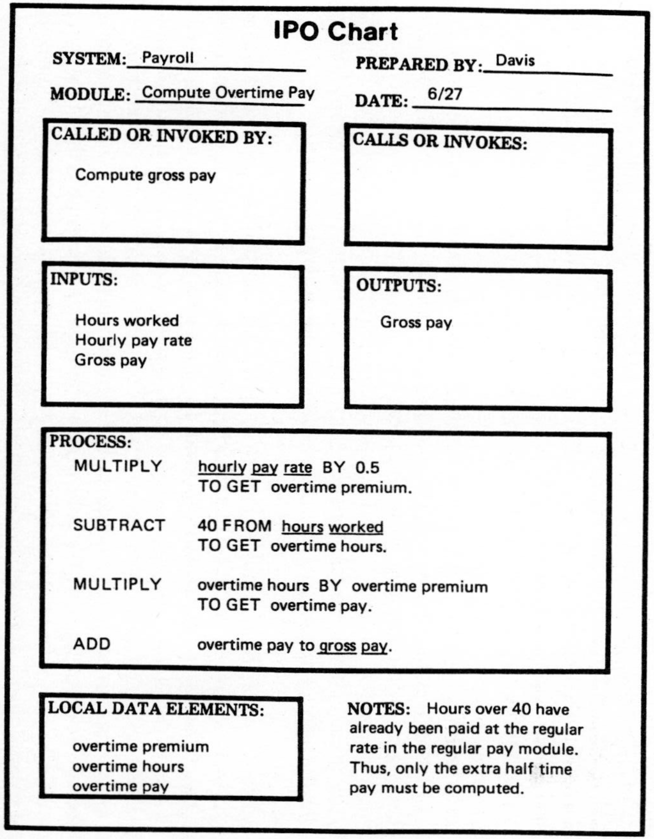

IPO Chart

SYSTEM: Payroll PREPARED BY: Davis

MODULE: Compute Overtime Pay DATE: 6/27

CALLED OR INVOKED BY:

Compute gross pay

CALLS OR INVOKES:

INPUTS:

Hours worked
Hourly pay rate
Gross pay

OUTPUTS:

Gross pay

PROCESS:

MULTIPLY	hourly pay rate BY 0.5 TO GET overtime premium.
SUBTRACT	40 FROM hours worked TO GET overtime hours.
MULTIPLY	overtime hours BY overtime premium TO GET overtime pay.
ADD	overtime pay to gross pay.

LOCAL DATA ELEMENTS:

overtime premium
overtime hours
overtime pay

NOTES: Hours over 40 have already been paid at the regular rate in the regular pay module. Thus, only the extra half time pay must be computed.

Fig. 6.9: *Compute overtime pay.*

Fig. 6.10: *Compute bonus pay.*

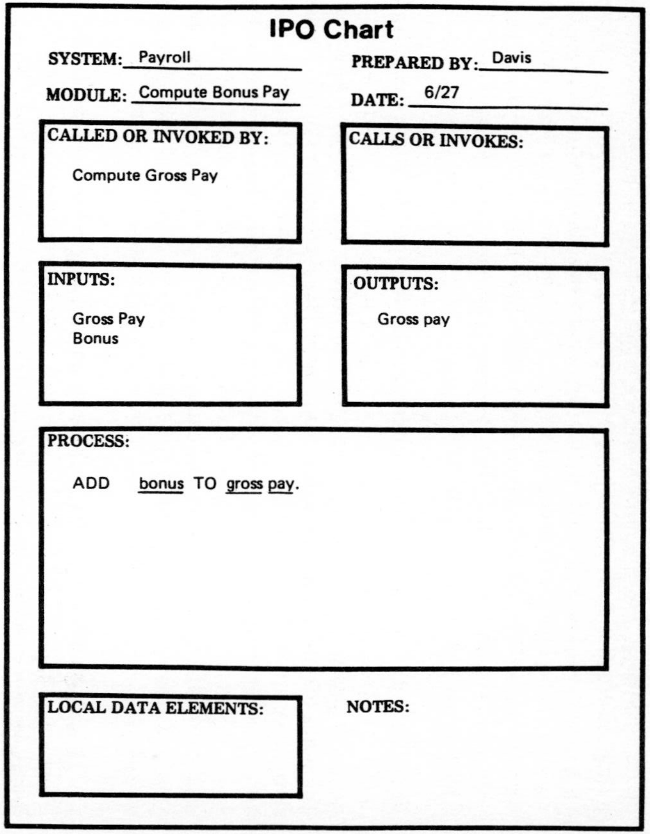

IPO Chart

SYSTEM: Payroll PREPARED BY: Davis

MODULE: Compute Bonus Pay DATE: 6/27

CALLED OR INVOKED BY:

Compute Gross Pay

CALLS OR INVOKES:

INPUTS:

Gross Pay
Bonus

OUTPUTS:

Gross pay

PROCESS:

ADD bonus TO gross pay.

LOCAL DATA ELEMENTS:

NOTES:

decomposition is probably a good idea. On the other hand, a module that consists of a single instruction or two might represent unnecessary decomposition, and moving the logic block to a higher level should be considered. However, the most important guideline is probably common sense. If decomposition makes the logic easier to follow, decompose. If decomposition complicates the logic, don't.

Functional Decomposition, Continued.

Refer back to Fig. 6.5. We have considered the decomposition of *Compute gross pay*, but what about the other modules? Let's move on to *Compute federal tax*. Should it be decomposed? The algorithm calls for two different tax tables, one for single and the other for married taxpayers. For any given taxpayer, one or the other will be used. Two independent functions suggest a need for two computational modules.

The tax tables are subject to frequent change. Good program design anticipates change. Coding the tax tables as constants would be a mistake; a better option is to create a table file, and initialize the tables as the program begins. Of course, this requires that an initialization routine be added to the payroll program.

Consider next the decomposition of another module, *Write reports* (see Fig. 6.3). What reports are to be written? The analyst has identified three: a detail report, a summary report for accounting, and an audit trail. The detail report contains data on each employee who is paid. The summary report and the audit trail, however, are written only once, at the end of the job. Thus the analyst adds a termination module to the program. Note that Fig. 6.3, the initial version of the hierarchy chart, did not anticipate a need for initialization or termination modules. As details become clear, plans must change.

The finished hierarchy chart is shown in Fig. 6.11. Start at the top with *Process payroll*. First, *Initialize* is invoked; it, in turn, invokes *Fill tables* and sends control back to *Process payroll*. Next, *Process one check* is invoked. It calls or invokes *Get data*. As a result, the three modules controlled by *Get data* are executed, and control is returned to *Process one check*, which invokes *Compute current pay*. You should be able to follow the rest of the logical flow on your own.

Why was *Process one check* added to the hierarchy chart? When one module directly controls too many lower level modules, the result can be confusion; the intermediate control module simplifies the program flow. (Note: a common standard is to limit a control module to 7±2 lower-level modules.) An IPO chart for Process payroll, the main control module, is shown in Fig. 6.12; note how it clearly distinguishes the initialize, process, and terminate phases of the program.

Each module on the hierarchy chart (Fig. 6.11) is described in detail on a single IPO chart. In THE PRINT SHOP, structured English is the standard for specifying each process, but flowcharts are sometimes used for high-level control functions, particularly when nested logic is involved. Other tools (tables, decision tables, and decision trees, for example) are used when appropriate. The objective of documentation is to explain a system, program, or module in the clearest possible way. There is a great deal to be said for consistency, for following a standard; if everyone follows the same documentation rules, everyone should be able to read the documentation. However, there are exceptions to any rule. Some types of logic are difficult to describe in struc-

Fig. 6.11: *A complete hierarchy chart for the payroll program. Note that this chart has been somewhat stylized for use in the textbook.*

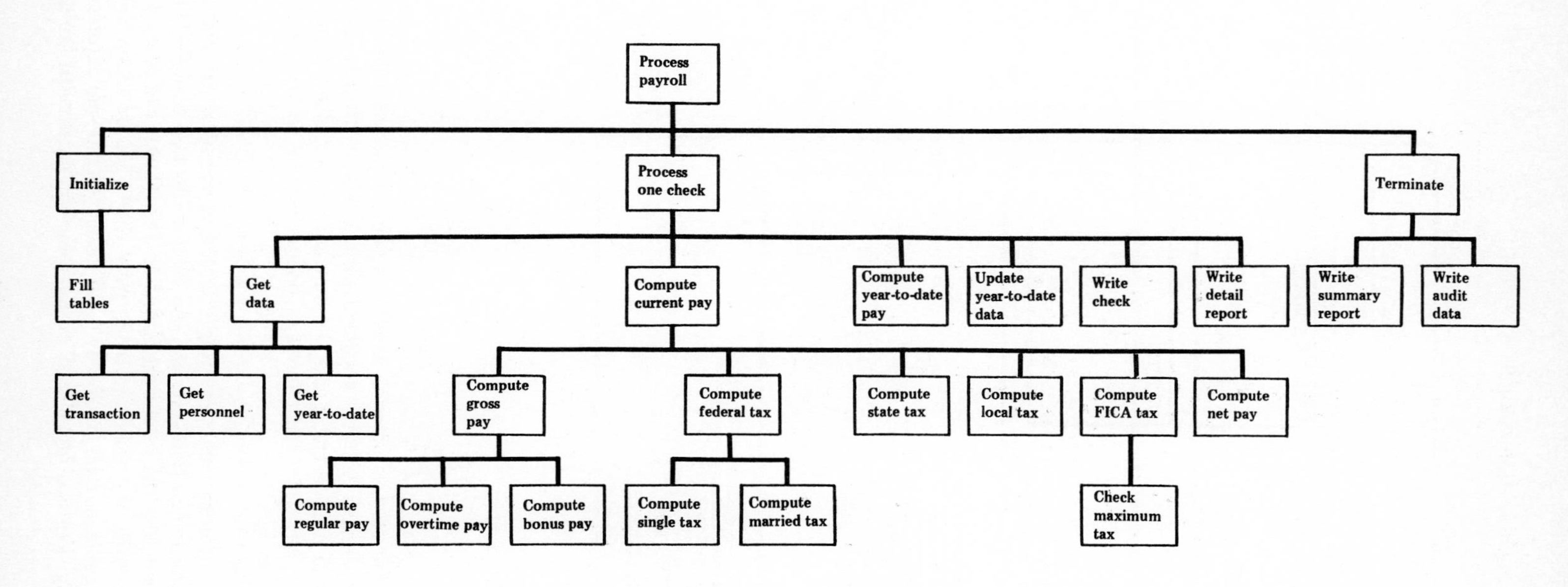

Fig. 6.12: *The main control module.*

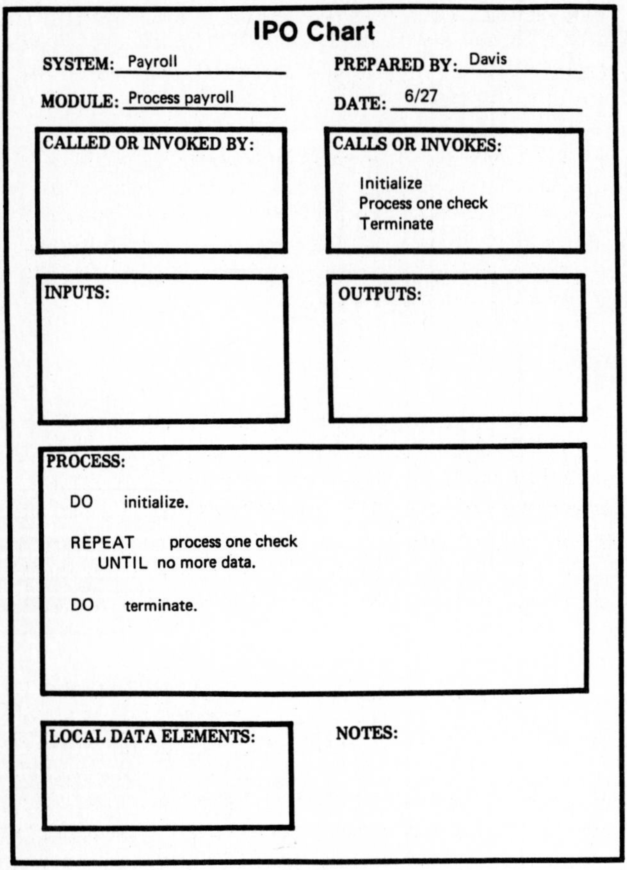

IPO Chart

SYSTEM: Payroll PREPARED BY: Davis

MODULE: Process payroll DATE: 6/27

CALLED OR INVOKED BY:

CALLS OR INVOKES:

Initialize
Process one check
Terminate

INPUTS:

OUTPUTS:

PROCESS:

```
DO      initialize.

REPEAT      process one check
    UNTIL  no more data.

DO      terminate.
```

LOCAL DATA ELEMENTS:

NOTES:

tured English, but easy to describe in another documentation technique. Insisting that the process be changed to fit the documentation is silly; it virtually guarantees that the programmers will ridicule and ignore the standard. A good idea is to use one technique as the standard, but allow other forms of documentation to be used if circumstances warrant.

In many organizations, preparing IPO charts has been automated to a degree. Using special documentation software, the analyst fills in a blank IPO chart displayed on a CRT screen, and the results are stored on disk. Charts can be modified simply by retrieving, changing, and restoring them. New charts can be added simply by adding them. Cross references (the "called by" and "calls" blocks) are automatically checked for consistency, and the inputs and outputs are compared with the data dictionary.

Once the charts have been completed, they can be printed and used as traditional documentation, but many such systems contain an additional feature. Rather than simply printing the IPO charts, the contents can be dumped as comments in the appropriate language, and subsequently included in the source code. Programmers tend to maintain source code. Comments buried in that code are more likely to be maintained than almost any other form of documentation. Incidently, this is one reason why we have used a simplified version of the IPO chart in this text.

Many analysts like to assign levels to the modules in an IPO chart. For example, in Fig. 6.11, *Process payroll* might occupy level A or level I, while *Initialize*, *Process one check*, and *Terminate* are on level B or level II; lower levels are identified similarly. A module's level is then added as a prefix to the module name; for example, *A-Process-payroll*, or *B-Initialize*. Later, when the module name is referenced in the program source code, the prefix clearly relates the name to the hierarchy chart.

Checking the Design

The hierarchy chart (Fig. 6.11) and its accompanying IPO charts represent a complete program design. Is it a good design? The only way to find out is to check it carefully. The analyst begins with the data flow diagram, concentrating on the automation boundary represented by the program. The logical flow is traced through the hierarchy chart. Are any processes missing? Are all data flows present? Are all the data stores accessed by the program? It is easy to overlook something when dealing with details; relating the data flow diagram to the hierarchy chart helps to ensure that all major functions have been included.

Next, the analyst cross-checks the hierarchy and IPO charts with the data dictionary. Missing elements might highlight an oversight; extra elements might be added to the data dictionary. The algorithm descriptions provide yet another check on the accuracy and completeness of the program plan. Are all algorithms accounted for? Can they be located on the hierarchy chart? Many analysts go a step further and cross-reference the hierarchy chart, the IPO charts, the data dictionary, and the algorithm descriptions, making sure that documentation is consistent on all.

As a final check, the analyst evaluates the modules for *independence*. An independent module is a true black box. Given a set of input parameters, it generates a set of output parameters, and has no other effect on the rest of the program. How can module independence be recognized? One key factor is the number of inputs to and outputs from the module; for example, a module with a single input and a single output tends to be more independent than a module with dozens of inputs and outputs. Look at it this way. Each parameter represents a potential error that can be transmitted to the rest of the program. The more parameters, the greater the chance that an error will occur. A module with numerous input and output parameters is *tightly coupled* to the rest of the program; a module with few parameters is *loosely coupled*. Loose *coupling* implies independence. Only those parameters essential to the function of a module should be passed to that module.

The first step in evaluating modules for coupling is to prepare a *structure chart* (see Module H). Starting with a copy of the hierarchy chart, the inputs and outputs listed on each IPO are written on the lines linking a routine to its control module; for example, the inputs to and outputs from *Get data* (Fig. 6.11) are written on the line linking *Get data* and *Process one check*. When all parameters have been listed, the analyst has an image of the data flows throughout the program. Now, the degree of coupling can be checked. Are all the parameters passed to a low-level computational module actually used by that module? Check the process block. Are all parameters passed to a control module needed by that module, or by a lower-level routine that it controls? Records and structures should generally be stored and controlled at a high level, and individual data elements should be passed to the lower levels. Do the data follow this pattern? Has a complete data structure been passed to a computational module simply because the analyst wanted to avoid writing the individual parameters? Unnecessary coupling should be eliminated before implementation begins.

Why is unnecessary coupling a problem? Why is it so important that the analyst take the time to check for it? Unnecessary coupling means unnecessary complexity, which increases the chances for coding or debug errors. In addition, excessive coupling can create a maintenance nightmare. If the value of a given parameter is wrong, any module that touches that parameter is suspect. Why should a maintenance programmer have to spend time checking a module because of an unnecessary data link? Another risk is the ripple effect. If a parameter is wrong, it can affect every module it touches. If a particular data link is not necessary to the function of a module, adding that parameter to the input or output list creates an unnecessary risk of contamination.

THE INSPECTION

An inspection follows completion of the HIPO documentation. The key question is: can code be written to the specifications? This is a technical question; thus the inspection team is composed of technical people. In a larger company, there are plenty of programmers and analysts to conduct an inspection, but THE PRINT SHOP employs a limited number of technical experts. Allowing the analyst and the programmer/operator to inspect the analyst's work is a bit dangerous, as they have a definite mentor/student relationship. While Carla Jones could certainly do the job, she is perceived as management, and an inspection is ideally a technical function. Thus THE PRINT SHOP has arranged to have a professor from the computer science department of a local university sit in on a consulting basis. Sometimes, an outside viewpoint can really help.

Note that this inspection concentrates on one of the programs described in the system plan. There are other programs; should they be inspected too? With the exception of the occasional simple program, yes they should; the basic sequence should be to design, inspect, and then implement each program in the system. Clearly, there will be several program design inspections.

Should each program design be subject to management review? Probably not. If a system were going to be developed in linear fashion, with analysis, followed by system design, followed by a complete detailed design before the first line of code was written, a management review of the detailed design might be in order, but frequent management reviews of pieces of the system are probably counterproductive. As the technical manager, Carla should certainly be aware of technical progress, but user and high-level management should become involved only on an exception basis. A schedule and a budget have been approved. As long as the project remains within the schedule and budget parameters, designing and implementing the system should be a technical concern. Of course, status reports might be scheduled at key points in the integrated design/implementation process, but management by exception seems to work best here.

SUMMARY

The chapter began with a discussion of how system design documentation facilitates subdivision of the remaining work. The data were seen as the framework that links the system components; thus the analyst planned the files and developed test data as the first step in detailed design. We then turned to the design of the payroll program. Starting with the data flow diagram, the analyst drew a high level hierarchy chart. Through functional decomposition, the hierarchy chart was expanded, with control functions at the top coordinating numerous detailed computational routines at the bottom. An IPO chart was prepared for each module on the hierarchy chart; structured English was used to define each process. A primary consideration was module cohesion.

The completed hierarchy chart was checked against the data flow diagram, the data dictionary, and the algorithm descriptions. A structure chart was then developed, and the modules checked for coupling. Finally, the completed payroll program documentation was inspected by technical personnel.

EXERCISES

1. How do the system flowchart and the project network help to control and coordinate the activities of several analysts and/or programmers?

2. Why does detailed design typically begin with the data?

3. Why is it so important that test data be designed to catch most errors before a project is released? What is wrong with debugging software after it is released? Note: this question is not explicitly answered in the text.

4. Test data should be realistic. Extreme conditions should be considered, and bad data included. Why?

5. In developing a hierarchy chart for the payroll program, the analyst started with the processes defined on the data flow diagram. Why? Don't look for an explicit answer in the text, because it isn't there.

6. What is functional decomposition?

7. Discuss the HIPO design and documentation technique. Explain the functions of the hierarchy chart and the IPO charts, and explain their relationship.

8. Detailed design is built on the framework of the earlier steps. Explain how. For example, explain how the HIPO documentation is related to the data dictionary, the algorithm descriptions, the data flow diagram, and the system flowchart.

9. Repeat the design of the payroll program using the Warnier-Orr technique (see Module K).

10. What is the difference between local and global data?

11. What is cohesion? Why is it important?

12. A good program is easy to follow. Do you agree or disagree? Why?

13. Why should good program design anticipate change?

14. A certain amount of flexibility is important in setting a documentation standard. Why? Why are standards needed?

15. What makes a module independent? Why is it important that a program be composed of independent modules?

16. What is coupling? How does coupling relate to module independence? The book says that loosely coupled modules are good and tightly coupled modules are bad. Do you agree? Why, or why not?

17. What is a structure chart? What is its purpose?

18. Why is it important that each program design be inspected?

19. In Fig. 6.11, the module *Compute state tax* was not decomposed; every state tax structure is different. Define the algorithm used by your state and, if appropriate, decompose this module and prepare the necessary IPO charts.

20. Define the algorithm used for computing local tax in your area. Modify the hierarchy chart of Fig. 6.11 to include the computation of local tax, and prepare necessary IPO charts.

21. Figure 6.11 shows the federal income tax computation broken into two lower level modules. Imagine that a flat tax proposal passes Congress, meaning that everyone pays the same rate. How would this change in the tax law affect the hierarchy chart?

22. In the chapter, the need to estimate the amount of space required by the various files was mentioned, but glossed over. Assume that THE PRINT SHOP currently employs fifty people, and expects to expand to one hundred within five years. Estimate the space required by each of the files in the payroll system. Assume that data will be stored on diskettes in 256-character sectors. A single diskette can hold about 6000 sectors. THE PRINT SHOP's minicomputer system has four diskette drives. Note: refer to Module M.

23. What file organization would you select for each of the payroll system's files? Why?

24. In the text, the payroll program was designed. Prepare a similar design for one or more of the other programs identified on the system flowchart (Fig. 6.1).

25. In the chapter, only selected IPO charts to accompany the hierarchy chart of Fig. 6.11 are shown. Prepare a complete set of IPO charts.

26. Complete the structure chart based on Fig. 6.11.

27. Continue with Chapter 5, exercise 13.

28. Continue with Chapter 5, exercise 14.

29. Continue with Chapter 5, exercise 15.

7

Case A Implementation And Maintenance

OVERVIEW

This chapter concludes the payroll system case study by discussing the implementation and maintenance steps. Implementation is concerned with coding, documenting, and debugging the programs. Since this is not a programming text, the discussion is general, focusing on principles rather than specific techniques. Basic structured programming concepts are presented, and a number of documentation guidelines suggested. The idea of top-down program implementation is then introduced, along with structured walk-throughs. The implementation step ends with a formal system test.

The need for training is an often overlooked aspect of system implementation. Operating procedures must be written, and the affected personnel trained in their use. Following the successful release of the system, maintenance begins; the chapter ends with a discussion of typical maintenance problems and techniques.

IMPLEMENTATION

The system has been designed. Specifications for the programs have been written. The time has come to code, debug, and document the programs. In addition, during implementation operating procedures must be written; affected personnel must be trained to use the new system; new hardware must be installed, tested, and debugged; and the system test must be conducted. It's a busy time.

Coding

This is not a programming text; thus we will not discuss coding in detail. We will, however, consider a few basic principles of structured programming, and discuss how they relate to structured systems analysis and design.

The economics of data processing are changing. Historically, hardware costs were so high that concern with hardware efficiency was not only justified but essential; thus programmers worried about the CPU time and main memory space their code used. Today, however, given the declining cost of hardware, developing and maintaining a program is often far more expensive than running it. Thus, emphasis is shifting from efficiency on the computer to controlling software costs. Debugging and maintenance are two of the more controllable elements of software cost. The objective of structured programming is to generate programs that are easy to understand, and hence easy to debug and maintain. This simplicity implies a lower cost. As a result, management has begun to take structured programming seriously.

The basic principle of structured programming is akin to the military strategy of divide and conquer. A program is broken into small, independent single function modules which, by their very nature, are clear and easy to follow. These modules are themselves composed of even smaller blocks: the sequence, decision, and repetition structures described in almost any structured programming text. The structured approach to systems analysis and design is directed at developing program specifications that can be implemented in structured fashion.

Documentation

Clarity is a primary objective; a good structured program must be easy to read. Thorough, up-to-date documentation is essential to program clarity. Like the code, the documentation must be maintained, and its value declines rapidly if it is not. Like the code, documentation must be planned with ease of maintenance in mind.

Clearly, there is a difference between the documentation required to support a commercial software package and that needed for THE PRINT SHOP's payroll program. When the documentation is for the benefit of programmers and not users or commercial customers, the best approach is often to bury much of it in the source code. Programmers will maintain the source code. As the code changes, the accompanying comments will, with minimum prompting, be updated to match. Programmers will not, at least willingly, redraw flowcharts or otherwise maintain independent documentation. Certainly they should, but realistically, the programmers will not maintain documentation unless it is both useful and easy to maintain. Documentation for the sake of documentation is a waste of time.

THE PRINT SHOP uses an interesting documentation technique. Assuming that the basic program structure will remain relatively constant, the analyst maintains a hierarchy chart in a file folder. The accompanying IPO charts, however, are coded as comments and incorporated directly in the source code. As the maintenance programmer changes the code, the accompanying IPO chart is changed with it. The result is a clear linkage of code and documentation maintenance.

Such tight linkage of the code and the documentation has other advantages as well. For example, consider the process of checking the code against the specifications developed in detailed design. The specifications are buried in the code. The IPO chart and the associated code can be viewed almost simultaneously, and often on adjacent pages of the program listing. Misinterpretations or coding errors should be obvious. It might be reasonable to expect the analyst to do a better job of program design, too. If documentation is viewed as merely a planning tool that will eventually be discarded or ignored, there is a natural tendency to take shortcuts. However, with the documentation permanently maintained in the code for everyone to see, the analyst is motivated to do it right.

Debug

During program debug, errors are removed from the code. Compiler errors are generally easy to find and correct. Logical errors, on the other hand, are often difficult to trace, and may even be overlooked. The program was designed in modular fashion, from the top, down. A similar, methodical, step by step strategy for implementing a program can greatly simplify the debug process.

The programmer begins by writing a skeleton version of the program, with the primary control structures in place, but with the lower level computational modules represented as stubs. The gross pay module, for example, might simply initialize a constant at this early stage. The skeleton version of the program is then tested, and the output compared to the expected values. Next, one of the detailed computational modules replaces its stub, and the test is repeated. If the results are acceptable, the next computation module is added to the developing program; however, if errors are found in the output, the most recently added module is the likely culprit. By focusing the search for errors on a single, one-page module, this top-down approach to program implementation can save a great deal of debug time.

Test data are the key to debugging. Initially, the programmer needs a small subset, yielding perhaps a page or two of output; many errors are obvious after a few lines have been printed or displayed, and waiting for volumes of output is an annoying waste of time. This subset of test data should be reasonably complete, exercising each possible alternative path through the program. Debugging with a limited subset of the test data is useful in the short run, but it does not constitute a real test of the logic. Each module should process a full complement of test data before being declared bug-free.

Structured Walkthroughs

It is often difficult for programmers to debug their own code. All too often, people read what they meant to write, and not what they really wrote. (Have you ever searched for hours in a vain attempt to find a bug, only to have a friend spot an obvious problem in seconds?) This problem is not unique to computer programming. Writers, for example, rely on copy editors to spot errors in spelling, punctuation, and sentence structure, and a programmer might consider using other programmers or analysts to perform a similar function. This is the essential idea behind a *structured walkthrough*.

THE PRINT SHOP uses structured walkthroughs on all programs. Following the first clean compilation of a module, the programmer and the analyst sit down for an hour or so, and read through the code together. The programmer reads aloud, paraphrasing the code and explaining its function. The analyst follows on a separate copy of the listing, and asks questions where appropriate. Errors or ambiguities are identified, and suggestions for improvement made. Since the analyst is writing some of the programs in the payroll system, the programmer will have an opportunity to play the role of the reviewer, too. In many larger organizations, much of the informality of THE PRINT SHOP's walkthroughs disappears; in fact, some organizations run formal inspections of all code as soon as a clean compilation is ready.

THE SYSTEM TEST

Implementation ends with a formal system test. The test data are crucial to this process. They must be realistic, and cover extreme conditions as well. Ideally, every alternative path through the program should be exercised at least once. Beyond the test data, the system test must involve all the elements that compose the system, including programs, files, forms, manual procedures, and personnel.

At THE PRINT SHOP, the initial phase of the system test involves the complete set of test data. Manual procedures are followed, with the people who will actually do the work participating. The real data entry programs are used. Test data are read from the actual files. The checks and reports are printed and carefully reviewed, line by line. Additionally, the contents of all files stored on secondary devices are dumped to the printer or to a screen, and reviewed just as meticulously.

Following a successful test, a *parallel run* is scheduled. First, the data for the most recent payroll are collected, and processed by the new system. Each line of the printed output is compared with the results generated by the old system. Once again, files are dumped and checked. Had the new system been used last week, the results would have been acceptable.

Friday is payday. As the data are collected, two copies are made, one for the bank and the other for THE PRINT SHOP. Both systems are used to prepare output, and both sets of output are compared. Ideally, they are identical; realistically, there will be some differences. Can these differences be explained? (Note: it is certainly possible for the old system to be wrong and the new system correct.) Significant errors could, of course, indicate a need for repeating the system test, but today the user is satisfied. The payroll system is finished! Development now ends, and productive use begins. Unfortunately, so does maintenance.

TRAINING

Before moving to maintenance, it is important that we briefly discuss training, as this is perhaps the most frequently overlooked aspect of system implementation. The analyst is responsible for training people to use the new system. People may hesitate when faced with something new, and many fear technology. User acceptance can make or break a system. No matter how good a system might be, it's useless if people won't use it. Good training can overcome such natural fears and hesitations.

Plan training carefully, starting at the very beginning of the project. The analyst should make it a point to interview the end users, elicit their ideas, and, where appropriate, incorporate these suggestions in the system. The users should feel that they have a part in designing the system, or at least that they have had an opportunity to be heard. Throughout the system development process, they should be kept informed of progress; inspections can help here. Finally, operating procedures should be inspected by the users, and they should be given adequate time, under the guidance of the analyst, to learn how to use the system. Above all, the analyst should remember two simple rules:

1. When it comes to the application, the user is the expert.

2. The user is the customer, and the customer is always right.

MAINTENANCE

Maintenance begins as soon as the system enters productive use. It *will be* required. It is easy to imagine that a just completed system is perfect, but no such thing exists. Little bugs will slip through the system test, and will show up weeks, months, or even years later. Algorithms will change; for example, the government will change its tax rates. Management or the user will ask for new or revised reports, and field formats will change. New systems will be developed and will require an interface with this one. Hardware will be upgraded, requiring a change in the code. The only program that does not require maintenance is the program that no one uses.

Maintenance is expensive. The best way to minimize this cost is to design the program with ease of maintenance in mind. One key is functional modularization. If each module is independent and performs a single, complete function, most changes will be limited to the logic contained in a single module. In addition, the ripple effect will be minimized, as changes made to an independent module should have little impact on the logic performed by other modules. The structured approach to systems analysis, system design, and programming is aimed at producing programs that consist of independent functional modules linked by a clear control structure.

Another valuable technique is to identify and further isolate likely changes. For example, in the payroll program, the federal, state, and local income tax rates will almost certainly change during the life of the payroll system. Thus the analyst created a table file, and designed the payroll program to read the tax rates from the file before starting processing. As a result, when tax rates change, the contents of the table file can be updated without affecting the payroll program. Another example is the independent program that maintains the personnel file. It might have been possible to include personnel data with the payroll transactions, but this would lock the personnel data to the payroll system. Isn't it possible that other systems—not yet designed—might need the personnel data? Certainly. In fact, the personnel data may well be more central to other applications than they are to payroll. An independent personnel file makes sense. An analyst is paid to look beyond "this" application.

Separating local and global variables also helps to simplify maintenance. Local data are defined, stored, and used strictly within a single module and, if properly identified, can be changed without affecting the rest of the program. Global data, on the other hand, are typically defined and stored at a high level, and used in more than one module. Changes to global data can create a ripple effect, causing unanticipated and often difficult to trace problems in other modules; thus any change involving global data must be much more carefully evaluated. If local and global data are separated, the programmer can quickly determine if the ripple effect is a possibility, and plan the maintenance work accordingly. Simply identifying a variable as local or global can also help. A common procedure is to insist that no module modify the value of any input global data element. If this procedure is followed consistently, the risk of a change to one module causing a ripple effect in another is greatly reduced.

In general, a program is easy to maintain if it is composed of well-documented, independent, complete, single-function modules.

SUMMARY

This chapter discussed the implementation and maintenance steps for THE PRINT SHOP's payroll system. Several basic structured programming concepts were briefly described. Next, a number of documentation pointers were presented, including the idea of burying much of the documentation directly in the source code. Program debug can be greatly simplified by using a top-down strategy for testing the code. First, a skeleton version of the program is written, with the detailed computational routines present as stubs. The detailed logic is introduced and tested, one module at a time. Structured walkthroughs were presented as a valuable debugging aid.

The chapter then turned to the system test. The test data were shown as crucial. Initially, all system components were tested using these data. Next, a simulated parallel run, using the input data from the prior week, was conducted. Finally, live input data were collected and processed by both the new and the old systems, and the results compared. Following successful completion of this system test, the payroll system was declared complete, and released to production. At this point, maintenance began.

Some of the problems associated with training were briefly discussed in the text. Training is often overlooked. It is a crucial part of the analyst's responsibility, however, and deserves careful attention.

The chapter ended with a discussion of maintenance. Maintenance will be required on almost every system, and a program should be designed with maintenance in mind. The functional modularity of a well structured program tends to help. Other design decisions that affect maintenance include the careful separation of local and global data, and the isolation of modules that are particularly prone to change. Clear documentation is another key.

EXERCISES

1. Relate the structured approach to systems analysis and design to the basic principles of structured programming.

2. Why must documentaion be maintained along with the code? Why does the idea of incorporating the documentation directly in the source code make sense?

3. Explain how the decision to bury the IPO documentation in the source code subtly forces the analyst to do a better job of design.

4. Explain top-down program implementation. How does it help simplify debug?

5. During debug, the programmer will often begin with a selected subset of the test data. Why? Why should a module be tested with a full set of test data before being declared bug free?

6. What is a structured walkthrough? Of what use is it? How does a structured walkthrough compare to a formal inspection?

7. Why is it sometimes difficult for programmers to debug their own code?

8. Why must the formal system test include all components of the system?

9. What is a parallel run? Of what value is it?

10. Training employees to use a new system is an extremely important implementation task. In fact, poor training can easily offset an excellent job of analysis and design, and make an otherwise excellent system ineffective. Why? Explain.

11. Why is maintenance necessary? Cite some examples of activities or occurences that require maintenance.

12. The best way to minimize maintenance cost is to design the system with maintenance in mind. Explain.

13. How does structured program design affect maintenance?

14. What is the ripple effect? Relate it to the difference between local and global data.

15. Discuss the training needs of THE PRINT SHOP's payroll system. Who should be trained? Who should do the training? What kinds of training might be needed? When should this training be conducted?

16. What operating procedures are needed in THE PRINT SHOP's payroll system? Prepare the operating procedures.

17. Continue with Chapter 6, exercise 27, concentrating on the system test, operating procedures, and training.

18. Continue with Chapter 6, exercise 28, concentrating on the system test, operating procedures, and training.

19. Continue with Chapter 6, exercise 29, concentrating on the system test, operating procedures, and training.

20. Prepare a report on the documentation standards used by your school's data processing center. If appropriate, any business data processing center can be substituted.

21. Prepare a report on maintenance programming. In addition to general concepts from the literature, cite specific examples from your school's data processing center, a business data processing center, or both.

PART II

A Small Business System

Case B Problem Definition

OVERVIEW

This chapter introduces the second case study, in which an analyst will be asked to plan and install a small business system in a doctor's office. We begin with an overview of the doctor's practice, and then discuss the problems that have led her to consider a computer as a possible solution to them. The physician's objectives, criteria, and expectations will be explored, and a statement of scope and objectives prepared.

DR. THERESA LOPEZ, M.D., INC.

For several years, Dr. Theresa Lopez has run a successful family practice in a suburban area just outside a major sun belt city. She has approximately five thousand patients, most of whom visit only occasionally. On an average work day, she sees between twenty-five and thirty-five patients (except during flu season, of course). Billings, including insurance and medicare claims, average about $20,000 per month.

To most people, $20,000 per month seems a great deal of money, but Dr. Lopez does not get to keep it all. A major expense is paying her office staff—an office manager, a receptionist, and two nurses. Several months ago, in an attempt to free more of her time for clinic and hospital work, she hired a recent medical school graduate as her assistant, and he is paid a salary plus a percentage of his billings. Additional expenses include office space, equipment and supplies, malpractice insurance, subscriptions to medical journals, and a variety of activities that might be grouped under continuing education.

Until a year ago, the office was run by Nancy Johnson, an old friend of Dr. Lopez. In fact, Nancy helped set up the practice, assuming responsibility for all financial matters, which allowed the doctor to concentrate on medicine. It was a very pleasant working relationship between two people who understood and trusted each other. Then, Nancy's husband was offered a job in another city. It was a difficult decision, but finally the Johnsons moved. Dr. Lopez has been trying to replace her ever since.

Her successor, a young man with some office management experience, just didn't work out. Part of the problem was that Nancy never established a system, for she knew the business and Dr. Lopez so well that she really didn't need one. In effect, Nancy *was* the system, the one indispensable employee. There were few, if any, written procedures for the new manager to follow, and he seemed to flounder. He didn't know a great deal about medicine, either, and the other employees quickly lost respect for him. The problems, of course, were not all his fault, but he just couldn't do the job. Everyone was relieved when he resigned after eight months.

Finding a replacement was to prove much more difficult than Dr. Lopez had imagined. For two months, she managed the office herself. It was both time consuming and frustrating—she did not go to medical school to become an accountant. There was one positive outcome from her experience, however; she really began to learn the business side of medicine. She saw, firsthand, how the lack of established procedures made the work so much more difficult than it had to be. Never again would she allow even a highly competent and completely trusted friend to become indispensable. She would control her own practice.

Given her new knowledge of the business, the doctor was able to redefine the office manager's position as more of a clerk than a manager, and was eventually able to hire a young community college graduate who showed promise of being able to grow into the job. Meanwhile, the doctor assumed much of the responsibility for managing the office herself. It was time consuming, and detracted from her real job, medicine. There had to be a way to allow others to manage the financial details of the medical practice.

She recalled a recent professional meeting where the speaker's topic was computers in medicine. It had sounded good; perhaps a computer could help solve her problems. Fortunately, Dr. Lopez was smart enough to realize that she didn't know everything. Computerese confused her, and she wasn't about to make an expensive decision on the basis of what might be little more than a sales pitch. One of her patients had recently opened an independent consulting practice called Systems Analysis, Inc.; she would ask a professional systems analyst for help and advice.

PROBLEM DEFINITION

Dr. Lopez and the analyst met over lunch one afternoon. She outlined the facts described above, and indicated that she needed help. Clearly, more time was needed to discuss the problem in detail, so they agreed to meet at the doctor's office the next afternoon. Let's join them as they begin to define the problem.

In their earlier meeting, Dr. Lopez talked about a number of general problems, such as a lack of precision in the office procedures, and the difficulty in training replacements. She also expressed a desire to be relieved of much of the financial management burden. What is this burden? The analyst begins the interview by attempting to define the problem more specifically.

The key problem, according to the doctor, is billing. Almost 75 percent of the doctor's earnings come from third party sources such as insurance carriers, medicare, medicade, welfare, and various other government programs; only 25% of her charges are paid directly by the patients. With the insurance companies and government agencies, the paperwork must be filled out precisely; little errors such as transposing two digits in an account number or a code can delay payment for as much as three months. This creates a cash flow problem, and the doctor has actually had to borrow money against anticipated insurance payments, just to keep her practice going. Lately, this "float" has amounted to as much as 10 percent of her cash flow; given the current interest that she might be able to earn on this money were she able to get it on time, she estimates that the cost to her exceeds $3000 per year.

Perhaps even more important to Dr. Lopez is the impact that such errors have on her professional image. Patients get upset when a bill is wrong, or when insurance payments are reported as unpaid, and they really become upset when their own payments are not credited promptly or correctly. The doctor is very sensitive to the possibility that some might consider her a "typical female" who just can't handle figures; she has always found that silly myth particularly infuriating, since she has never had a problem with math. Above all, however, Dr. Lopez is a perfectionist; she wants the job done right simply because she wants the job done right, and the profit and loss statement has very little to do with it.

Another problem is with patient recalls. When a patient visits the office, she often suggests a follow-up visit in a month, six months, or some other time interval. She likes to note such recalls and have the receptionist call the patient a week or so before the recall date to schedule an appointment, but the office has not done a very good job on this lately. It is professionally embarrassing to have a patient call and ask, "Wasn't I supposed to come back for a follow-up visit?" Again, there is the professional image problem; she is a well-organized, highly professional person, and resents anything that makes her seem less than that.

Drug inventories are a major concern. Every doctor's office maintains a limited supply of injections, various medications, and drugs to treat patients. Recently, the office of another doctor in town was burglarized, and drugs were taken. Because he didn't know exactly how much was stolen, he had a difficult time collecting from his insurance company, and almost got into trouble with the police. Dr. Lopez had to admit that she would have faced a similar problem had she been the victim. Perhaps better control of the drug inventory could save a little money, but even if it didn't, she wants this aspect of her business brought under control.

Payroll is another problem. When Nancy Johnson was in charge of the office, it was no problem at all, but the somewhat negative relationship that has developed between the office manager and the rest of the staff has led to some grumbling. They trusted Nancy, but are not sure that they want the new employee to be quite so familiar with their finances. Ideally, payroll should not involve the office manager so directly as does the present calculator, scratchpad, and typewriter system. In response to the concerns of the nurses, Dr. Lopez has assumed responsibility for the payroll herself, but she'd rather not do it.

The analyst then asks about accounts payable, another common small business application. The bills are paid on time; accounts payable can wait. What about patient medical histories; are there any problems in maintaining them? "No," replies the doctor. In fact, she rather likes the informal, personal touch of handwritten medical histories, and so do the patients. The schedule isn't a problem either; Dr. Lopez established clear patient scheduling guidelines when she first opened her practice, and they have been working well ever since. The concerns are basically limited to billing, the drug inventory, payroll, and patient recall.

The Doctor's Objectives and Criteria

As the interview draws to a close, the analyst takes time to summarize the problems. Basically, Dr. Lopez wants to define her office procedures so that they are consistent, easy to use, and independent of the office manager's personality. Another concern is with training. She has accepted a certain level of job turnover, and wants to minimize training for the inevitable new employees. Financially, she wants to improve her cash flow by processing paperwork more quickly and accurately. An improved system might save some money too, but financial concerns are not really at the top of her list. Dr. Lopez is a professional, and is proud of that fact. Lately, her office management has been handled in a manner that frankly embarrasses her, and she doesn't want this to continue. She views an improved system as an insurance policy to protect her practice from employee turnover.

Dr. Lopez is satisfied with the analyst's apparent understanding of her concerns, but she has some strong opinions about computers, too. Frankly, she doesn't feel comfortable with the machines, and is more than a bit hesitant to buy one. She realizes that a computer could be the solution to many of her problems, but she wants to hear that from someone who does not sell them for a living (which is why she called Systems Analysis, Inc.). She has heard enough horror stories to have formed some strong opinions on what a computer must do, and wants to make sure that the analyst knows how she feels.

A key criterion for any system must be ease of use. Dr. Lopez is not a computer expert, and does not wish to become one; she is a physician. She does not want to find herself in the position of replacing a hard-to-find office manager with an even harder-to-find computer person. She wants to avoid the essential employee problem, not exacerbate it. Above all, she doesn't want any system that she can't understand. She sees herself as a doctor, and resents anything that takes time from her practice of medicine. She certainly does not want to become a computer expert.

Training support is essential. She has heard of other doctors who purchased computer systems that were basically dumped on their doorsteps, and she doesn't want that. Before she purchases a computer, or any new system for that matter, she wants to see easy-to-read documentation and on-line training aids, and she wants to be assured that continuing support will be available.

Perhaps reliability and dependability are the real keys. She does not want down-time taking up her time or threatening to put her out of business, and she does not want to become a slave to some computer system. Another doctor she knows recently experienced a problem with a computer supplier going out of business; Dr. Lopez does not want to get caught in that same trap. She will insist on a maintenance contract. She wants guarantees on maximum downtime. She wants some form of backup to keep her going in the event that downtime does occur. Dr. Lopez may not be a computer expert, but she has certainly thought about computers in her practice.

Her final criterion is cost; she does not want to spend more than $20,000 on a new office management system.

The Statement of Scope and Objectives

Shortly after the interview, the analyst takes the time to summarize the results. The next morning, this summary serves as a base for preparing a statement of scope and objectives (Fig. 8.1), which is made part of the agreement between Dr. Lopez and Systems Analysis, Inc.. The agreement calls for a feasibility study of two weeks' duration for which the analyst will be paid no more than $2000. The doctor agrees that the statement accurately reflects the problem, and signs the contract. Work can now begin on the new office management system project.

SUMMARY

This chapter discussed the problem definition stage for a project to develop a new office management system for a physician. The doctor's problem was described in general terms. During an interview, the analyst's questions help to focus on four specific problem areas: billing, patient recalls, payroll, and drug inventory control. The doctor stated her concerns with computer equipment, and a set of system objectives and criteria was discussed. Finally, the analyst prepared a statement of scope and objectives.

Fig. 8.1: *The statement of scope and objectives.*

STATEMENT OF SCOPE AND OBJECTIVES: *July 15, 1983*

BETWEEN: *Theresa Lopez, M.D., Inc.*

AND: *Systems Analysis, Inc.*

THE PROJECT: *OFFICE MANAGEMENT SYSTEM*

PROJECT OBJECTIVES:

1. *To define office procedures more clearly.*
   a. *To simplify training of office personnel.*
2. *To improve cash flows.*
3. *To reduce office management cost.*
4. *To improve record keeping accuracy.*
   a. *Billing or accounts receivable.*
   b. *Patient recalls.*
   c. *Drug inventory.*
   d. *Payroll.*

KEY SELECTION CRITERIA: *Ease of use, availability of training support, reliability, dependability, and cost.*

PROJECT SCOPE: *The cost of the project will not exceed $20,000.*

PRELIMINARY IDEAS: *A small business computer system. Improved manual procedures.*

THE FEASIBILITY STUDY: *In order to investigate the potential for this project more fully, a feasibility study lasting approximately two weeks is suggested. The cost of this study will not exceed $2000. The cost of the feasibility study is included in the project scope; in other words it is part of the project's $20,000 limit.*

EXERCISES

1. What is wrong with an indispensable employee?

2. Why, do you suppose, does a lack of clear procedures make the job of managing an office so much more difficult?

3. Why is it important that a professional faced with a technical decision get advice from a specialist?

4. Why is it important that an analyst attempt to define a problem in concrete terms? Can you see any danger in this approach?

5. Outline the key problem areas in the doctor's office, and briefly describe how each might result from poorly defined procedures.

6. Many of Dr. Lopez's concerns were nonfinancial; for example, she worried a great deal about her professional image. Contrast this case study with THE PRINT SHOP.

7. Dr. Lopez clearly stated her concerns about computers. Why was this important?

8. Perhaps your instructor can arrange an out-of-class project with a local professional (doctor, lawyer, insurance broker). If so, prepare a statement of scope and objectives for a new office management system.

9. Discuss with a doctor the statement of scope and objectives presented as Fig. 8.1. Is it realistic? What factors might be added or deleted?

10. In the chapter, such tasks as accounts payable, patient record keeping (medical history), and patient scheduling were ignored—Dr. Lopez did not consider them problems. In many practices, however, they are. Discuss the nature of these potential problems, and then add them to the statement of scope and objectives.

11. Perhaps your instructor can arrange an out-of-class project with a local professional office. If so, prepare a statement of scope and objectives for a proposed new office management system.

9

Case B
The Feasibility Study

OVERVIEW

In this chapter, a feasibility study of the proposed office management system for Dr. Lopez will be conducted. After the project's scope and objectives are verified, the existing system will be studied, and problems noted. Next, the analyst will develop a high level model of the proposed system, and discuss it with the doctor. Alternative solutions will be defined, and a course of action recommended. Finally, the feasibility study report will be written and presented to the client, Dr. Theresa Lopez.

THE FEASIBILITY STUDY

The statement of scope and objectives presented at the end of problem definition represents the analyst's understanding of the problem. Is there a solution? The feasibility study's objective is to find out. The analyst will be spending much of the next two weeks in the doctor's office, methodically performing a feasibility study as outlined in Module C. Let's follow the study, step by step.

Define the Project's Scope and Objectives

Is the statement of scope and objectives correct? Does Dr. Lopez really understand what is to be done? It is important that both she and the analyst have the same image of the project, so there are no misunderstandings or unrealistic expectations. Thus, the analyst explains that a feasibility study is not intended to produce a solution to the problem, but merely to determine if a solution is possible. The doctor should not expect miracles or quick cures. At the end of two weeks, she will be given an opportunity to choose between continuing work on the new system or terminating the project; at that time she will have a much clearer picture of costs and benefits. Dr. Lopez understands the purpose of this first phase in the system development process, and appreciates the analyst's openness; a sense of trust is beginning to develop. Their discussion concluded, Dr. Lopez introduces the analyst to her staff, and the real work begins.

Study the Existing System

In order to determine the actual problems affecting Dr. Lopez's practice, the existing system must be studied. The first step is to interview the office manager. Most interviews begin with an open question such as, "Tell me what you do." This case is a bit different, however; the office manager is a new and relatively inexperienced employee, who may not know how to respond. Thus, the analyst decides to quickly focus the interview. Billing is a key problem, and occupies a great deal of the office manager's time, so she is asked to walk through the steps involved in billing a single patient.

Following each patient visit, the doctor attaches a note that lists the services performed to the patient's medical folder, and gives it to the office manager. An invoice (Fig. 9.1) is prepared using this information. Each procedure or service is itemized. Next, the office manager looks up the charges in a reference book. Some items will be paid by such third parties as an insurance company, medicade, medicare, or welfare, and the manager must identify them; this involves cross checking the patient's insurance coverage (from the medical records) with information from the reference book. After each service is listed and the charges noted, the manager adds the columns to get the total amounts to be billed.

If any services are billed to a third party, the office fills out yet another form. Each insurance company seems to have its own, as does each of the government programs; the analyst is able to count over twenty different third party forms on file! All the forms seem to use essentially the same information copied directly from the invoice, an obvious duplication of effort.

Next, the manager turns to the ledgers (Fig. 9.2). Each patient has a ledger card stored in a large filing cabinet. The patient's card is pulled, and information from the

Fig. 9.1: *The invoice used in the existing billing system.*

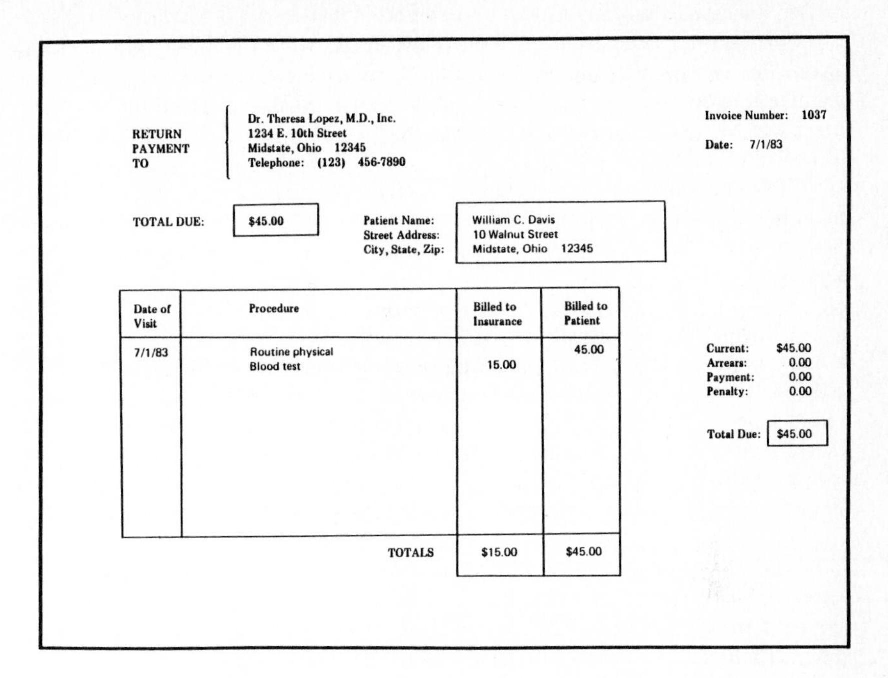

RETURN PAYMENT TO

Dr. Theresa Lopez, M.D., Inc.
1234 E. 10th Street
Midstate, Ohio 12345
Telephone: (123) 456-7890

Invoice Number: 1037

Date: 7/1/83

TOTAL DUE: $45.00

Patient Name: William C. Davis
Street Address: 10 Walnut Street
City, State, Zip: Midstate, Ohio 12345

Date of Visit	Procedure	Billed to Insurance	Billed to Patient
7/1/83	Routine physical		45.00
	Blood test	15.00	
	TOTALS	$15.00	$45.00

Current: $45.00
Arrears: 0.00
Payment: 0.00
Penalty: 0.00

Total Due: $45.00

Fig. 9.2: *A patient ledger card.*

PATIENT: William C. Davis
10 Walnut Street
Midstate, Ohio 12345
(123) 456-1234

PATIENT LEDGER
No. 3718

Date	Invoice	Billed to Insurance	Billed to Patient	Payment Received	Balance Due
6/05/83	Carry forward				$30.00
6/15/83	0935			30.00	0.00
7/01/83	1037	15.00	45.00		45.00

current invoice is copied to a new line. Similar ledgers are kept for each insurance carrier, medicare, medicade, and the welfare department; if services are charged to a third party, the carrier ledger is pulled, and information is copied from the invoice yet again. The ledgers are updated by adding current charges to the balance due, but this arithmetic step does not mask the duplication of effort. Data should enter a system once, and be used again and again without reentry.

"What happens when a payment is received?" asks the analyst. In response, the manager explains how the patient's ledger card is pulled, the payment is entered on a new line, and the amount due is recalculated. Payments from third parties are handled similarly.

"How often are patients billed?" Bills are sent out monthly. The information for a bill is copied directly from the ledger card, and the invoices from each office visit are attached to explain the charges. The office manager considers the process of generating bills highly inefficient, as less than ten percent of the patients are active in any one month, and yet each ledger card must be checked. She has an idea for marking active ledger cards with colored tabs, and thinks it might save a few hours every month. The analyst agrees that the idea has merit, and notes it for future consideration.

The next question is an open one: "Is there anything else you think I should know?" Yes, there are some billing problems that trouble the office manager. One is the need to age accounts. The doctor gives a two percent discount to invoice amounts paid within thirty days of the billing date. Between thirty and sixty days, the full amount of the invoice is charged. After sixty days, interest is charged at a rate of 1.5 percent per month. These calculations are very time consuming, and the office manager admits to numerous mistakes. Once, she tried to charge the government interest on a delayed medicare payment, and learned the hard way that this policy applies only to patients.

Another problem is actually showing the insurance payments on the patient's bill. Technically, a patient is responsible for all charges the insurance company doesn't pay, and most people like to know the status of their claims. It is difficult to cross reference the payment to the patient ledger; in fact, the office manager is still not clear on exactly how this is done.

To summarize what has been learned about the present billing system, the analyst prepares a system flowchart (Fig. 9.3). To begin with, there is a great deal of duplication of effort, with the same information copied over and over again onto slightly different forms. Quality control is lacking, as no one (but the patient) really checks on the accuracy of the output. The procedures are fuzzy in places, with too much left to chance; for example, the office manager could not really explain how insurance payments are credited to a patient's ledger. Finally, there is a clear violation of an elementary business principle: One person both writes the bills and collects the payments, and, without good auditing procedures, that's just asking for trouble.

The office manager also worked with payroll. The nurses and the receptionist are paid an hourly wage with time and a half for overtime. The other doctor receives a salary plus fifty percent of his billings, and this has been a minor problem, as billings are sometimes difficult to accumulate. The analyst notes that until recently, when the

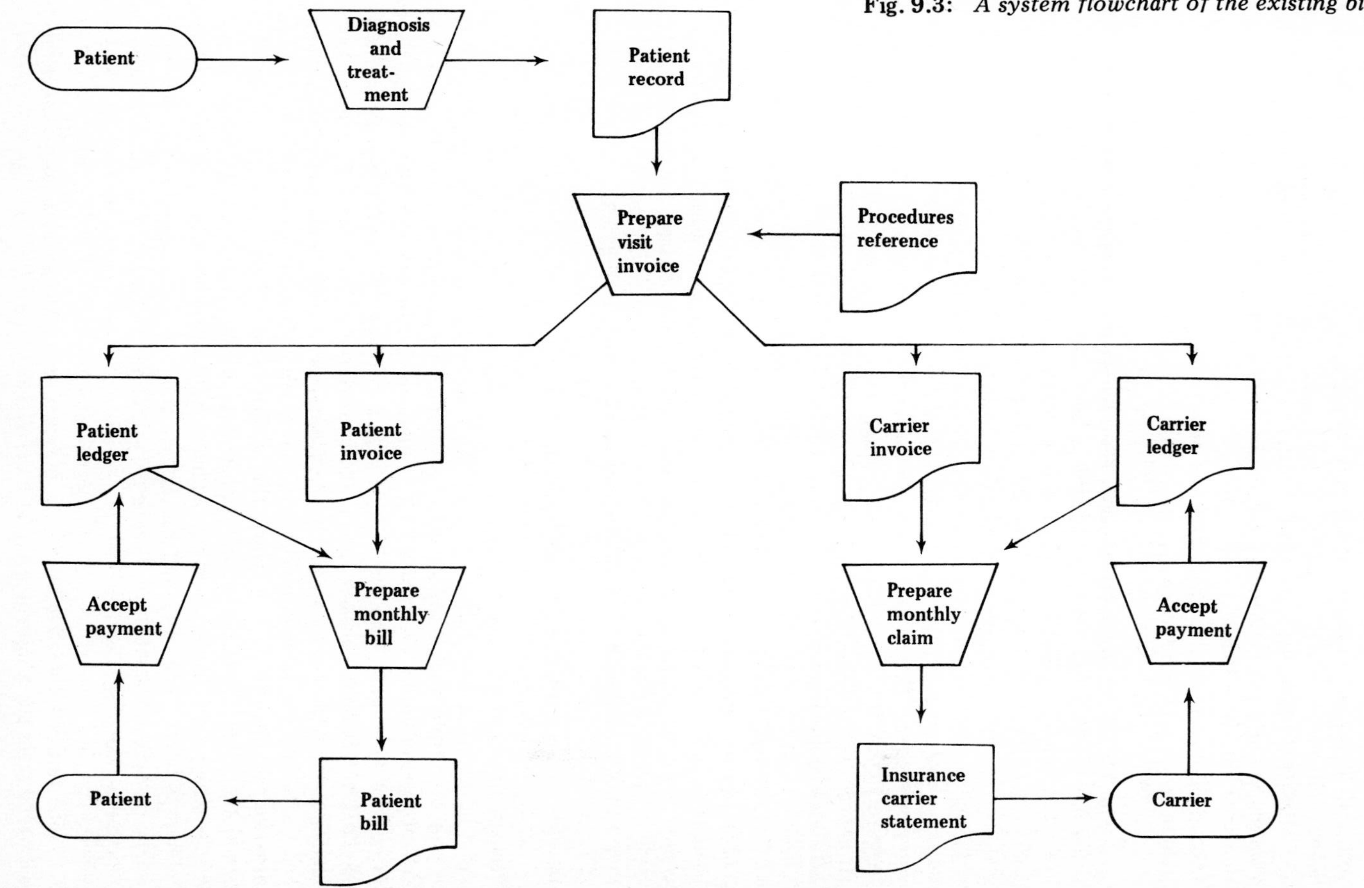

Fig. 9.3: *A system flowchart of the existing billing system.*

doctor assumed responsibility for payroll, the office manager paid herself. No employee should ever be allowed to prepare his or her own paycheck!

After lunch, the analyst interviews the receptionist; she is responsible for the recall system. Once the office manager finishes with the patient's medical folder, it is returned to the receptionist, who files it. First, however, she reads the doctor's comments, and notes recalls on a desk calendar. Lately, she has had some problems with the new doctor; frankly, she can't read his handwriting. Assuming that the need for recall is noted on her calendar, she almost always remembers to call the patient in plenty of time to schedule an appointment. The only trouble is that she can't always reach each patient, and occasionally one or two are never called. She could use some help in keeping track of follow-up calls.

The nurses are interviewed next; the drug inventory is their responsibility. As drugs are received, they are logged into a note book. As they are dispensed, entries are made on the patient's medical folder (drug charges are added to the invoice) and in another note book. No one ever cross checks the two books, however. The problem is that drugs are received in large quantities and dispensed one at a time; for example, how could anyone possibly keep track of the number of aspirin tablets left in a 500 count bottle? As a result, except for an annual physical inventory that is conducted every March, one one really knows exactly what is in the storeroom. Occasionally, the office has run out of a drug, and been forced to buy it at retail cost, an unnecessary expense. There is a problem, and the nurses admit it. They just aren't sure what to do about it.

As a check on the interviews, the analyst collects samples of each of the forms and documents handled by the office staff, and studies them. Added questions arise, leading to brief, informal followup interviews. The present system is indeed a mess, and it might be necessary to start from scratch. It's not all bad, however. The office staff knows that improvements are needed, and will probably not defend the status quo with too much energy; positive change will be welcomed by all.

Develop a High-level Model of the Proposed System

Perhaps the most obvious problem with Dr. Lopez's system is the need to copy the same information over and over again. What essential data must be processed? The copies of various forms collected while studying the present system provide an excellent starting point for answering this question. Extracting the data elements from these forms one at a time, the analyst begins to prepare a data dictionary (see Module E). For example, consider the invoice shown in Fig. 9.1. One data element on the invoice is the patient's name. A data dictionary record is created for "patient name," and the fact that it is found on the invoice is noted. A similar record is created for each data element on the invoice. Next, the analyst turns to the ledger (Fig. 9.2). Once again, the patient name appears. Rather than create a new data dictionary record, a reference to the ledger is simply added to the old one; "patient name" now cross references both the invoice and the ledger. This process is repeated for each of the fields on the ledger, and then the analyst moves on to yet another form, extracting and cross referencing its data elements.

The finished data dictionary contains one entry for each unique data element, with references showing where the element is used. Given the long lists of references, it is clear that most data elements are used in a variety of places; the duplication of effort needed to copy the same element over and over again is obvious. The analyst's suspicions are confirmed; most of the paperwork simply recopies existing data.

Are there other sources that might suggest either the nature of the problem or a possible solution? Surely other physicians have struggled with their paperwork systems; why reinvent the wheel? Thus the analyst schedules interviews with several doctors and their office managers. The insurance companies might be able to help, too. Through these interviews, the analyst is able to identify a standard invoice acceptable to almost all insurance companies; by itself, this could greatly simplify paperwork flow.

Ignoring the largely artificial distinctions between different patients and different third parties, what are the basic functions that must be performed by the billing system? The analyst can identify four. Diagnosis and treatment defines the services to be charged. Given these data, the office manager can generate the charges. Next, the invoices are written, and eventually sent to the patients and to the insurance carriers—the distinction between writing the invoice and writing the bill is largely an accident of timing, as exactly the same information is involved. Finally, payments are accepted and recorded on the appropriate ledger. Figure 9.4 shows the analyst's view of this logical system; note how much of the physical detail has been removed.

Fig. 9.4: *A data flow diagram of the logical billing system.*

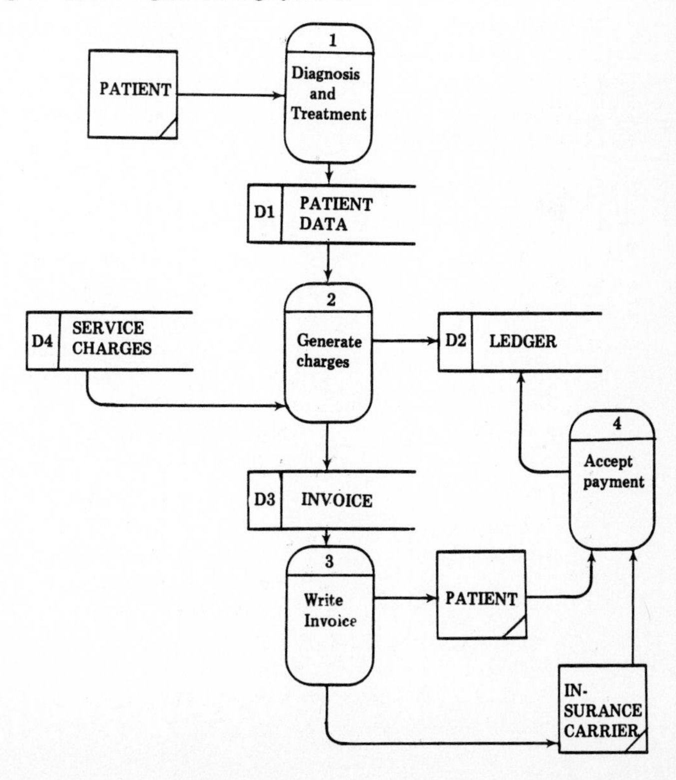

Payroll was covered in Part I of this text. The doctor's logical payroll system (Fig. 9.5) is even simplier than THE PRINT SHOP's, so we won't discuss the details. Patient recalls (Fig. 9.6) are equally simple, with the diagnosis and treatment process triggering a recall notation, which eventually leads to the patient recall.

Drug control is a bit more difficult to define, simply because there is no present system. Borrowing from prior experience in other settings, the analyst develops a logical drug inventory system (Fig. 9.7). When a supplier ships drugs, they are accepted and noted on a drug inventory data store. Diagnosis and treatment is the process that consumes the drugs, updates the drug inventory, and generates reorders when necessary. The office manager uses the reorders as a trigger to order new drugs from the supplier, and the cycle starts again. It's a very basic inventory control system.

Fig. 9.5: *The data flow of the payroll system.*

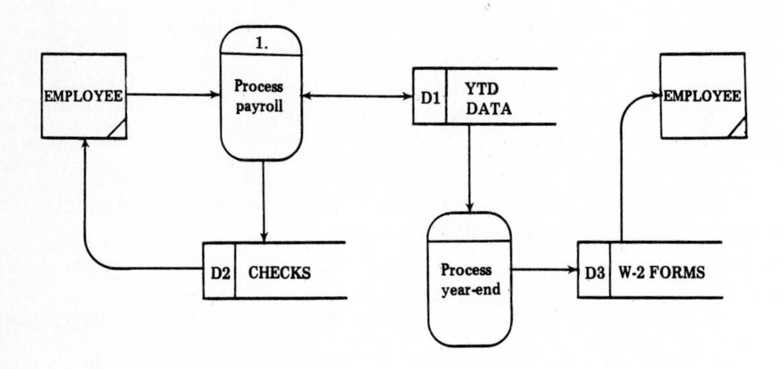

Fig. 9.6: *The data flow of the patient recall system.*

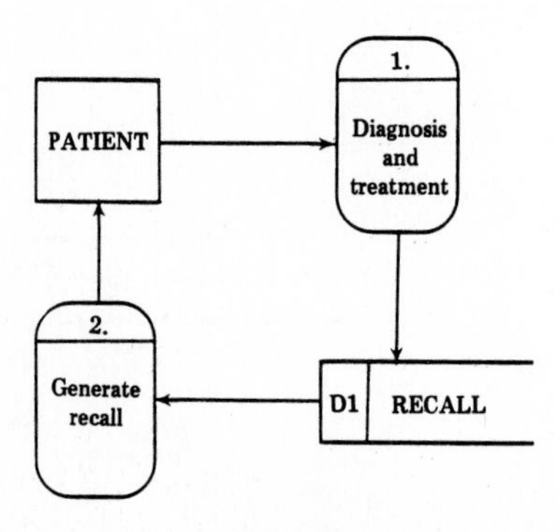

Fig. 9.7: *The data flow for the proposed drug inventory system.*

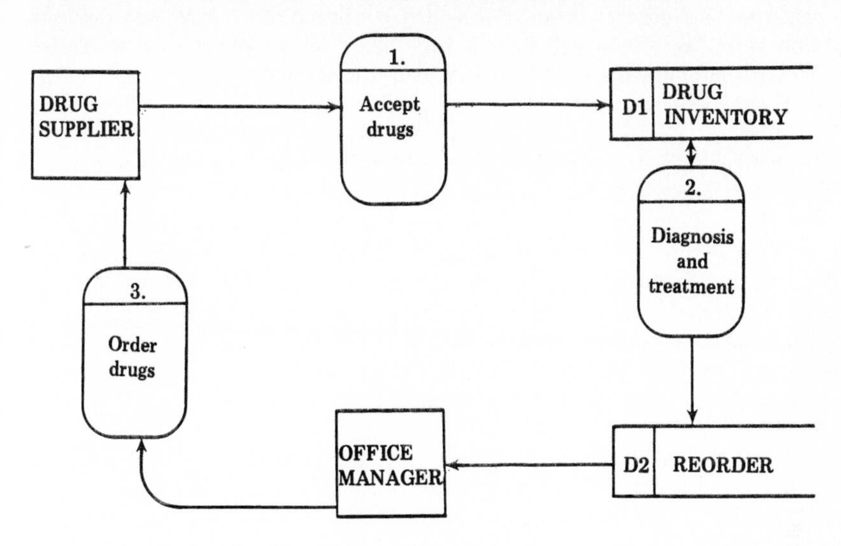

Define the Problem in Light of New Knowledge

Dr. Lopez has, of course, received daily updates on the analyst's progress, and she is pleased. Still, clear communication is crucial if a successful system is to be designed and installed; nothing should be left to chance, and approval should never be assumed. Thus, a more formal meeting is scheduled. The analyst describes the logical system, and makes sure that the doctor really understands it. Operational feasibility is every bit as important as technical and economic feasibility, so the doctor is asked if she objects to any preliminary ideas. The use of a common form for all (or almost all) third party billings is greeted enthusiastically. The way the analyst has simplified the data flow through the office makes sense, too. The doctor is satisfied that the analyst's view of the problem and ideas for solving it are compatible with hers. The time has come to begin evaluating alternative physical solutions for feasibility.

Develop and Evaluate Alternative Solutions

To an experienced analyst, it is obvious that a computer or even an improved set of manual procedures could solve many of the doctor's problems; technical feasibility is almost a given. The analyst decides to concentrate on a small computer system first, as it seems intuitively the best choice. Such operational criteria as ease of use are difficult to judge at this stage, but the analyst has enough experience to know that easy-to-use, turnkey systems are commercially available. What about economic feasibility, however? Can a user-friendly computer system be installed for less than $20,000?

To find out, the analyst begins by compiling a list of the system's primary functions: invoicing or billing, payroll, patient recalls, and drug inventory. Most technical libraries have reference manuals that describe commercially available computer systems, listing hardware features, software, and prices. A few hours of research quickly reveals that systems containing what seem to be the necessary elements can be purchased for between $8,000 and $15,000. Advertising for turnkey systems in medical journals confirms this general range. For added confirmation, the analyst contacts a friend who works as a salesperson for a large computer supplier. Their complete physician's office system sells for $16,000. Clearly, a computer system can be purchased for less than $20,000.

An alternative might be to purchase a low cost microcomputer system and develop original programs custom designed for Dr. Lopez. The idea is intriguing to the analyst, as a medical practice package might be sold to other doctors. Perhaps a special price could be arranged, with Dr. Lopez paying only part of the development cost in exchange for granting Systems Analysis, Inc. a right to market the system. There is, however, the question of operational feasibility. Dr. Lopez wants a system that is easy to use, reliable, and dependable. Systems Analysis, Inc. is not equipped to handle a long-term maintenance committment. While this option certainly should be discussed, the analyst cannot in good conscience recommend it.

Streamlining the manual procedures is a very real possibility; no one said that Dr. Lopez had to use a computer. The data flow diagram has already suggested areas where effort might be concentrated and significant improvements are possible for an investment of perhaps $10,000. This is an attractive low cost option.

Decide on a Recommended Course of Action

The analyst believes that Dr. Lopez should purchase a small computer system. During analysis, a set of functional specifications will be developed. During system design, general physical specifications will be written and sent to a number of potential suppliers for bid. The various alternatives will then be compared with the doctor's actual requirements during detailed design, and the best fit system selected. Finally, during implementation, the system will be ordered, installed, customized, and tested. The total cost, including the analyst's fees, should not exceed the $20,000 target, although it could go as high as $25,000.

By accepting the analyst's recommendation, the doctor will essentially be committing funds to pay for an in-depth problem analysis. It appears that there is a feasible solution, but some questions still remain unanswered. The doctor will face at least two more decision points before being asked to commit the bulk of her investment, but additional work by the analyst will still be expensive, and she should know exactly what she is paying for. Thus, one more step must preceed the final presentation of the results of the feasibility study.

Rough Out a Development Plan

The two key questions are: how long will it take, and how much will it cost? Thus the analyst prepares a rough schedule and a cost/benefit analysis. Consider the schedule (Fig. 9.8) first. It shows the estimated elapsed time and cost for each stage in the process. Note that the two week feasibility study cost $2000, while two weeks of analysis will cost only $1000. Why? The analyst will be working less than full time on the doctor's system, and she should know this. The plan calls for completing and installing the system in three months.

Fig. 9.8: *A rough implementation schedule.*

Step	Elapsed time	Cost
Feasibility study	completed	$2000
Analysis	2 weeks	1000
System design	2 weeks	1000
Detailed design	1 month	1000
Implementation	1 month	1000
TOTALS	**3 months**	**$6000**

Next consider the cost/benefit analysis (Fig. 9.9). The system will cost $20,000. Operating expenses should be about $3500 per year, and quantifiable savings will be perhaps $4000 per year, yielding a net annual benefit of only $500. A $500 annual return on a $20,000 investment is equivalent to 2.5 percent simple interest, and no one would invest money at that rate. The cost/benefit analysis also shows a computation of the net present value using an 18 percent discounting rate (see Module G). The net present value is negative; the cost of the system exceeds the tangible benefits by $18,436.42. Dr. Lopez has indicated that a number of intangible factors such as control of her practice and professional image are the real concerns. The cost/benefit analysis shows that these intangibles will cost her the equivalent of $18,436.42. Are they worth that much? Only Dr. Lopez can answer this question. The cost/benefit analysis gives her a basis for making the decision.

Write and Present the Feasibility Study Report

Finally, the feasibility study is finished. The results are compiled, and a report (see Module C) written. A formal meeting with Dr. Lopez is scheduled. She is surprised by the negative cash flow shown on the cost/benefit analysis, and thinks long and hard before reaching a decision: yes, she will purchase a computer system. Thus, the analysis stage begins.

SUMMARY

This chapter discussed the feasibility study for a proposed office management system. After clarifying the scope and objectives, the analyst began to study the existing system. It was a mess, with duplication of effort and an almost complete lack of controls. Creating a preliminary data dictionary containing the data elements found on the various forms used by the office helped to reenforce the analyst's view that the present system was inefficient. Logical models of the proposed new system elements were prepared using data flow diagrams. The doctor approved these preliminary ideas, so the task of generating alternatives began. Of three alternatives considered, a small computer system was the analyst's recommended choice. After preparing a rough implementation plan and a cost/benefit analysis, the feasibility report was written and

Fig. 9.9: *An estimated cost/benefit analysis.*

Development cost	
Analysis	$ 6,000
Hardware & software	12,000
Customizing	2,000
TOTAL	**$20,000**

Operating costs	
Maintenance	$ 2,500
Utilities	500
Supplies	500
TOTAL	**$ 3,500**

Cost Savings	
Cash flows	$ 2,000
Reduced overtime	1,000
Inventory control	1,000
TOTAL	**$ 4,000**

Net Present Value Analysis (18% discounting)

Year	Net benefit	Present value	Cumulative present value
1	500	423.73	423.73
2	500	359.09	782.82
3	500	304.32	1087.14
4	500	257.89	1345.03
5	500	218.55	1563.58

Net present value = [1563.58 − 20,000.00] = − [18,436.42]

presented to the doctor. Her decision, in spite of a poor return on investment, was to accept the analyst's recommendation.

EXERCISES

1. Why is it so important that the analyst and the client agree on a project's scope and objectives?

2. Develop a system flowchart to describe how your major department updates your advisor's student records.

3. Develop a logic flowchart to describe the algorithm for aging accounts in Dr. Lopez's existing billing system. Use structured English to describe the same algorithm.

4. In the text, the office manager was not clear on how insurance payments are credited to patient accounts. Think about this problem. How would you do it?

5. Using the data elements on the invoice (Fig. 9.1) and the ledger (Fig. 9.2), prepare a complete data dictionary (see Module E).

6. Assume that accounts payable is another problem area. When the office manager or one of the nurses orders supplies, the order is dated and entered in a log book. When the order arrives, this fact is recorded on the same log book line. Later, when the bill arrives, the office manager checks the log book to see if the order has arrived before paying for it. Draw a system flowchart to describe this process. Draw a data flow diagram that reduces the process to its logical elements.

7. Distinguish, once again, between technical, operational, and economic feasibility.

8. In developing alternative solutions, the analyst spent a few hours researching technical literature, and then asked a friend for an opinion. Was this adequate? What do you think? What exactly is the purpose of a feasibility study?

9. Why is it important that the analyst recommend a course of action?

10. Use the data of Fig. 9.9 to compute a payback period. Explain what the payback period means. (See Module G.)

11. Chapter 8, exercise 10 suggested other tasks that might be problems in a doctor's office. Select one or more of these tasks, and incorporate them into the feasibility study.

12. Continue with Chapter 8, exercise 11, and conduct a feasibility study.

10

Case B Analysis

OVERVIEW

In this chapter, we will consider the analysis step in the office management system project. We'll start with the data flow diagram from the feasibility study, and design a new logical system. Next, the data flows will be analyzed using a data dictionary, and algorithm descriptions will be developed. Exploding the data flow diagram will help us define additional details about the new logical system. Analysis will end with a presentation to the client, Dr. Lopez.

ANALYSIS: OBJECTIVES AND CRITERIA

Analysis follows the feasibility study. We know that the problem can be solved. The objective is to determine, at a logical level, exactly what the proposed system must do. The exit criteria include a complete data flow diagram, a data dictionary, and a set of preliminary algorithm descriptions.

The existing office management system is clearly not adequate. It was never really planned; it just evolved. Because so much of the system reflects the personality and style of a former employee the problem is compounded. The existing system will prove of little use in designing a new one; the analyst will be starting almost from scratch.

ANALYSIS AND INTUITION

Dr. Lopez's existing office management system is unacceptable, and automating a bad system will not make it better. In developing a basic design for the new system, the analyst will be forced to rely on experience and intuition. The feasibility study provides a starting point. Consider the four data flow diagrams from Chapter 9 (Figs. 9.4 through 9.7). One process, *Diagnosis and treatment* (the central task of the office), is common to three of the four major systems: billing, patient recalls, and drug inventory. Perhaps a system can be designed with this process as the driving mechanism. Thus the analyst prepares a new data flow diagram (Fig. 10.1) with *Diagnosis and treatment* as the central process.

Follow the data flow through the logical system. A patient is treated. As part of the treatment, injections and/or other drugs may be administered, so DRUG INVENTORY is accessed. Services performed are noted on PATIENT DATA. After the patient has been treated, the next process, *Generate charges*, accesses PATIENT DATA and DRUG INVENTORY and makes entries on LEDGER and INVOICE. Additionally, a notation (if necessary) can be made on the RECALL data store. Finally, after checking the drug inventory, necessary reorders can be generated. Eventually, the contents of the various data stores can be used to write invoices, note payments, generate recalls, and order drugs.

Note how the data generated by one process, *Diagnosis and treatment*, are used to update every data store; can you see how this process drives the system? *Generate charges* is responsible for processing all these data. Notice, however, that one data store, DRUG INVENTORY, can be updated by two different processes. Remember the doctor's concern about a lack of control over drug inventory? With the proposed system, responsibilities for withdrawing items from inventory and tracking inventory levels are separate. In effect, the drug inventory is double checked, an important auditing consideration.

Pause briefly and consider the source of the initial design of Fig. 10.1. Where did it come from? What was the analyst's model? It is easy to attribute a design's origin to experience and intuition, but what do these terms mean?

Experience comes with time. Often, a young systems analyst is teamed with an older analyst, in hopes that some will rub off, but experience is also a function of

Fig. 10.1: *A data flow diagram of the proposed logical patient system.*

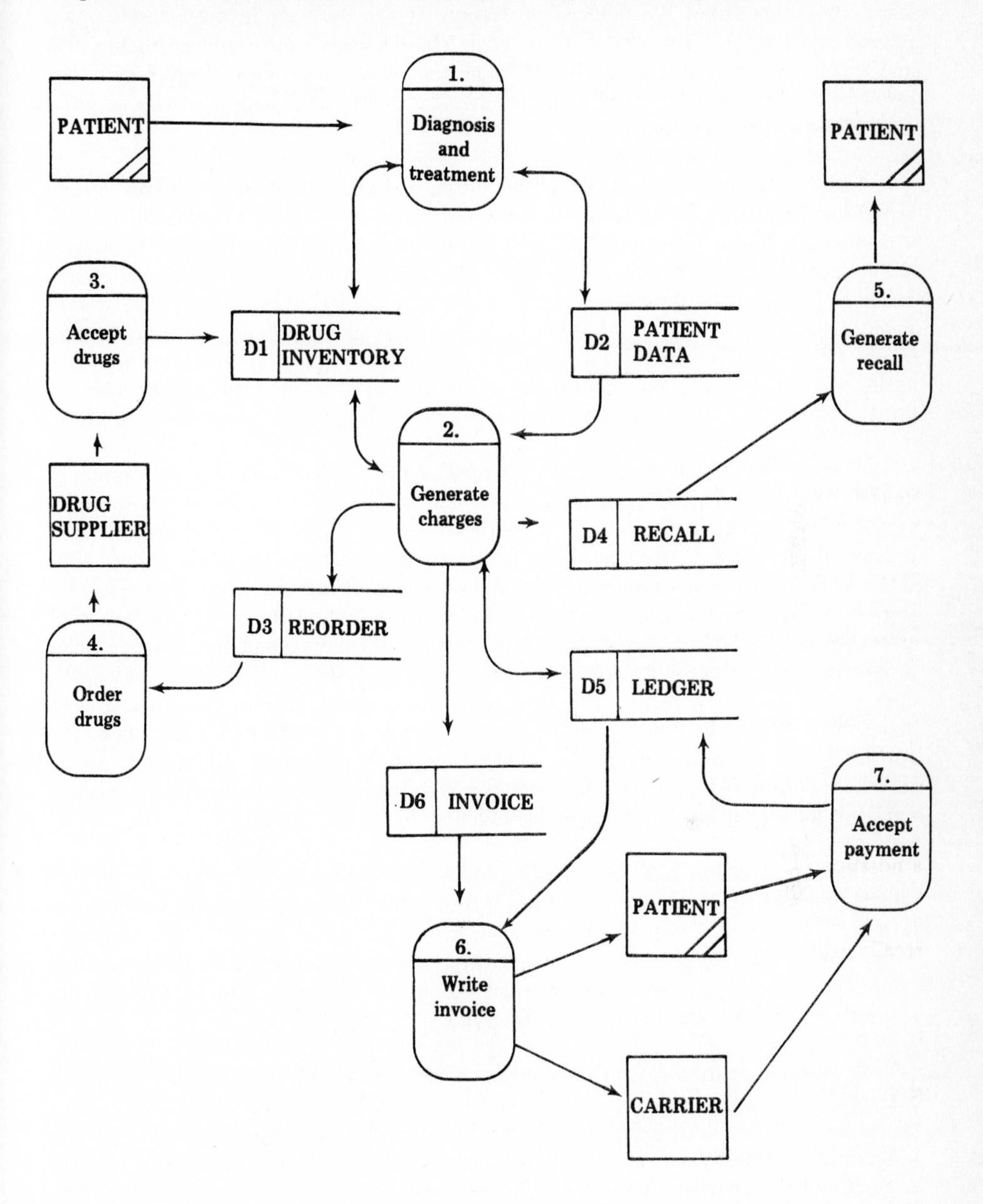

preparation. A good systems analyst working for a physician will take the time to learn something about the medical profession. If the next job involves developing a payroll system for a print shop, the analyst should study both the printing business and the payroll application. Let's put it another way. Experience is not merely the result of the passage of time; it results from hard work and intelligent study. A good analyst never stops learning.

Intuition is more difficult to define. Often, the intuitive element involves sensing a pattern. In THE PRINT SHOP, the old payroll system provided a reasonable pattern or model for the new one, but Dr. Lopez's system is unacceptable. Perhaps the analyst might spot similarities between her needs and an office management system previously installed for another physician, or even for a non-medical office; if so, that system might serve as a starting point. There is a danger in this approach, however. All too often, a less capable analyst will continue to install the same system, over and over again, whether it fits or not, because that system is all he or she really knows. There is no such thing as a universal system. Find an *appropriate* pattern, not merely a pattern.

The analyst of our case study developed a new logical system from scratch. A key element was identifying a *driving mechanism*. Most systems are driven by some event or group of events, usually a particular set of input data. For example, the process of updating checking accounts is driven by a batch of checks and deposit slips that can be sorted by account number and processed sequentially. In contrast, consider a 24-hour bank teller system. This system is driven by customer transactions that arrive in random order. While both the master file update and the 24-hour teller programs may reduce the current balance of a customer's checking account, the designs of these two programs will be radically different, simply because their driving mechanisms are so dissimilar. Identifying the driving mechanism of a system gives the analyst a starting point. From this starting point, a variety of alternatives can be sketched, considered, and, perhaps rejected. Given the driving mechanism as an anchor, a reasonable logical system can often be designed.

In part, intuition is a natural skill; some people have it, and others don't. The creative individual does have an advantage, but intuition is no substitute for preparation and experience. For example, the analyst working on Dr. Lopez's office management system identified a driving mechanism, the data generated by *Diagnosis and treatment*. An analyst, even a highly creative one, who lacked an understanding of the medical profession might have missed this point.

A systems analyst is a professional problem solver, whose work is difficult to categorize. At times, applied mathematics will be enough to solve the problem. Other tasks may call for a knowledge of computer programming or computer hardware, while others may demand personal communication skills, or flashes of insight and creativity. A good systems analyst is prepared to do whatever is necessary to solve the problem. Often, this involves learning a new body of material quickly. Systems analysis is such a broad profession that technical training, by itself, is not enough. In addition to knowing how to apply certain tools and techniques, the analyst must also know how to learn. The ability to study and absorb new information quickly may well be the key to gaining experience rapidly, and preparation is certainly an element of what we call intutition.

ANALYZING THE DATA FLOWS

The high-level data flow diagram represents a broad overview of the proposed logical system. Is the design reasonable? Will it do the job? The best way to find out is to trace the data flows through the system. During the feasibility study, the analyst started a data dictionary. The various forms processed or prepared in the doctor's office were, as you may recall, collected, and the data elements extracted, one at a time; for example, Fig. 10.2 shows entries for three elements: *patient name*, *current charges*, and *balance due*. The first step in analyzing the data flows is to complete the data dictionary.

Note (Fig. 10.2) that *patient name* appears on both the invoice and the ledger. Note also that *current charge*, in addition to appearing on both these documents, is known by two other names: *billed to patient* and *current*. One objective of a data dictionary is to define and cross reference unique data elements, and to highlight those redundant elements that appear, perhaps in different guises, in several places. At this stage, we will not consider the data structures; thus, the fact that the amount "billed to patient" seems a repeating element on the ledger is temporarily ignored.

Once a data dictionary has been developed, we can begin to trace the data flows. Start with the data destinations. Identify the elements flowing to a destination and trace the flow back, element by element, through the data flow diagram. For example, consider the balance or total due appearing on the invoice and on the ledger (Fig. 10.2). Near the lower right of Fig. 10.1, start with the patient. The balance due flows to the patient; what is the source of this data element? It is computed in process 6, *Write invoice*. The total amount due is computed (using information from the INVOICE and LEDGER data stores) by adding current charges, any prior balance due, and a service charge (if appropriate); an algorithm description in IPO diagram form is shown as Fig. 10.3. Consider the inputs to this algorithm: current charges, the prior balance due, and the dates associated with prior amounts due (for aging the account). Where do they come from? The current charges (there may be several) come from process 2, *Generate charges*. Another algorithm is needed; perhaps a list of standard charges by service performed. Working methodically, the analyst can trace each data element from its destination back to its ultimate source, highlighting the necessary algorithms and defining new data elements along the way.

Let's assume that, as part of this process, a new, independent data store was identified. We won't add it to Fig. 10.1, but look for data store D7, MEDICAL HISTORY, in subsequent data flow diagrams.

EXPLODING THE DATA FLOW DIAGRAM

Let's assume that, by tracing the data flows, the analyst is able to account for each data element processed by the system. The preliminary design makes sense. Before we move to system design, however, it should be analyzed in greater detail. An excellent approach is to explode each process on the data flow diagram.

We'll begin with *Diagnosis and treatment*. What functions must be performed by this process? The analyst can identify three:

Fig. 10.2: *A portion of the data dictionary for the office management system.*

Name: Patient name
Aliases: Patient; Name
Description: The name of a patient in the form:
Last, First Initial.
Format: Character
Location: Invoice
Ledger

Name: Current charge
Aliases: Billed to patient; Current
Description: The charge to the patient for a single office visit.
Format: Numeric; maximum value = 999.99.
Location: Invoice
Ledger

Name: Balance due
Aliases: Total due
Description: The total amount that a given patient owes the doctor. The sum of current charges, the prior balance due, and any service charge.
Format: Numeric; maximum value = 9999.99.
Location: Invoice
Ledger

Fig. 10.3: *An algorithm description for computing the balance due.*

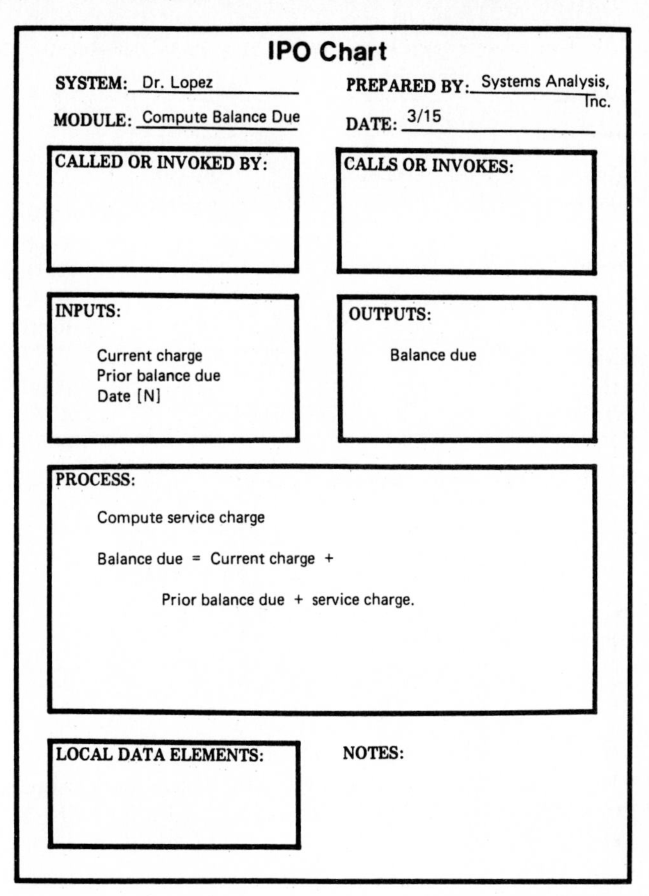

IPO Chart

SYSTEM: Dr. Lopez

PREPARED BY: Systems Analysis, Inc.

MODULE: Compute Balance Due

DATE: 3/15

CALLED OR INVOKED BY:

CALLS OR INVOKES:

INPUTS:

Current charge
Prior balance due
Date [N]

OUTPUTS:

Balance due

PROCESS:

Compute service charge

Balance due = Current charge +
Prior balance due + service charge.

LOCAL DATA ELEMENTS:

NOTES:

1. Get the patient.

2. Treat the patient.

3. Record the services performed.

An explosion of *Diagnosis and treatment* is shown in Fig. 10.4; note how the relationships between the various subprocesses and the data stores is clearly illustrated. Can any of these lower level processes be further decomposed? Perhaps treating a patient (1.2) can be broken into components such as a nurse's preparatory work, the doctor's investigation and diagnosis, and the subsequent treatment (an injection, for example), but these steps are a bit too detailed for analysis, identifying implementation details such as who does what.

Fig. 10.4: *An explosion of the Diagnosis and treatment process.*

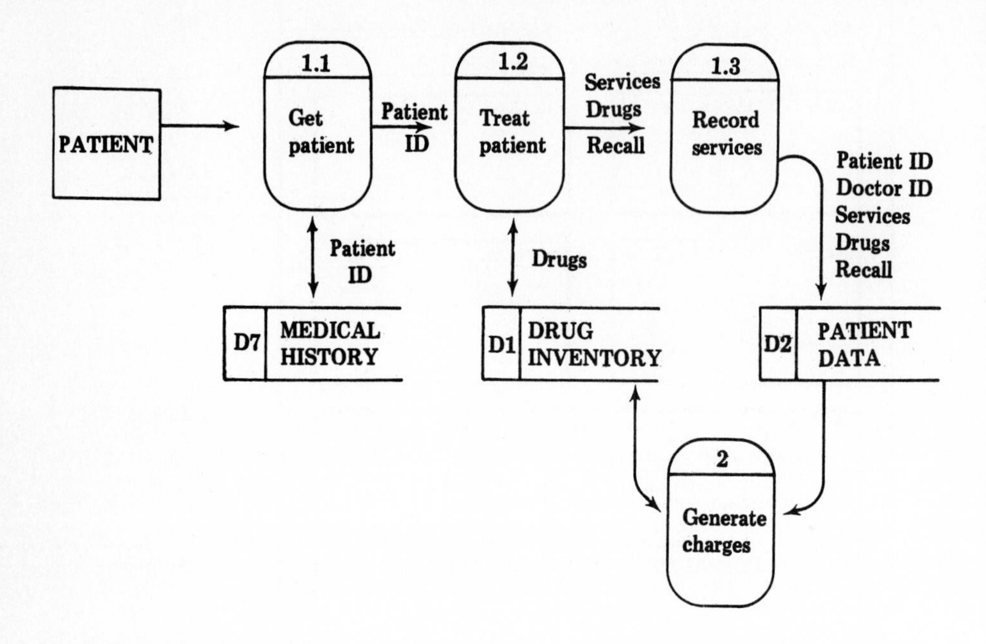

Consider next process 2, *Generate charges* (from Fig. 10.1 again). What functions must be performed in generating charges? Let's assume that, based on prior studies of the system and follow-up interviews, the analyst can identify the following seven:

1. Check the drug inventory for charges.
2. Check the drug inventory for a possible reorder.
3. Accumulate the current visit charges.
4. Update the patient ledger.
5. Update the third party (or carrier) ledger.
6. Prepare an invoice for the current visit.
7. Check to see if a recall is necessary.

An explosion is shown as Fig. 10.5. Can any of these processes be decomposed further? Certainly. For example, what steps are involved in accumulating the current charges (process 2.3)? Charges for drugs, services, injections, laboratory work, and other medical functions are all computed using different algorithms. (We won't show processes 2.31 through 2.35, but expect an end-of chapter exercise.)

Fig. 10.5: *An explosion of the Generate charges process.*

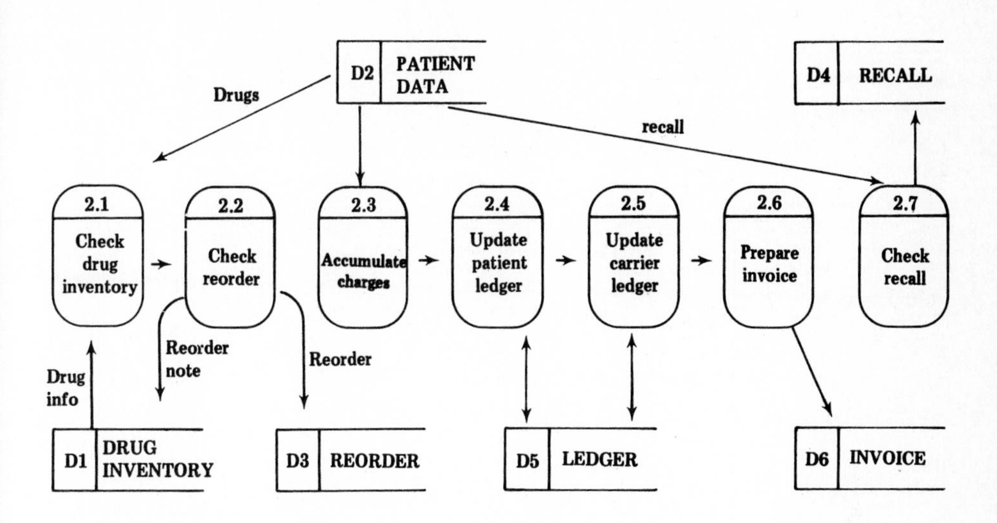

What about process 6, *Write invoice*? What steps are involved? Figure 10.6 shows a functional decomposition of the process. Similar decompositions (not shown in the text) can be developed for the other processes as well. Later, when presenting the proposed logical system to Dr. Lopez and her employees, the analyst will use Fig. 10.1 as a system overview, and then turn to the exploded data flows to explain each process, one at a time. (Note the payroll system has not been discussed. Don't forget that payroll will be included in the doctor's system.)

THE PLAN

The data flow diagrams represent a logical plan for the office management system. The data dictionary and algorithm descriptions provide some detail to support the plan. Next is system design, where the key objective will be to convert the logical system into physical specifications against which the various physical alternatives can be measured. Of primary concern is the driving mechanism, the diagnosis and treatment process. This key to Dr. Lopez's system provides the data used by almost all the other processes. It is essential that this process be designed to simplify the doctor's work while still allowing her to control the practice. The analyst will spend considerable time on the question of how to best collect the source data, and this factor will be a key to selecting a system. Ease of use is, after all, a primary criterion.

Fig. 10.6: *An explosion of the Write invoice process.*

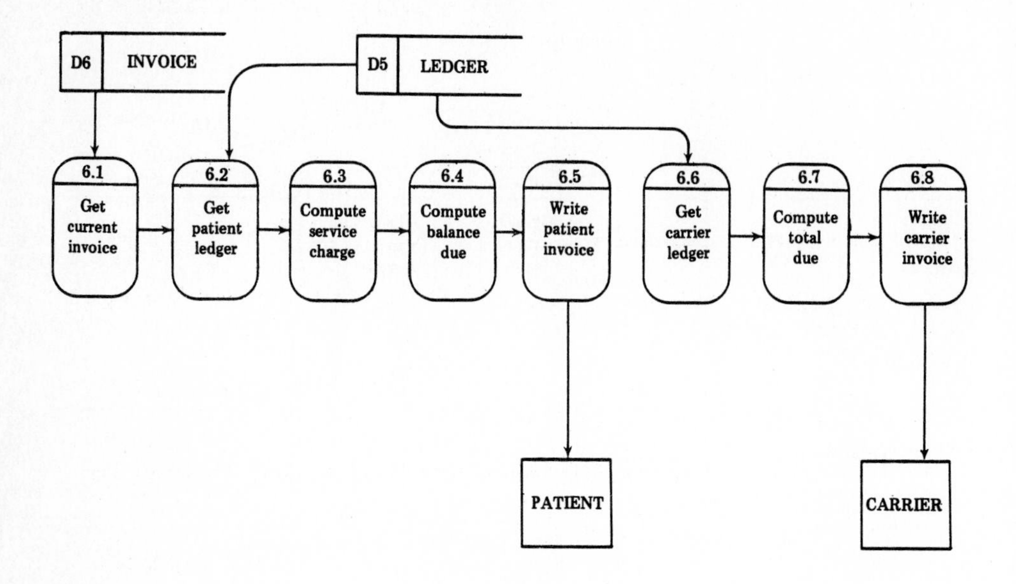

THE PRESENTATION

Dr. Lopez and her employees are not technical professionals, so no inspection is planned. This does not, however, relieve the analyst of the responsibility to present the proposed system clearly. Thus, independent interviews with the office manager, the receptionist, and the nurses are scheduled, and each employee is given an in-depth explanation of her piece of the system. Next, a group meeting involving the doctor and all four employees is scheduled, and the entire system is overviewed. In the interviews and in the group meeting, the analyst is asking, indirectly, a key question: *Do I understand your business*? If the new system is faulty, now is the time to find out.

Finally, the analyst meets with Dr. Lopez. She is pleased with the logical plan, and has little trouble following it. The preliminary schedule and cost estimate prepared during the feasibility study still seem reasonable. Thus the next stage, system design, can begin.

SUMMARY

This chapter discussed the analysis step in Dr. Theresa Lopez's office management system project. Starting with the data flow diagrams from the feasibility study, the analyst envisioned an integrated system driven by the diagnosis and treatment process, and sketched a data flow diagram incorporating the functions of what were previously three separate systems. A data dictionary was developed, and, element by element, the data flows were traced through the logical system. Needed algorithms and additional data elements were identified. By exploding the data flow diagram to a lower level, the logical system was defined in greater detail. Finally, the analyst's plan was presented to the employees and to the doctor, and funds to pay for system design were authorized.

EXERCISES

1. What is the objective of analysis?

2. In designing a new office management system for Dr. Lopez, the analyst was not able to use the old system as a model. Why?

3. Why is it important that the analyst define or identify the driving mechanism of a system? What is a driving mechanism?

4. Why wasn't the payroll system included on the data flow diagram of Fig. 10.1?

5. The text states that experience results from more than just the passage of time. Explain.

6. Intuition, at least in part, involves recognizing patterns. Explain.

7. Explain the process of tracing data flows through a data flow diagram. How is the data dictionary involved? How are algorithms highlighted by this process?

8. Explain how a data dictionary can be used to identify redundant data elements. Why are aliases maintained on a data dictionary?

9. The process *Generate charges* was decomposed into seven lower-level elements in the text. Further decomposition is possible. Starting with Fig. 10.5, decompose each process to an appropriate level.

10. In the text, decompositions were not shown for processes 3, 4, 5, or 7 of Fig. 10.1. Based on what you know about the doctor's practice (from Chapters 8 and 9), decompose these processes.

11. Develop a logical plan for the doctor's payroll system. You may want to refer back to the first case study. Additional details can be found in Chapters 8 and 9.

12. Why is it so important that the analyst clearly present all work to the client?

13. Continue with Chapter 9, exercise 11.

14. Continue with Chapter 9, exercise 12.

11

Case B System Design

OVERVIEW

Continuing with the office management system example, this chapter considers the system design step. A series of automation boundaries is assumed, and alternative implementation strategies identified. Three reasonable alternatives are selected and evaluated, and a recommendation is made to the client.

THE OBJECTIVES OF SYSTEM DESIGN

Analysis is now complete. The next step is system design, where the analyst will determine, at a black-box level, how the system should be implemented; physical design begins here. During this step, several alternative strategies will be considered and evaluated; the objective is to select one as a blueprint for the system. The exit criteria include a system flowchart and a cost/benefit analysis for each of several reasonable options.

Note, however, that Dr. Lopez is not a technical professional, does not intend to become a technical professional, and does not even plan to hire one. She needs a system that can be used by people with little or no technical training. Almost certainly, her system will be purchased, not custom designed. Physical specifications are still needed; in other words, the necessary programs, files, and forms will be identified. However, no vendor can realistically be expected to have precisely what Dr. Lopez needs. The best the analyst can hope to do is obtain a reasonable balance between the ideal system and what is commercially available. The specifications will be used as a standard against which alternatives can be measured.

GENERATING ALTERNATIVES

The first step in system design is to generate alternative physical strategies for solving the problem. Let's assume that the analyst is able to identify three:

1. An expensive, fully automated system.

2. An intermediate cost, partially automated system.

3. An inexpensive, improved manual system.

Note that the general guideline of expensive, intermediate, and inexpensive options has been followed.

EVALUATING THE ALTERNATIVES

The Fully Automated System

The fully automated system (Fig. 11.1) places the doctor on-line. Data concerned with the patients, the insurance carriers, the drug inventory, payroll, and standard service charges are maintained in an integrated data base that can be accessed through a CRT terminal. As services are performed, the appropriate data are entered into the system; charges are automatically added to the patient's account, and a notation made on the medical history. Also, using these same data, the system checks to see if the service is covered by a third party such as medicare or an insurance company and, if it is, the charge is debited to the appropriate carrier. When drugs or injections are administered, data are once again entered through the CRT. In response, the system automatically updates the inventory, and charges the service to the patient and/or the

Fig. 11.1: *A fully automated system for the doctor's office.*

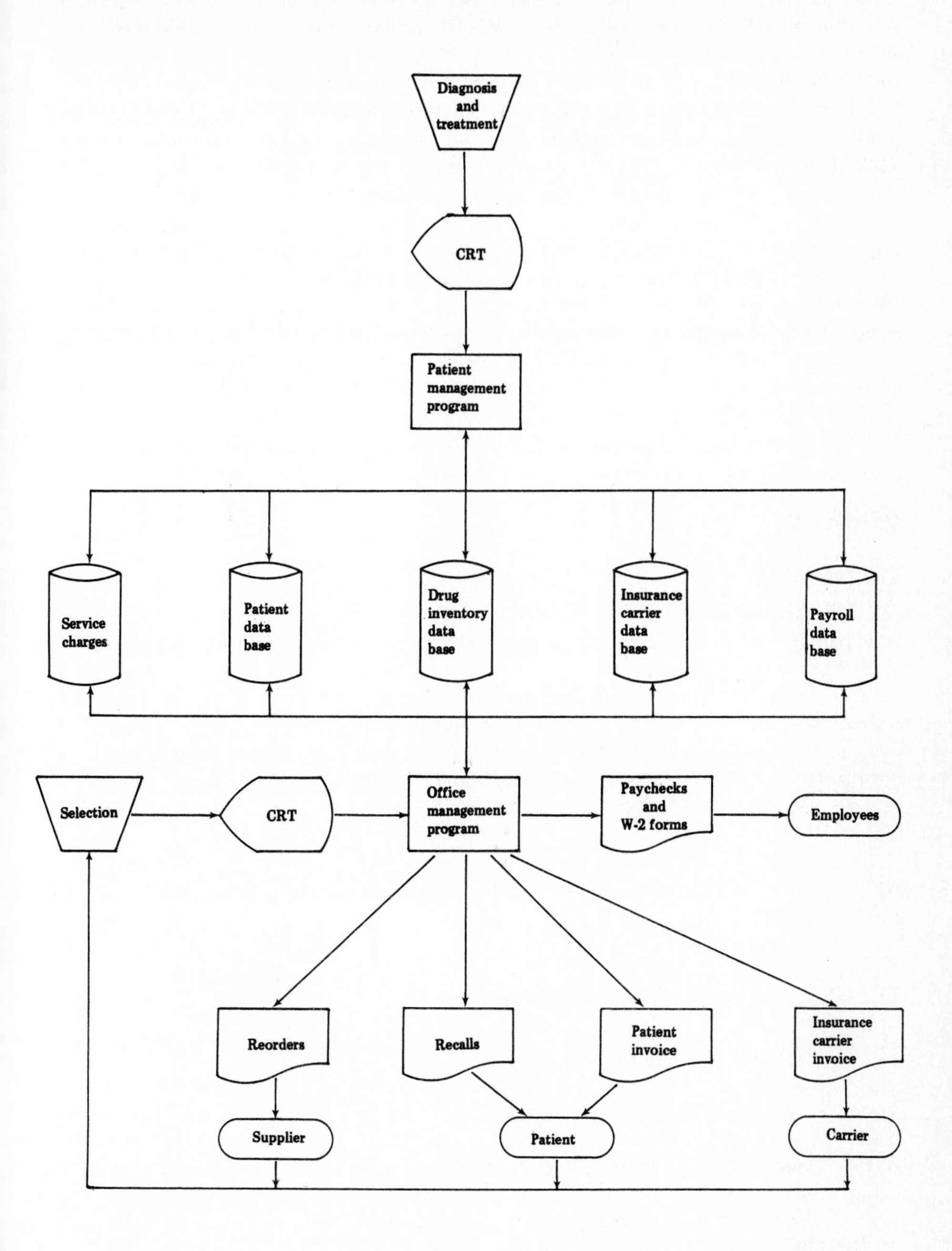

insurance carrier. If a recall is necessary, the doctor simply enters the fact, and the system notes it for later use. In other words, all functions associated with recording and charging for services performed are automatically handled by the system as a direct result of entering the data.

Not all activities take place as a patient is being treated, of course. Issuing patient recalls, reordering drugs and supplies, and posting payments received are daily tasks, while payroll is processed weekly, and invoices mailed monthly. To deal with the batch functions, the analyst has envisioned an office management program (Fig. 11.1 again). A nonmedical employee, such as the office manager or the receptionist, accesses this program via CRT terminal, and selects from a menu of options (a tentative list is shown in Fig. 11.2). For example, consider "Recall patients." If this option is selected, the clerk receives (based on some algorithm) a list of patients to be recalled today. In response, telephone calls are made, and appointments are scheduled and noted on the system. Patients not contacted—for whatever reason—are added to tommorrow's recall list.

Invoices are normally generated monthly. To perform this task, the clerk starts with the standard menu and selects "Generate patient invoices" (see Fig. 11.2). In response, clear explicit, step-by-step instructions are displayed, walking the clerk through the process. For all practical purposes, the task of preparing invoices is completely automated.

With this system a patient can be given an up-to-date invoice at the time of the visit; perhaps the doctor can offer a small discount for prompt payment. Additionally, insurance company invoices might be accelerated. Many carriers accept charges either once a month or whenever the total claimed exceeds a minimum level. With the new system, it will be possible to monitor each insurance carrier for the appropriate claim level, and automatically generate an invoice as soon as the critical point is reached. Both features should help the doctor's cash flow.

Fig. 11.2: *Tentative options for the office management program menu.*

1. RECALL PATIENTS
2. WRITE REORDER FORMS
3. POST PAYMENT RECEIVED
4. GENERATE PATIENT INVOICES
5. GENERATE CARRIER INVOICES
6. PROCESS WEEKLY PAYROLL
7. PROCESS YEAR-END PAYROLL

Note that the office manager isn't mentioned in the above discussion. With a fully automated system, this position could be eliminated, and the receptionist trained to run the clerical access program. Considering the present office manager's salary and benefits, and a possible salary increase for the receptionist, the doctor should save about $18,000 per year in personnel costs. Improving the cash flow should save about $3500 annually, while better inventory control could account for another $1000, yielding an estimated annual benefit of $22,500 (Fig. 11.3).

Of course, there will be additional costs. The analyst estimates a development cost of $35,000 to $40,000, considerably above the projected scope of the system. Maintenance will be expensive; $6000 per year seems reasonable. Supplies should run about $1500. Subtracting the estimated operating costs from the tangible benefits yields

Fig. 11.3: *A cost/benefit analysis for the fully automated option.*

Development Cost: $35,000 to $40,000

Annual Cost Savings		Annual Cost Increases	
Personnel	$18,000	Maintenance	$6000
Cash flow	3,500	Supplies	1500
Inventory	1,000		
Total	**$22,500**	**Total**	**$7500**

Net Annual Savings = $15,000

Assuming an 18% discounting factor

Year	Net Benefit	Present Value	Cumulative present value
1	15,000	12,711.86	12,711.86
2	15,000	10,772.77	23,484.63
3	15,000	9,129.46	32,614.09
4	15,000	7,736.83	40,350.92
5	15,000	6,556.64	46,907.56

Net Present value = [$46,907.56 − 40,000] = $6,907.56
Return on investment = 25% to 26% Payback period = 3.95 years

a net annual savings of $15,000 on an investment of $40,000. Although the initial cost seems high, there is a significant return on this investment (see Fig. 11.3).

Intangible factors are also important. A fully automated system would eliminate the key employee problem, and greatly simplify training new employees. However, the doctor would be almost completely dependent on an office management system that she understands only superficially. As a result, she would be at the mercy of maintenance services completely beyond her control, trading one dependency problem for another.

The Intermediate Cost Alternative

Partial automation represents another option (Fig. 11.4). The system begins with improved manual data collection procedures. After a patient is treated, the new forms are given to the office manager who enters the data via CRT terminal. Several different programs are used to update the drug file, the patient ledger, the insurance carrier file, and the patient recall file. Options of these same programs are also used to generate reorders, patient invoices, and third party invoices. Additionally, the same CRT terminal can access programs to accept drugs, accept payments, and process both the weekly and year-end payroll. Finally (Fig. 11.4), another program reads the recall file and prints a list of patients for the receptionist to contact.

The total cost of this system should be about $20,000, which is within the initial scope of the project. Cost savings will be limited, however. Earlier, during the feasibility study, the analyst prepared a cost/benefit analysis for this system (Fig. 9.9), and sees no reason to change the estimates. For all practical purposes, there is no return on this investment.

There are some benefits. The cost falls within the estimated scope of the project, and Dr. Lopez can afford it. The key employee problem is minimized. Since the system now explicitly defines each process, training is greatly simplified. Additionally, the system will be far more consistent and predictable than the existing one, so embarassing errors should be less common. However, the bottom line is still dollars and cents. To get a return on her investment, Dr. Lopez must spend more than she planned. If she spends only what she has budgeted, she will not get her money back. Perhaps she should seriously consider a manual system.

The Manual System

The flow of paperwork through Dr. Lopez's office is highly inefficient. Simply cleaning up this flow might lead to significant improvements; in fact, it's common for an office of this size to achieve 50 percent of the benefits of a computer without actually installing one. The process involves developing better forms and simplified, more explicit procedures. As a result, data accuracy can be improved, the key employee problem reduced (but not eliminated), training simplified, and the inventory brought under control.

The analyst believes that, for an investment of about $10,000, half the cost of the intermediate system, Dr. Lopez can expect to reap about half the benefits, roughly $2000 per year. With such a system, there would be no increase in maintenance or supplies, so the estimated $2000 would be a net benefit. Over five years, the dis-

Fig. 11.4: *A less automated system requiring more manual intervention.*

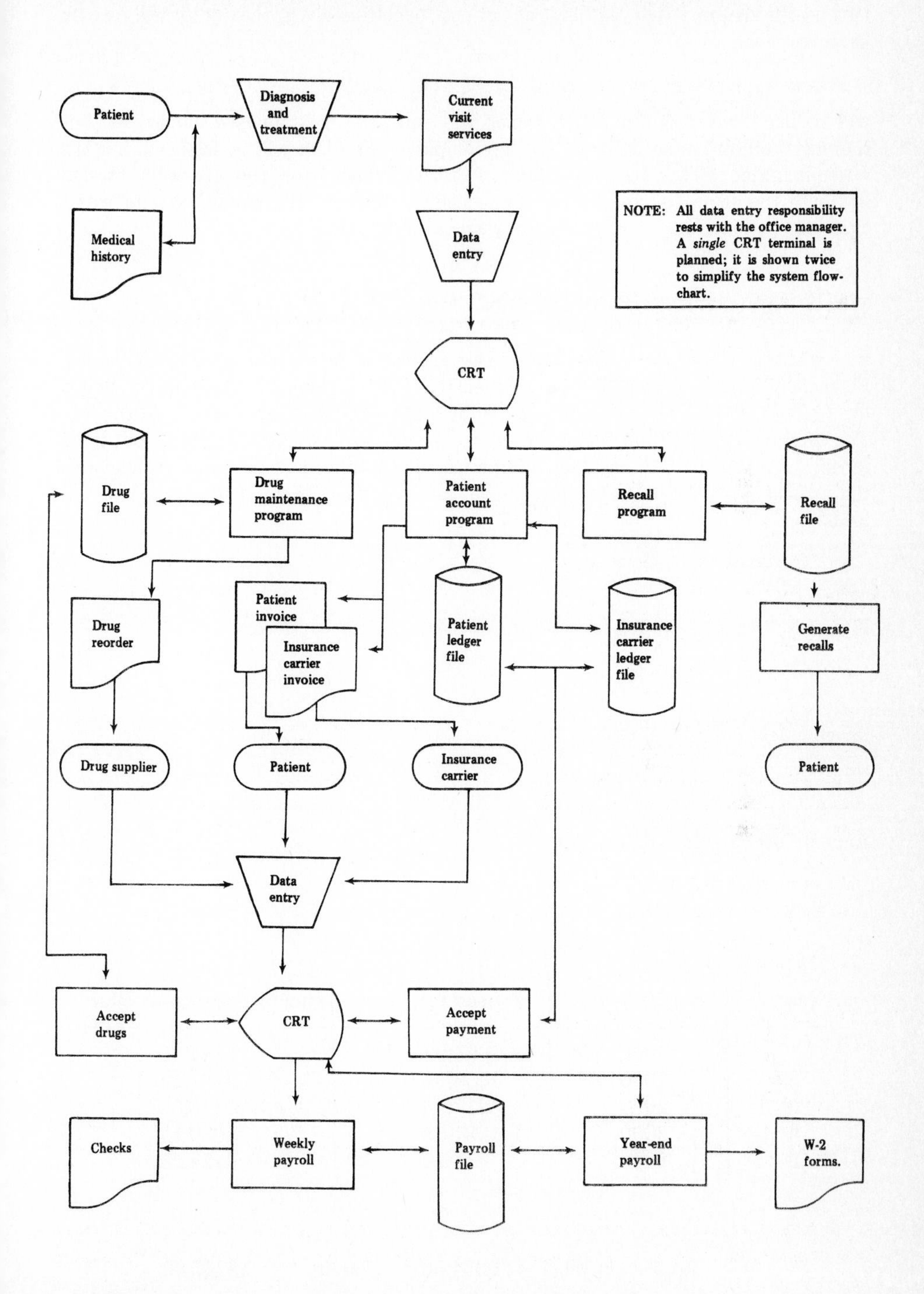

counted net present value (Fig. 11.5) is minus $3745.65, hardly an attractive return, but considerably better than the proposed intermediate system. Eventually, looking beyond the arbitrary five year limit of our cost/benefit analysis, the doctor will recover her investment.

The manual system will not really solve the office management problems, but it will reduce their magnitude. Dr. Lopez will still have to train new employees, but training shouldn't take as long. The key employee problem will be lessened, but the office manager will still be at the center of things. Human errors will still occur, though not with the same frequency. To cite a medical analogy, the manual system won't cure the "patient," but it will make the pain easier to bear.

THE ANALYST'S RECOMMENDATION

In evaluating these three alternatives, the analyst is faced with a difficult choice. Financially, the best system is the expensive, fully automated one, but Dr. Lopez can't afford it. The manual system is a poor investment that really won't do the job. The intermediate solution seems closest to the doctor's intangible criteria, but it's a terrible investment. An unstated alternative, do nothing and continue to live with the present system, is even less acceptable. In short, there is no clear solution to this problem.

Fig. 11.5: *A cost/benefit analysis of the proposed manual system.*

Development cost: $10,000

Annual Cost Savings: $2000/year

Assuming an 18% discounting factor

Year	Net Benefit	Present Value	Cumulative present value
1	2000	1694.92	1694.92
2	2000	1436.37	3131.29
3	2000	1217.26	4348.55
4	2000	1031.58	5380.13
5	2000	874.22	6254.35

Net present value = (6254.35 − 10,000) = − $3745.65

Is this situation unusual? Not at all. The idea that the methodical application of the proper technique leads to *the* answer is a textbook fiction. More often, the analyst is faced with a choice between several less than ideal options. Imagine if, shortly after the office management project began, the analyst decided that the problem was to *install* (rather than investigate) a small business system. By now, a computer may well have been selected and contracts signed; the doctor would be committed to a project with *no possible return* on her investment. Shouldn't the client (or the user) have a choice? Shouldn't those who will invest the money be given an opportunity to access realistically what they will get for that money? Of course. The fact that our methodical approach to systems analysis and design has led us to this point is an argument *for*, not against the methodology. So far, the doctor has invested perhaps $4000. To invest more must be her decision. The analyst has given her the facts she needs to make that decision, and that is the responsibility of a professional technical advisor.

Of course, the analyst should make a recommendation; the doctor is looking for technical advice. We'll assume that the recommended option is the intermediate cost system; the fully automated system is simply too expensive, and the manual system will almost certainly disappoint Dr. Lopez, who expects (and needs) some very real support. While there will be no tangible return on her investment, the scope of the system is sufficiently limited so that there should be no surprises, either; in other words, the risk of actual development costs substantially exceeding $20,000 is slight.

Assuming that Dr. Lopez agrees, the analyst intends to prepare a set of technical specifications and ask a number of potential suppliers to submit bids. The intent will be to purchase a complete system from a single source, as multiple-source systems can be a maintenance headache. The potential supplier's reputation for service and support will be a primary factor in choosing a system. Under no circumstances will a custom designed system be selected; writing custom programs can turn into a bottomless financial pit.

It is unlikely, of course, that any supplier will be able to match the doctor's needs exactly, so some compromise will be necessary. The technical specifications will be written to emphasize functions that must be performed, rather than the implementation techniques. In an attempt to find the best fit, the analyst intends to study proposed alternative systems, using such criteria as cost, apparent quality, maintenance support, training support, system specifications as developed to this point, and a variety of other factors. It won't be easy, but this is the technical advice that Dr. Lopez is paying for.

THE PRESENTATION

The results of the system design phase are presented to Dr. Lopez. All three alternatives are explained, and nothing is withheld; the fact that the results of the feasibility study predicted a limited return on investment lessens the impact. Finally, the analyst recommends the intermediate cost solution as the "lesser evil." Dr. Lopez agrees, because an improved office management system is worth $18,000 to her. Now, detailed design can begin.

SUMMARY

In this chapter, we considered the system design step in the office management system project. Starting with the data flow diagram from analysis, the analyst developed three alternative solutions: an expensive, fully automated system, an intermediate, partially computerized system, and improved manual procedures. Cost/benefit analysis favored the fully automated system, but it was too expensive. The intermediate system promised little or no return on an investment of $20,000. The manual system failed to meet the doctor's intangible requirements. Faced with three less than acceptable alternatives, and knowing that the existing system was unacceptable, the analyst recommended the intermediate system, and Dr. Lopez agreed.

EXERCISES

1. Discuss the objectives of the system design phase for this project.

2. "No vendor can realistically be expected to have precisely what Dr. Lopez needs." Why not?

3. This chapter skips over the process of generating alternative physical strategies, assuming that you have read the first case study and the appropriate modules. To be sure that you understand this process, briefly explain it.

4. In describing the fully automated system, the chapter glossed over the problem of recording (at the time of the initial visit) a need for patient recall. Where would you expect recall data to be noted? How? Why?

5. In the chapter, issuing patient invoices at the time of a visit and accelerating insurance company invoices were said to have a potentially positive impact on the doctor's cash flow. What does this mean? Why is it important? No, your text does not explicitly answer these questions.

6. Figure 11.3 shows a net present value of $6,907.56 and a payback period of 3.95 years for the fully automated system. Both computations assume a discounting factor of 18 percent. Repeat the calculations using a discounting factor of 12 percent.

7. Why were intangible factors so important in ultimately rejecting the fully automated system? Or were they?

8. Explain, in your own words, the differences between the fully automated and the partially automated systems. Why, for example, is a reasonably skilled office manager still needed in the latter? What is the significance of using separate programs and separate files rather than an integrated system with a data base? Why would the partially automated system be less expensive? Don't look for explicit answers in the text.

9. Figure 11.5 shows a cost/benefit analysis for the manual system. Recompute the net present value using a 12 percent discounting rate. Next, extend the estimated life of the system by assuming savings of $2000 for each of ten years. Using ten years of benefits and a 12 percent discount rate, compute the net present value, the payback period, and the estimated internal rate of return.

10. In evaluating alternatives, the analyst discovered that there was no clear solution to the problem. Explain how this situation was predicted by the feasibility study. Why is it important (in arguing for a methodology) that early steps predict the results of later steps? Explain how the analyst resolved this problem; in other words, why was the intermediate solution recommended to the doctor?

11. Continue with Chapter 10, exercise 13.

12. Continue with Chapter 10, exercise 14.

12

Case B Detailed Design

OVERVIEW

In this chapter we will consider the detailed design phase of the office management system project. The analyst will develop physical specifications and submit them to a variety of potential vendors for bids. The bids will then be screened, and the available alternatives compared with the doctor's needs. Finally, a system will be selected and ordered.

THE OBJECTIVES OF DETAILED DESIGN

Generally, detailed design's objective is an exact description of how the system should be implemented. During system design, the necessary programs, files, data bases, hardware, and procedures were identified at a black box level. During detailed design, the contents of these black boxes are specified; for example, in Chapter 6, we used the HIPO technique to develop a detailed plan for a computer program.

Dr. Lopez's office management system is different, however. Rather than designing programs, files, or hardware, the analyst will be selecting an already existing system. The first step will be to prepare reasonably detailed but still flexible technical specifications, and send them to a number of possible suppliers for bids. The available alternatives will then be compared to the doctor's needs, and the system that best fits those needs identified. Finally, the selected system will be ordered.

PREPARING PHYSICAL SPECIFICATIONS

How should the analyst prepare physical specifications? The exit criteria from system design serve as a base, representing an ideal. Perhaps there is a turnkey system close to this ideal; to find it, the specifications must be precise. Overspecification should be avoided, however; inappropriate requirements can easily lock out a potential supplier or unnecessarily tie the system to a single source. For example, consider the problem of selecting a device to play movies on a television set. Three distinctly different technologies are available. If, for whatever reason, the specifications call for video disk, only one of these technologies will do; the available choices will effectively be cut to one third. The doctor needs a system to solve a logical problem; the physical structure of the solution is irrelevant. The analyst should specify *what* the system must do, rather than *how* it must do it. Physical details should be included only when absolutely necessary.

The system must perform certain minimum functions, including billing and accounts receivable, patient recalls, inventory control, and payroll. There are many ways to implement these functions. One possibility is a series of independent programs. Another option is a menu driven system of programs and modules that appears to the user as a single routine. Data base and query techniques might be used. The analyst should state the functional requirements so no reasonable alternative is excluded.

Data are an important element in system selection, too. The specifications should clearly identify types and volumes of data; the data dictionary is the source of this information. Using the data stores or the files defined in earlier steps as a guide, the analyst collects the elements in the dictionary into a series of data structures, computes the number of characters or storage locations in each structure, and estimates the number of times a structure occurs; for example, as many as 500 characters might be stored for each of an estimated 5000 patients, yielding a volume of 2.5 million characters. Given such a number, the potential supplier knows that the system must be capable of storing up to 2.5 million characters, and this logical requirement, in turn, limits the physical options. Data should be linked to the minimum system functions. Clearly, patient data is associated with accounts receivable and patient recalls. Additionally, data are needed to describe the insurance carriers, the inventory, and payroll details.

The physical devices and/or logical organizations assumed during system design should not be identified in the specifications. When details are specified, they become requirements. For example, consider the estimate of 2.5 million characters of patient data from the prior paragraph. This is a real requirement; the system won't work if these 2.5 million characters are not stored. Given this data volume, hard disk is the likely storage medium. It is possible, however, to devise a system with storage capacity for 2.5 million characters that does not require hard disk. The functional specification allows for such options. If the analyst were to go a step farther and state the "obvious," hard disk required, this potential option would be eliminated. The key is that data be stored; how to store them has yet to be decided. If hard disk is specified, the decision is made. Specify technical details only when necessary.

Support aids are particularly important to the doctor. She requires a turnkey system with nontechnical documentation and extensive training support, and will reject any system that does not provide these services. She will insist on single-source maintenance support, and would like assurances on minimum downtime and system backup. In general, factors considered important by the client should appear in the specifications. Potential suppliers should know the criteria that will be used in selecting a system.

In her practice, Dr. Lopez uses highly sensitive equipment, such as an electrocardiograph. Electrical interference from a computer is possible, and cannot be tolerated, so the computer must be shielded. Another problem is occasional power surges. Equipment to protect a computer is available, and should be part of the system. These details would not have occurred to the doctor; they are one reason why technically competent, professional advice should be sought before computer equipment is selected.

Needed reports and other forms should be described at a functional level. Avoid defining precise formats, as they do little more than lock out potential bidders. Don't ignore the possibility of meeting the doctor's report requirements through a query feature.

Finally, the analyst should clearly indicate an acceptable price range or upper limit, and let it be known that bids will be sought from numerous potential suppliers.

The specifications should tell a potential supplier what the system must do and identify any restrictions or limitations. They must be flexible enough to encourage a variety of bids, including solutions the analyst does not anticipate. At the same time, they should be precise enough to discourage inappropriate bidders. The technical specifications for the doctor's system are shown as Fig. 12.1.

IDENTIFYING OPTIONS

Once the specifications are complete, bids should be requested from a variety of sources, including nationally known firms, local franchise holders, and local independents. How should the analyst identify potential bidders? A good technical professional should know several possibilities. Contacts made while attending professional society meetings might suggest others. During the feasibility study, several medical offices were visited, and the systems these doctors use might be worth considering.

Fig. 12.1: *A set of technical specifications for the office management system.*

SYSTEM PROSPECTUS

The Business

Dr. Theresa Lopez, M.D., Inc.. Fewer than ten employees. Estimated 5000 patients.

Minimum System Requirments

1. Patient billing and accounts receivable.
2. Patient recall scheduling.
3. Drug and supplies inventory control.
4. Payroll.

Optional Features

Accounts payable, patient scheduling, financial analysis, word processing.

Minimum Data Requirements

Medical history and financial data are maintained on all patients. A maximum of 500 characters are maintained per patient, giving an estimated maximum total of 2.5 million characters. The inventory is limited to roughly 500 items. Allowance should be made for up to 50 insurance companies, and for up to 25 concurrent employees.

Required Support Features

Dr. Lopez requires a true turnkey system. The following support features are essential:

1. Nontechnical, user-level documentation.
2. Training support.
3. Single-source maintenance support.
   a. Minimum downtime assurances.
   b. System backup.

Special Features

All computer hardware must be shielded so as to minimize interference with sensitive medical equipment. Protection from power surges is also required.

Reports and Forms

Patient invoices, third party invoices, inventory reorders, payroll forms, and a variety of patient and financial reports are required; the precise formats are open. Query capability could meet the doctor's needs for report generation.

System Structure

It is the doctor's intent to purchase or lease a complete system from a single source. Ideally, you have a package that matches the doctor's needs. Realistically, some compromise will be necessary. Bids will be requested from a number of potential suppliers, and the best fit system will be selected.

Estimated Budget

The cost of the system, installed, should not exceed $16,000. Please include estimates for such continuing costs as maintenance, training, and software updates. Indicate any custom programming that might be needed. If a lease is available, indicate lease terms.

Installation

Please estimate your delivery time from the date an order is placed.

Demonstrations

Dr. Lopez would like to see a demonstration of your system before making her selection. Please state your procedures for conducting such demonstrations.

Pre-Acceptance Testing

Please indicate if your firm is willing to allow Dr. Lopez to test your system in her office for a period of time before making a final decision to accept or reject it.

Advertising in technical publications and in medical journals might suggest additional sources. Finally, the analyst might consult a technical information service such as the *Datapro Reports* or the *Auerbach Reports*. As an added check on the bids, the analyst might also use one of these information services to scope a custom, off-the-shelf system; while certainly not exact, the cost of such a system should be roughly comparable with the bids that are received.

After the requests for bids are mailed or delivered to potential suppliers, there will be a delay. Some vendors will respond very quickly; others will not. How long should the analyst wait before beginning to evaluate the bids? In most cases, a few weeks should be enough. The intent is to purchase a turnkey system; the expectation is that the system already exists. Vendors who cannot reply quickly either do not have an appropriate system, or really don't want the doctor's business.

As bids are received, they must be screened. First, proposed systems costing more than the stated limit should be set aside; they will not be given serious consideration unless no one submits an acceptable bid under the established limit. (Possibly, of course, even at this late date, the doctor may be forced to choose between increasing her budget or cancelling the project.) Concentrating on those bids that have passed the financial screen, the analyst compares each with the doctor's criteria: ease of use, dependability, reliability, backup, and maintenance. This is not an easy task. Several of the criteria are difficult to quantify, calling for value judgements. Additionally, there are questions concerning how the various criteria should be weighted; for example, is an extremely easy to use system with slightly below par maintenance support preferable to a moderately easy to use system with very good maintenance support? Clearly, a system rated as unacceptable on one or more of the criteria can be eliminated, but the best system will rarely be obvious.

The need to balance the actual facilities of a given option with the doctor's ideal system complicates the selection process. The analyst would like to come as close as possible, but compromise will be necessary. Is an acceptable system that comes quite close to the tentative plan better than an excellent system that will require substantial changes in the doctor's business practices? Maybe. Unfortunately, "maybe" may well be the best the analyst can do during this initial screening process.

SELECTING A SYSTEM

Screening should narrow the choices to a handful of options. How should the final choice be made? Remember that the system is being selected for Dr. Lopez. She should play a part in the final selection process. Clearly, with no technical training, she could not have fully understood the variety of options available. The analyst has, however, narrowed the choice to a reasonable number of acceptable alternatives. She can certainly evaluate these limited options using her own (perhaps unstated) criteria.

An important step is to arrange for a field test or a live demonstration of the finalists. Many organizations have data centers specifically to support such demonstrations. Others may be able to arrange a visit to a medical practice that recently installed a similar system, or to a nonmedical installation where the hardware can be demonstrated on an unrelated application. Field tests and demonstrations are accepted practice among suppliers of small business systems, and a vendor who cannot or will not accomodate such a reasonable request is suspect.

Ideally, the doctor should be able to touch and control the demonstration system. She should load programs, enter data, and access files. She should feel the keyboard and read the display and the printed output. She will have to live with her choice, and the importance of the impossible-to-quantify "chemistry" factor should not be ignored. Different systems *feel* different. Technical people may ridicule such emotional responses to electronic devices, but Dr. Lopez, a user, is not a technical person. Ease of use is a key criterion. Simply, people find systems they like easy to use, and may not be able to explain why.

Additionally, potential suppliers should make a list of local users of their products available. The analyst should check these references. Often, the telephone is both the quickest and most effective medium. In these days of instant litigation, many people hesitate to write information that might negatively affect another party, but will willingly offer criticisms and candid comments informally and "off the record." A call to the local *Better Business Bureau* is another good idea; it maintains records on customer complaints and questionable business practices. One final comment: don't trust advertising unless you are able to verify the claims personally. No one tells you about bad features willingly, and almost everyone tends to accentuate (and perhaps inflate) the positive. While not as bad as ads for many consumer products, technical advertising does suffer the same deficiencies.

Try to negotiate a pre-acceptance test; ideally, the doctor should be given the opportunity to "live with" the system for a time before making a final decision to buy. Many firms cannot or will not accept such arrangements. A vendor's refusal should *not* be seen as a reason to reject a product, as pre-acceptance testing is not yet standard in the microcomputer marketplace. Rather, a willingness to accept such testing should be viewed as an inducement to do business. One possibility is a lease/buy option. Under such a contract, the doctor agrees to lease the system for a specified period of time, perhaps six months. If, during this period, the system proves unacceptable, it is returned. Should the doctor decide to keep it, most of the money paid under the lease is applied to the purchase price. Such lease/buy arrangements are a form of insurance policy, limiting the customer's risk to the terms of the lease.

The analyst should, of course, recommend a specific system; the final choice, however, is the doctor's.

SUMMARY

This chapter discussed the detailed design phase of the office management system project. Using the tentative system design as a reference, the analyst prepared a set of physical specifications. The specifications were then submitted to a variety of potential suppliers for bids, which were screened, and narrowed down to a handful of viable options to be studied in depth. The doctor was personally involved in this final selection process, attending several system demonstrations. References were checked and a preacceptance test negotiated. Finally, the analyst recommended and the doctor selected a system.

EXERCISES

1. In general, what are the objectives of detailed design? What makes this case study different from THE PRINT SHOP's payroll system?

2. Why should the technical specifications written to support the selection of a turnkey system be more logical than physical?

3. In Chapter 11, preliminary planning for a physical office management system was performed. Here in Chapter 12, we seem to be returning to the logical system in order to write technical specifications. What purpose is served by the tentative physical system design?

4. Why are the data so important in developing system specifications?

5. Why did the analyst avoid adding physical storage media and logical data organizations to the technical specifications?

6. Explain the process of screening bids as described in this chapter. Why was system cost used as an initial gross screen? Why was the process of weighting the criteria so difficult?

7. Why is it so important that the client participate actively in the final selection process?

8. Why is it so important that the client have an opportunity to actually use a system before making a selection?

9. What is a pre-acceptance test? What is the value of such a test?

10. Continue with Chapter 11, exercise 11.

11. Continue with Chapter 11, exercise 12.

13

Case B Implementation And Maintenance

OVERVIEW

In this chapter, we consider the implementation and maintenance steps for Dr. Lopez's office management system. Prior to installation, the physical site must be prepared and documentation, operating procedures, an installation plan, and a training plan written. Following implementation, the system must be tested and the employees trained to use it. Long-term considerations include establishing preventive maintenance, emergency, and backup procedures, and agreeing on the analyst's continuing relationship with the system.

PLANNING FOR IMPLEMENTATION

A system has been selected and ordered. The job may seem complete, but it isn't. Advertising in mass media suggests that a computer is "just another household appliance," but installing a system is considerably more complex than plugging in a new television set. There may be a time delay of days, weeks, or even months, before the new hardware arrives; the analyst must use this time to plan for a smooth installation. Specifically, before the equipment actually arrives, the analyst will:

1. Prepare the physical site.
2. Prepare documentation and operating procedures.
3. Prepare a test plan.
4. Prepare a training plan.
5. Handle other implementation details.
6. Prepare backup procedures.

Let's consider each of these pre-installation tasks.

Prepare the Physical Site

We'll begin with the problem of physical space. A computer system consists of a number of components—the computer itself, a disk drive or two, a printer, a CRT terminal, and possibly more—linked by cables. People use this equipment, and must have enough space to work. Also, such things as printer paper, extra disk packs, diskettes, backup media, various forms, and documentation or procedure manuals must be stored, and this takes even more space. The doctor, we'll assume, plans to place the computer in the room presently occupied by the office manager. Is the space adequate? The only way to find out is to measure the room and the hardware, and to compare these measurements. If the planned location will not do, find out now.

One useful technique can be borrowed from the field of interior decorating. A schematic of the room is drawn to scale on a sheet of graph paper. Using this same scale, cut out small pieces of cardboard or paper to represent the components. Given this simple model, alternative room arrangements can be evaluated simply by moving the "components" around on the schematic.

Equipment to support a large computer center is often free-standing, but the hardware associated with a small computer is not. Desks, tables, work surfaces, book stands, storage cabinets, and terminal pedestals may be required. Perhaps existing furnishings will do; perhaps new ones will be needed. Decisions should be made and necessary furnishings purchased and installed before the computer arrives. Since the furniture selected impacts physical space planning, these two tasks should be carefully coordinated.

Utilities represent another potential problem. For example, most people tend to assume that electricity is a given; new appliances can simply be plugged in and used. Although small computer systems typically run on standard household current, the amount of equipment involved can easily tax the limits of existing wiring. The analyst should obtain a copy of the technical specifications for the system, and determine how much current each component requires. Given this information, a qualified electrician will be able to determine if the existing wiring is adequate and, if not, to install necessary new wiring.

With most household appliances, a blackout is merely inconvenient—the television won't work until the power comes back on. With a computer, however, an interruption in the power supply can be disastrous. Main memory is volatile; it loses its contents when power is lost. Additionally, other components such as disk drives can be damaged if power is cut at the wrong time. If such problems are anticipated, an *uninterruptable power supply (UPS)* can be placed between the electric company's lines and the computer. Basically, a UPS consists of a switch and a battery. As long as power is available from the usual source, it flows through the UPS and into the computer, charging the battery in the process. The instant that power is interrupted, the switch closes and the battery takes over. This backup source gives the operator sufficient time to bring the system down in an orderly fashion. If such equipment is deemed necessary, its installation should be completed before the computer arrives.

Inconsistent power can cause more subtle problems. For example, a brownout occurs when power is supplied continuously, but at a lower than normal voltage. This can affect a computer, causing it to lose precision or to "forget" things. A UPS can help, with a battery providing supplemental power to drive the system at a constant voltage. Power surges are an entirely different matter, however. A surge might occur when lightning strikes a line, or when heavy equipment located somewhere on the line is suddenly turned off or on. A small surge can change the setting of a few bits; a large surge can destroy the almost microscopic connections between components inside the computer. The solution is often an inexpensive *surge arrestor* placed somewhere between the power supply and the computer.

Another consideration is air conditioning. Computer equipment, in particular disk drives and printers, doesn't work well when the temperature or the humidity soars. The technical specifications for the computer hardware probably lists temperature extremes and a suggested normal operating range; a reasonable range might be between 65 and 80 degrees Fahrenheit (roughly 18 to 27 degrees Celsius). Complicating the problem is the fact that the computer equipment itself generates heat, adding to the air conditioning load. If there is any question about the adequacy of the existing climate control equipment, the advice of a competent heating and air conditioning consultant should be sought. The cost of inadequate air conditioning is often measured in excess downtime and maintenance.

Some systems involve teleprocessing. It is a good idea to test the telecommunication link. Borrow a terminal and establish communications. If problems are encountered, ask the telephone company for help. It is possible that rewiring the telephones may be necessary; it is even possible that a special, private, conditioned line may be needed (at a substantial extra charge). Again, the time to identify problems with the physical site is now, while there is still time to correct them before the equipment arrives.

Nothing is more annoying than having expensive equipment sitting around, unusable, because someone forgot the plug or a telephone. Careful physical site planning is essential to a smooth installation.

Prepare Documentation and Operating Procedures

If Dr. Lopez's employees are to be trained to use the system, they must be given clear documentation and operating procedures. System documentation should be obtained from the supplier and carefully read by the analyst. Although much of this material is excellent, it tends to be general purpose, and Dr. Lopez has certain specific needs. Thus, the documentation should be carefully related to the application, and necessary supplements ("How we do it here") prepared.

Operating procedures are a bit different. They should start with the doctor's application, and state exactly when and how each task is to be performed. Ideally, the operating procedures are brief and precise, explaining each discrete task on a single page. Rather than incorporating details in the procedures, the appropriate documentation should be cross referenced.

Prepare Installation and Test Plans

When the hardware arrives, it must be installed. Who does it? When? What checks must be performed? Who is responsible for these checks? The answers should not come as a surprise; an implementation plan is necessary.

How should the system be tested? Ideally, testing should involve all system components—the hardware, the software, the data, and the people. To get the people involved, integrate testing and training. Schedule a parallel run for each major system function, starting with the easiest ones. Have the employees use the hardware and the software to process the data by following the procedures. Then have them repeat the job the old way, pointing out the parallels. Compare the results of the two processes; if they differ, find out why. If the existing system is bad or inconvenient to use, the employees should notice that the new system is, in fact, better, and be anxious to phase over; success is more likely if people accept change instead of having it forced on them. This general idea must, of course, be translated into a concrete test plan. Activities must be scheduled, and the rules for conducting each parallel run clearly defined.

Prepare a Training Plan

Given the documentation, operating procedures, implementation plan, and test plan, a specific training plan can now be devised. Selected elements of the documentation can be identified and assigned to the employees before the equipment arrives. Perhaps an overview of general computer concepts might be in order. The intent of this initial training phase is to familiarize the employees with general computer concepts, the doctor's system, and operating procedures. Technical details should be avoided until the equipment is available.

Once the equipment arrives, each employee must be trained to use his or her piece. Installation, training, and testing should proceed function by function, one step at a time. Initially, the analyst should show the employee how to perform a task, and then gradually turn over responsibility. This should be followed by a brief period of "hand holding," where the analyst or another technical expert is available to answer questions and deal with problems, if needed. Gradually, the direct technical support should fade away.

To help minimize the key employee problem, every individual using the system should have a backup; for example, the receptionist might be trained to take over when the office manager is unavailable. Training backups should begin a week or two after the primary training is complete. Ideally, the person responsible for a given function should train his or her backup, with the analyst supervising the process.

The doctor should, of course, be included in both the planning and the training. It is assumed that she will be able to perform all functions, serving as a third-level backup to everyone.

Plan Implementation Details

Invariably, there are little details that tend to be overlooked. For example, consider file initialization. Dr. Lopez has roughly 5000 patients. How long will it take to enter the initial data for 5000 patients? Who is responsible for verifying the accuracy of these data? These questions must be answered before the hardware arrives. Perhaps temporary help can be used. The vendor might be able to provide access to an equivalent system, or even personnel to help. A good idea is to link file initialization and the test plan. Begin by creating the files in the week or two immediately preceeding installation, and then accumulate the transactions that change the contents of the files. Once the system is installed, use the accumulated transactions to update the files, and compare the results to the output from the old system. Discrepancies might highlight file creation errors, program bugs, or system bugs.

Supplies must be ordered, too; forms, diskettes, disk packs, printer ribbons, and paper will be needed, and may involve a lead time of a week or more. It would be a shame to be unable to test a system because there was no ribbon for the printer.

Plan Backup Procedures

The final element of preinstallation planning is to develop backup procedures. Why is backup necessary? Without adequate backup, there may be no way to recover from a disaster. For example, program errors, human errors, hardware failures, fires, floods, and natural disasters can destroy the data on disk; thus, the contents of disk are regularly copied to another medium such as tape or diskette so that, in the event of a disaster, the information can be recovered. What medium will be used? How often should disks be backed up? Where will the backup media be stored? Who will be responsible? The plan should answer these questions.

Another concern is with backing up the entire system. For example, what happens if the doctor's office is struck by fire? Temporarily, at least, the computer may be unusable. Are there other systems that can be borrowed or rented for a time? If so, they should be identified, and procedures for contacting this backup source written. Extensive maintenance is another possible source of lengthy downtime, and many suppliers will provide a loaner system under certain circumstances.

THE IMPLEMENTATION PLAN

In planning for system implementation, the analyst's responsibility is to anticipate everything that could possibly go wrong, and then to devise non-technical procedures for dealing with each problem. In general, the purpose of preinstallation planning is to ensure that there are no surprises. It is important that a complete plan exist. The steps should be clearly defined and scheduled; if the installation is at all complex, a project network or at least a Gantt chart should be prepared. A smooth installation is no accident; it is the result of meticulous planning.

POST INSTALLATION TASKS

When the hardware arrives, it must be installed. Components must be placed on the appropriate work surfaces, and linked by bus lines or cables. Diagnostic tests must be performed to ensure that the hardware is working properly. Often the supplier is responsible for setting up and checking out the hardware. As the doctor's technical advisor, the analyst should be present. Typically, the initial setup and checkout of a small computer system takes less than a day.

What happens after installation? The employees must be trained and this takes time. Initially, the analyst and (if possible) a vendor representative should be deeply involved. Most people are anxious (perhaps even a little frightened) when first faced with a need to actually use computer equipment. On one level, they know the equipment is expensive, and are afraid of breaking it—the old "Don't push *that* button!" syndrome. On another level, they are afraid of personal failure. Perhaps they will not be able to deal with this new technology, and find themselves unemployed. Maybe they will make fools of themselves. People who understand computers know that a well-designed system is difficult to "break" and easy to use, and as a consequence tend to overlook the untrained person's concerns. In addition to checking the technical accuracy of the system, a good test and training plan must consider this factor. If the people can't use the system, it is useless, no matter how technically outstanding it might be.

Picture yourself as a systems analyst. As training for a given task first begins, you may want to sit at the console and demonstrate the job for a time. However, the best way to learn to use a computer is to use it; thus, the trainee should take over the console and begin pushing the buttons or inserting diskettes as quickly as possible. Instead of demonstrating the system, the analyst should now begin to talk the trainee through the process, providing step by step instructions. After a few iterations, guidance should no longer be volunteered; instead the analyst should stand by, observe, and wait for the trainee to ask for help. Gradually, the level of "hand holding" should diminish, with help available only on call. Legitimate problems will occur, and, at least for the first day or two, the analyst must be available to help smooth over the rough spots, but the employees must learn to use the system on their own. While the analyst should not seem in a hurry to leave, he or she should not remain too long, either.

What if something is wrong with the system? Modern computer equipment follows a predictable reliability curve. Initially, start up errors will be encountered; it is not uncommon to have a defective card or board fail within the first several hours of use. Once the initial bugs are corrected, the system enters a period of stability, with occasional, widely spaced failures, but, generally, it maintains good, solid, dependable performance. After several years, components (in particular the mechanical components such as disk drives and printers) begin to wear out, and failures become more common. Basically, this means that if anything is going to fail, it is likely to fail quickly, which is another argument for technical support during the first few days.

Remember that training involves more than merely showing someone what buttons to push. Training is psychological, too; a well trained person uses a machine confidently. If the initial results generated by a system are unacceptable, user confidence in the system is eroded. If, time after time, a system generates strange errors that only the local guru can even understand, much less fix, a person's confidence in his or her ability to use the system is eroded. During the initial shakedown period, it is essential that quick, accurate technical support be available.

LONG-TERM FOLLOW-UP

What happens after the initial shakedown period is completed? Dr. Lopez's employees know how to use the system, and no longer need the direct, continuous technical support of the analyst. The system now enters a maintenance phase; clear maintenance procedures are essential. The employees should know how to identify and document a system problem, and who to contact for help. Often, a checklist is prepared. By following the step by step procedures on the checklist, the employee can investigate several possible solutions before documenting the problem and calling the maintenance service. Often, the maintenance supplier can provide a standard checklist.

Preventive maintenance is perhaps the best way to avoid excessive downtime due to equipment failure; for example, disk and diskette drives should be cleaned regularly, and a printer cared for much like a typewriter. Some tasks might be performed by the doctor's staff; others may require regular visits from the maintenance service. Clear preventive maintenance procedures are a must.

Backup procedures are equally important. Unfortunately, backup is easy to put off. Copying data from disk to tape seems a waste of time, and, all too often, becomes a victim of other, apparently more important work demands. Suddenly, a disk pack is dropped or a hardware failure destroys the contents of a few tracks. Without backup, the lost data may be impossible to recover. (Imagine, for example, how you would feel if your school lost all your academic records.) Backup is a form of insurance; it must be done in case the unlikely ever happens.

Following implementation, what is the analyst's relationship with Dr. Lopez's system? In many cases, there is none; the job is done. In others, a continuing relationship is established, with the analyst stopping in occasionally to check on how things are going, and the doctor calling for unscheduled help when a new release of a program is received or a new employee is hired and must be trained. Often, a retainer is paid, with the analyst agreeing to be available, on-call, for a limited amount of time in exchange for a fixed fee. Numerous arrangements are possible; the important thing is that all parties agree.

A key element of the structured approach to systems analysis and design is avoiding surprises; thus it is fitting that we end this case study with all parties agreeing.

SUMMARY

This chapter discussed the implementation and maintenance steps in the office management system case study. We began with such preinstallation tasks as preparing the physical site, preparing documentation and operating procedures, developing installation and training plans, preparing backup procedures, and handling a variety of other details. We briefly considered what happens during installation, and then turned to post-installation problems, particularly employee training. The chapter ended with a brief overview of the maintenance phase.

EXERCISES

1. Why is it so important that the physical site be prepared before the hardware arrives?

2. Obtain, from a local computer store or from such sources as the *Datapro* or *Auerbach* reports, technical specifications for each of the components in a system consisting of a processor, 128K of main memory, two diskette drives, one hard disk drive, a CRT terminal, and a printer. Given this information, determine how you would install it in a 10'x10' room.

3. Why is the risk of a power interruption such a concern on a computer system? What is a UPS? Explain how a UPS helps to minimize the danger of a power interruption.

4. Why is proper air conditioning so important?

5. In the text, it was suggested that operating procedures be "brief and to the point, explaining each discrete task on a single page." Why? The answer is not in the text.

6. Why should testing involve all the system components, including the procedures and the people?

7. Once again, what is a parallel run? Why is this such a valuable testing technique?

8. "The person responsible for a given function should train his or her backup." Why?

9. Why is backup necessary?

10. Why must nontechnical, psychological factors be considered during training?

11. Describe the typical reliability curve of a computer system. (You might want to draw it.) How does this curve relate to the need for technical support during the early phases of training.

12. Distinguish between maintenance and preventive maintenance.

13. "Backup is a form of insurance." Explain.

14. Continue with Chapter 12, exercise 10.

15. Continue with Chapter 12, exercise 11.

PART III

An On-Line System

14

Case C Problem Definition

OVERVIEW

This chapter introduces the third case study, beginning with a brief overview of the organization. Next, the problem is introduced. Management has recently identified a potential new market for a computer games service, and a team of analysts is assigned to investigate. The first step is to define the problem. At the end of this phase, a statement of scope and objectives is prepared.

THINK, INC.

Eight years ago, Tom Bowa and Mary Lewis graduated from college, married, and began working for different companies located in the high technology area just outside Boston. They shared two hobbies: personal computing and sports. Seldom do people have an opportunity to transform their hobbies into a full time job, but it happened. Tom and Mary were attending a party, and the subject turned to baseball. A good natured argument ensued and, in mock frustration, one of the participants commented that perhaps "the computer should be asked to settle the issue." Why not? Tom and Mary knew both baseball and computers; developing a computerized baseball information system might be fun. So they did. Much to their surprise, it became a commercial success. Then, success itself became a problem; the baseball information service was consuming so much time, that they had to make a decision—abandon their hobby, or devote full time to the system. They chose the latter option, and decided to strike out on their own.

The number of baseball enthusiasts around a major league city may seem endless, but the market for a baseball information service is limited. If Tom and Mary were to make a real success of their idea, they would have to broaden its appeal. Thus, football, basketball, and ice hockey were added to the service. This helped, but not enough.

Then Mary had a brilliant idea. Why not add ski information? Skiers are concerned about the condition of the slopes. Bad conditions can destroy a ski vacation or weekend, and are difficult to predict in advance. Why not list up-to-date ski conditions and allow subscribers to make last minute reservations through their personal computers or terminals? With such a system, a skier could decide after work on Friday to schedule a mini-vacation for that same weekend! Subscribers would pay for that service, and resorts, lodges, and motels might be willing to pay for the right to be listed. With a computer-based service, even last minute cancellations could be converted into an occupied room. The ski information service was an immediate success. In fact, it was so successful that skiing lodges from other parts of the country began to enquire about the possibility of expanding the service. For this, Tom and Mary would need help, and their first thought was of an acknowledged leader in the field, THINK, INC.

Tom and Mary could not have approached THINK, INC. at a better time. While the company was certainly successful, management had just begun to question the growth potential of its products: legal, medical, and news information services. All were highly specialized. There are advantages to specialization. Because such products appeal to a well-defined group of people, it is easy to estimate the market potential, which reduces risk. Specialization limits the potential market, however; there are only so many doctors, lawyers, or newspapers. The real potential for generating revenue lies in mass marketing. The Bowa/Lewis Ski Service had mass market potential. THINK, INC. would purchase rights to the ski and the sports information services, and hire both Tom and Mary as technical advisors. Soon, the ski service went national, and was an almost immediate hit.

What next? Now that the ski information system was an established success, what new products might be offered? Management was in a mood to gamble a bit. The established products had created a consistent, dependable revenue base. Money was available to support a risky venture with the potential for a large payoff, and management was actively looking for such a project. The idea for a computer games service was born against this background.

A NEW MARKET—GAMES

What is the source of new, innovative ideas? People. Sometimes, there is a flash of insight or intuition; at other times, the germination of a new idea can be traced to an accident. That is how the games and information system got its start.

Tom and Mary were having lunch with two of their colleagues, and the conversation turned to video game parlors. They are obvious money makers. Perhaps THINK, INC. should take a serious look at that market. One of the technical people, a highly skilled analyst, was skeptical. He found playing the games boring, and preferred writing them. In his opinion, many "real" programmers felt the same way, and "real" programmers, as disproportionate buyers of personal computer systems, represented a significant part of the potential market for an at-home games service.

Why not develop a system that would give the subscriber a choice of playing or *writing* computer games? Perhaps royalties might be paid to successful game authors. Quickly, the conversation became animated, with everyone throwing out thoughts and suggestions. The idea sounded good. It might work! Later that afternoon, Tom and Mary outlined the concept to the marketing manager, who was equally excited by the prospects. The idea began to snowball. With Mary as the sponsor and driving force, meetings were held and memos written. Gradually, some sense of a viable project began to emerge. Finally came a decision: let's look into it. The video games system was no longer just an idea, it was a project.

THE ANALYSIS TEAM

Given the complexity of its products, THINK, INC. has long followed a structured approach to systems development. Let's assume that the games system project has just been established, and follow the analysis and design process. We'll assume that a team of three members is assigned to work on the project. The leader is an experienced analyst. He will work with two recent college graduates, both of whom own personal computers and enjoy video games. Hopefully, the team leader will provide a sense of stability, and the young analysts will contribute fresh and creative ideas. Following a briefing by Mary, the project leader is handed a stack of memos, notes, and other references, and told to "look into this." The project has begun.

PROBLEM DEFINITION

Where should the project leader begin? First, define the problem. The notes and memos represent the thoughts of numerous people; perhaps by summarizing this information, the analyst can get a better sense of what is needed. Working alone, he begins. The material of the next several paragraphs represents the project leader's summary of the existing technical information. Note that it is presented as an abstract of material that may not have been very well organized. It should be read as a compilation of ideas, not as a coherent plan.

A Summary of Preliminary Ideas

THINK, INC. would like to develop a computerized games and recreation system, perhaps as an added feature of the existing sports, ski, and concert information services. Subscribers will be able to access the system via standard telephone lines or through a cable television network. Once on-line, they will be able to select a game or an information service from a standard menu. An initial fee will be charged for accessing the system; a basic hourly usage fee will also be charged. Each time a user plays a game, an extra charge (perhaps 25 cents per play) will be assessed. There will be a one-time membership fee of about $25 before a user is issued a user number and password. Of course, the traditional information services will continue to be charged at the present rates. Customers will be billed at the end of each month.

As an added feature, subscribers will be encouraged to develop original games and add them to the system library. The system will support BASIC and Pascal; there will be no charge—beyond the standard hourly usage fee—for subscribers using an interpreter or a compiler. When a subscriber completes an original game, a note will be placed in a "customer comments queue." Our personnel will evaluate the game and accept or reject it. If acceptable, the game will be added to the library and made available to other subscribers. Authors will be paid a royalty (perhaps 5 cents) each time their games are played. The royalty will appear as a credit on the subscriber's system usage bill; net credit bills will be accompanied by a check for the net royalty amount. According to the legal department, however, the entire royalty, not just the net royalty, must be reported to the internal revenue service as annual income.

We will need an information module containing brief descriptions of the games, a clear statement of the charges, and similar material. This module will be part of the system overhead; there will be no charge, other than the standard hourly usage charge, for accessing it. Each subscriber will be assigned a standard disk workspace (perhaps 32K); subscribers who need more will be assigned more space, but at an extra charge.

Management will require a variety of reports, including: usage by game and by service, usage by customer, royalties by subscriber, and a series of more general usage, royalty, and exception reports. Popular games should be identified for possible purchase, and unpopular games for purging. Credit decisions must be made on customers, and those who do not pay their bills must be denied access to the system. Subscribers who earn high royalties should be identified, as they might be potential employees. A standard monthly billing program will be needed; all charges and all royalties must be itemized. (Note that THINK, INC. may eventually owe royalties to non-subscribers.) Finally, federal law requires us to report, to the internal revenue service, all royalties paid our subscribers; thus we must store each author's social security number.

Security and backup are important concerns. Access to user number and password lists must be strictly controlled, and we must be able to quickly change a user number, password, or both, if security is violated. Each transaction processed by the system is to be logged. System areas, standard services, and games are to be backed up on-line; user work areas should be backed up daily. Users should not be allowed to access material in another user's work area without permission. To help avoid program theft, it should be impossible to obtain a source listing of any game or program on the system library.

The customer comments queue is an important feature of this system. Subscribers will be encouraged to write, on the comments queue, suggestions, complaints, original game notices and similar messages to our staff and to other customers. By using their name or a user number, they should be able to retrieve messages from the same queue; what is envisioned is almost an electronic mail network, limited to our subscribers. Messages should be free form, and will be limited to a single 80x20 CRT screen. System messages might also be stored on this queue.

Customers will be responsible for maintaining their own terminals or personal computer systems, and for paying their own telephone or cable television bills; our responsibility begins and ends at the front end. Finally, if this system does as well as is hoped, it will grow rapidly. Provision must be made for rapid growth.

Problem Definition, Continued

How long might it take to summarize a stack of memos, notes, reports, and other preliminary documentation? A few days might be reasonable. The project leader might do this work alone, or assign portions of it to the younger team members. Many young analysts, after reading only the first paragraph of the summary presented above, will be tempted to sit down and begin designing programs, or even writing code. It is essential to overcome this temptation. Before the project team begins its job, it must define that job. The very nature of the summary makes this difficult. Notice, for example, that a number of specific suggestions for implementing the system are mixed in with the general parameters. This is normal. Given an idea, technically trained people just naturally think of how they might implement it, in effect conducting a quick technical feasibility study in their own minds. The good analyst must recognize these preliminary thoughts for what they are: preliminary thoughts. The fact that Tom and Mary are visualizing user number/password security or a message system based on an 80x20 character CRT screen does not necessarily mean that these options are the best choices available. It merely means that they were able to envision these solutions; they are presented more to illustrate meaning than to limit choice. The analyst must carefully separate the real problem from the implied solutions. Young analysts find this difficult, which is why the project leader may decide to work alone during this early stage.

What should the project leader do with the summary? An excellent idea is to cycle it back through management. Mary is the official sponsor of this project. By discussing the summary with her, the analyst is essentially saying, "This is what I think you want; it is accurate?" Given written documentation, misunderstandings can often be corrected.

Initially, this process of defining, questioning, and redefining may be quite informal. Eventually, however, a formal statement of scope and objectives (Fig. 14.1) will be needed. This document is a one-page summary of the problem definition. It outlines key functions that the system will be expected to perform. At this stage, the project's scope is difficult to estimate, and this is a potentially serious problem. Technical feasibility can almost be assumed on this system; the people at THINK, INC. have developed similar systems before, and there is little question that they can do it again. Operational feasibility is another given; an on-line games and recreation system clearly fits the company's image. The only real concern is with economic feasibility. How much will the system cost? How much will it return? Note that the statement of

Fig. 14.1: *A statement of scope and objectives.*

STATEMENT OF SCOPE AND OBJECTIVES: *July 18, 1984*

PROJECT: *GAMES AND RECREATION SYSTEM*

THE PROBLEM: *To provide THINK, INC. with a new, mass market source of revenue.*

THE OBJECTIVES: *To investigate a proposed new on-line system to support:*

1. *Playing video games.*
2. *Writing original video games.*
3. *Providing system information to the customer.*
4. *Providing a communication link with customer.*
5. *Accounting for customer usage.*
6. *Accounting for author royalties.*
7. *Generating appropriate management reports.*

The proposed system must be designed for rapid growth. It may be integrated with the present information services. It may require access via telephone line or cable television network.

PROJECT SCOPE: *Difficult to estimate at present, but it will certainly exceed one million dollars.*

FEASIBILITY STUDY: *A preliminary feasibility study is recommended. The objective will be to establish a realistic scope for the project. The feasibility study will cost no more than $10,000, and will be completed within one month from the date of formal authorization.*

scope and objectives recommends a preliminary feasibility study aimed at establishing a realistic scope for the project. This reflects the analyst's concern for economic feasibility. In effect, the statement of scope and objectives is incomplete, and a preliminary feasibility study costing roughly $10,000 will be needed to complete it.

Mary agrees. The games and recreation system project can now begin.

SUMMARY

In this chapter, a problem definition for a proposed games and recreation system was prepared. The idea for the system was born during a lunchtime conversation, and quickly sparked the interest of a number of people. A team of three systems analysts was formed to investigate the potential for such a system. Existing documentation was summarized, and the summary shared with management. Eventually, both the project leader and management agreed on the general structure of the problem, and a statement of scope and objectives was written. The analyst recommended a preliminary feasibility study to more accurately estimate the scope of the project. Management agreed, and the project was officially funded.

EXERCISES

1. In the text, the origin of the idea for a games and recreation system was a lunchtime conversation. Is this realistic? What do you think? How do you think innovative ideas originate?

2. Distinguish between an idea and a project.

3. Why should an analyst take the time to summarize existing documentation during problem definition? Why is this particularly important on a completely new (in contrast with improved) system?

4. Reread the summary of preliminary ideas from the chapter. Identify the implied solutions buried in this material. Pick out the real system requirements. Relate them to the statement of scope and objectives.

5. Have you ever attempted to define a technical problem by citing a possible solution as an example? Why, do you suppose, do analysts so often read such implied solutions as system requirements? Why is this a problem?

6. Why is it so important that the analyst share a written summary of his or her understanding of the problem with management?

7. Explain, once again, the function of a statement of scope and objectives.

15

Case C
The Feasibility Study

OVERVIEW

In this chapter, we consider a preliminary feasibility study of the proposed games and recreation system at THINK, INC. During the problem definition stage, it became apparent that the scope of the proposed system was not adequately defined, so the first task is to estimate the system's cost. To do this, the analysts develop a high level data flow diagram, and use it to help prepare a revised statement of scope and objectives. Now, the question becomes estimating revenues. Marketing feels it does not have sufficient information, and suggests a prototype system; management agrees. Given this clear direction, preliminary plans are made, and analysis begins.

CLARIFY THE SCOPE AND OBJECTIVES

The statement of scope and objectives is incomplete; thus the analysis team's first task is to complete it. Earlier, the project leader summarized information from a number of documents, and used the summary to prepare a problem definition. As a check, the other members of the team independently repeat the process. Working alone, they review the documents, and extract system objectives. The next morning, they meet to discuss their findings. The team leader's version serves as an agenda; it is discussed point by point. Is anything missing? Has anything been misinterpreted or misunderstood? Using their own notes as a reference, the two younger analysts suggest additions, deletions, and changes to the preliminary summary. Eventually, they have a document with which all can agree. The system's scope is still missing, however; how might a reasonable cost estimate be prepared?

DEVELOP A HIGH-LEVEL MODEL

While useful, a verbal description of a system leaves a great deal to be desired. For example, try describing a room in enough detail to allow a contractor or an interior decorator to reproduce it precisely. Now, compare your verbal description to a photograph or to a blueprint. If you were the contractor or the decorator, which would you prefer? Simply, graphic techniques convey significant detail better than written descriptions.

Data flow diagrams are excellent for summarizing a system graphically. Working independently, the three analysts extract lists of sources and destinations, processes, data stores, and data flows from the verbal summary. Again, a meeting is scheduled. One of the analysts reads the verbal description aloud, while the other two identify and list data flow diagram elements (Fig. 15.1). Finally, just to ensure that nothing has been skipped, each individual's list is checked.

Given this list, a data flow diagram can be prepared. Students tend to think that professionals simply sit down and generate perfect documentation. They don't; data flow diagrams are usually drawn by trial and error. To simplify the task, the team might begin by ignoring the processes, viewing the entire system as a single process in the center of a chalkboard. The sources and destinations are added first. What data flows link a source or destination to the system? To answer this question, the list of data flows is referenced, and more detail added to the diagram. Next, the data stores are incorporated. What data flows link a data store to the system? Again, the list of data flows is referenced. Step by step, the diagram evolves, with the freehand sketches erased, redrawn, and erased again and again. Eventually an acceptable version is produced (Fig. 15.2); now, a more permanent copy can be drawn using a pen and a template.

To simplify drawing the data flow diagram, the processes were initially ignored; the next step is to integrate them into the system. Figure 15.1 lists quite a number; a data flow diagram showing all of them will be quite complex. Is there any way to simplify the list? Can any of the processes be grouped to form larger processes? One possibility is arranging them by their timing requirements. For example, some tasks such as responding to a customer request or accessing the games file must be performed on-line, while others, such as preparing bills, are performed monthly. To get a broad

Fig. 15.1: *A list of data flow diagram elements.*

SOURCE/DESTINATION
subscriber
customer (same as subscriber)
author
internal revenue service (IRS)
management
system manager

PROCESSES
access game
access information
access system information
record charges
access comments queue
check security
log activity
generate royalty report
generate management reports
compile charges
compile royalties
prepare bills
assign work space
assign security parameters
backup system files
backup user files

DATA STORES
sports information
games
ski information
concert/entertainment information
customer charges
bills
customer work space
comments queue
royalty
YTD royalty (for IRS)
system information
management reports
credit information
security information
usage log

DATA FLOWS
customer access information
game information
information services information
bill
new game
messages
royalty information
income statement (to IRS)
reports (to management)
usage information (to log)

sense of the system, it might be possible to hypothesize an on-line process, a daily process, a monthly process, and an "as required" process (Fig. 15.3), and use them to design a high-level, logical system. Note that these timings are not arbitrary; they are legitimate system requirements.

The process timing breakdown makes sense. Rather than fifteen to twenty detailed processes, the analysts must now cope with only five, which greatly simplifies their task. Figure 15.2, represents a starting point. By exploding the single process into its functional components, a more precise logical design can be developed. Once again, a sketch is prepared, erased, revised, and erased again, until finally an acceptable version evolves (Fig. 15.4). One young analyst is assigned responsibility for documenting the logical solution, and, working with a template and a pen, copies the diagram from the chalkboard to paper. This formal version is reproduced and distributed to all team members; it will be used as the basis for the next step in the feasibility study.

Fig. 15.2: *An initial data flow diagram.*

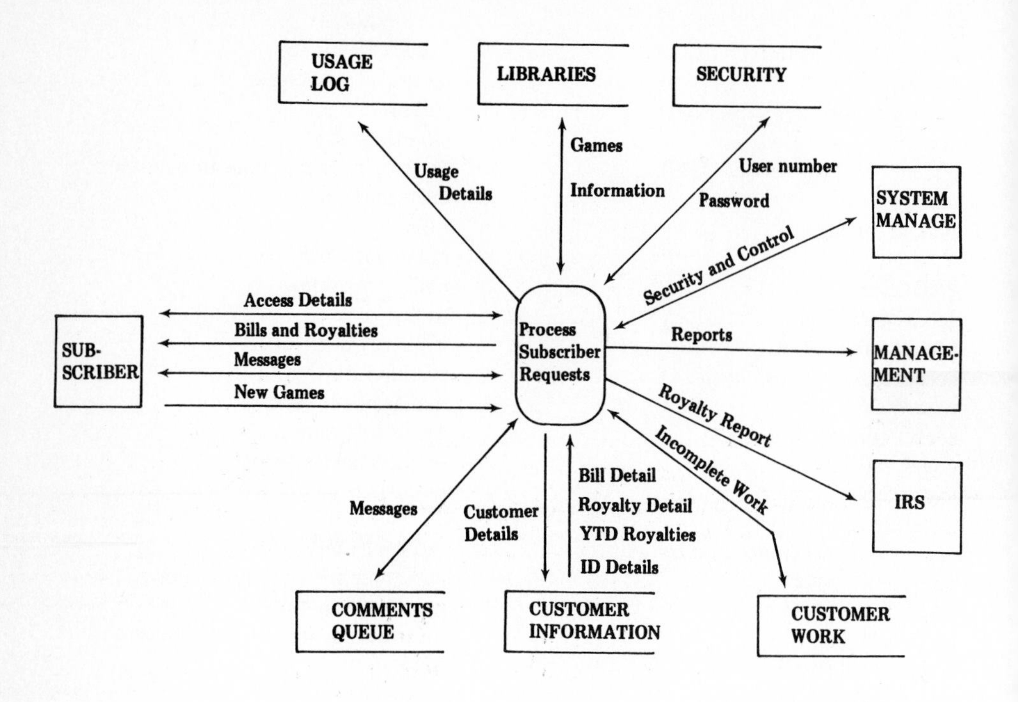

Fig. 15.3: *Processes can be grouped by their timing requirements.*

SYSTEM OVERHEAD—IMMEDIATE
access games
access information
access system information
record charges
record royalties
access comments queue
check security
log activity

IRS REPORT—ANNUAL
generate royalty report

MANAGEMENT REPORTS—DAILY
generate management reports

MONTHLY
compile charges
compile royalties
prepare bills

LIBRARY MAINT.—AS REQUIRED
assign work space
assign security controls

BACKUP—AS REQUIRED
backup system files
backup user files

Fig. 15.4: *A high-level data flow diagram of the system.*

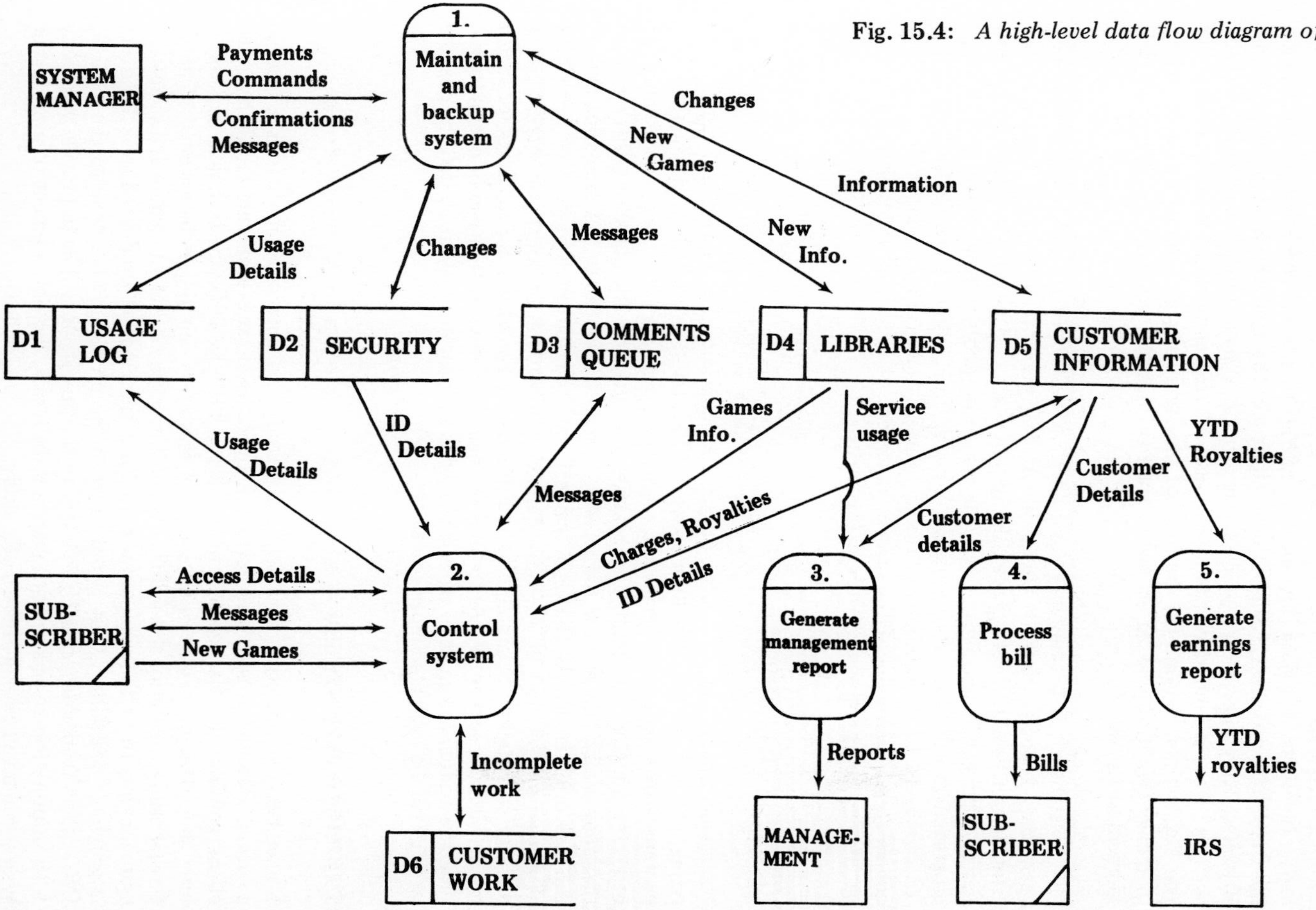

REDEFINE THE SCOPE AND OBJECTIVES IN THE LIGHT OF NEW INFORMATION

Estimating the cost of a system is difficult. Estimating the cost of a single process within that system is much easier, however, as each process represents a set of concrete logical functions. The data flow diagram of Fig. 15.4 is a logical model that highlights the individual processes; using it as a reference, the analysts develop ballpark cost estimates. They discuss each process, one at a time. What functions must this module perform? Does it remind you of a previous job? How long would it take you to do it? The key is time, as time (in the form of programmer years) can easily be converted to money. We'll assume an estimated total of eighteen programmer years; Fig. 15.5 summarizes the details.

What next? A programmer costs about $30,000 per year; eighteen programmer years thus becomes roughly $540,000. At ±50 percent (which is as precise as any estimate can be at this early stage), the system could cost as much as $8000,000. Additionally, there will be management, coordination, and overhead costs, and all those programmers will consume significant computer time. Adding these "extras" brings the total estimate to $1,200,000. In round figures, twenty programmers and 1.2 million dollars seems reasonable. Now, a new statement of scope and objectives can be prepared (Fig. 15.6) and shared with management.

REDEFINE THE PROBLEM

Management now has a concrete proposal to consider. The system still looks interesting, but is it worth 1.2 million dollars? If the system is created, what benefits might be expected? Clearly, the games and recreation system is seen as a potential source of new revenue; thus estimating the benefits is a question for marketing. Let's assume that they really don't know the answer. The proposed games and recreation system is a new product. Unlike the legal, medical, and news information services, it is aimed at a mass market, not at an easily defined group of professionals, so market size is difficult to estimate. Compounding the problem is knowing what to charge for such a loosely defined service. There are just too many variables; without additional information, the system is a gamble. What is needed is a *prototype* system, so that the possible benefits can be estimated more accurately.

THE PROTOTYPE SYSTEM

Remember the objective of a feasibility study: to determine if there is a feasible solution. Remember also that technical, operational, and economic feasibility must all be considered. The people who work for THINK, INC. are technical experts; if they think the job can be done, technical feasibility can almost be assumed. Pending further study, operational feasibility is another given; an on-line games and recreation system seems to fit the company's need for a mass market product. The real concern is economic feasibility. Cost is only part of the problem; what benefits can THINK, INC. expect from the system? The intent of the prototype will be to provide marketing with information to answer this question. In effect, the development of a prototype system will be an extended feasibility study.

Fig. 15.5: *The following estimates were generated for the processes of Fig. 15.4.*

PROCESS	ESTIMATED TIME (programmer years)
system control	10
report generator	1
billing	2
earnings report	2
maintenance & backup	1
overhead	2
TOTAL	**18 programmer years**

Fig. 15.6: *A revised statement of scope and objectives reflecting the estimated project scope.*

STATEMENT OF SCOPE AND OBJECTIVES (VERSION II): *August 1, 1984*

PROJECT: *GAMES AND RECREATION SYSTEM*

THE PROBLEM: *To provide THINK, INC. with a new, mass market source of revenue.*

THE OBJECTIVES: *To investigate a proposed new on-line system to support:*

1. *Playing video games.*
2. *Writing original video games.*
3. *Providing system information to the customer.*
4. *Providing a communication link with the customer.*
5. *Accounting for customer usage.*
6. *Accounting for author royalties.*
7. *Generating appropriate management reports.*

The proposed system must be designed for rapid growth. It may be integrated with the present information services. It may require access via telephone line or cable television network.

PROJECT SCOPE: *The estimated system cost is $1,200,000. This is a preliminary cost estimate with a precision of ±50 percent. The system will require the efforts of roughly twenty programmers for one full year.*

How much is THINK, INC. willing to gamble to find out if the proposed system is economically feasible? Obviously, 1.2 million is excessive, but management still believes in the system, and is reluctant to drop it. Let's assume that they are willing to risk $250,000. What might be done for a quarter million? How much of the proposed system could be written? What limitations would this lower investment entail? Would the reduced system provide a realistic model for test marketing? Once again, the analysts are asked to investigate. The prototype system will be limited to $250,000, which will be used to pay for the three analysts, no more than five programmers, and necessary hardware support. Additionally, the prototype must be designed so that the complete system might be constructed on its framework. Given a revised statement of scope and objectives (we won't present it, but look for an end of chapter exercise), the analysts go back to work.

CLARIFY THE SYSTEM'S SCOPE AND OBJECTIVES (ONCE AGAIN)

We have essentially returned to the beginning of the feasibility study. The analysts have a new problem definition. It cites new limitations, and accurately defines the prototype system's scope, but the objectives are now subject to question. Clearly, a $250,000 system differs from a $1,200,000 system. What might the new, smaller system reasonably be expected to do? More precisely, if the prototype system is to provide the information needed to estimate market potential accurately, what functions *must be* included? Can these functions be performed on a $250,000 system? To answer these questions, the analysts prepare, with marketing input, a prioritized list of system functions.

Given such a list, the next step is mechanical. A cost for each function is estimated, and then the analysts run down the list, item by item, verifying and accumulating the costs until the $250,000 limit is reached. Functions above the cutoff will be included in the prototype system; functions below the line will not. Now, it's back to marketing for some fine tuning. Do they agree with the new list of included functions? Can they live with the excluded functions? Should partial functions be implemented? If a given function must be included, what might be dropped? Gradually, after numerous discussions and meetings, the analysts and marketing reach agreement on the general structure of the prototype system. A new statement of scope and objectives is prepared (Fig. 15.7), clearly highlighting the limitations of the prototype. Again, management can respond.

THE MANAGEMENT REVIEW

Is the scaled-down, prototype version of the system worth doing? Can feasibility be demonstrated with such a system? These are the questions management must answer, and several more iterations might be required. We'll assume, however, that both management and marketing are satisfied, and the analysts are told to continue.

ROUGH OUT A DEVELOPMENT SCHEDULE

Normally, the next steps in a feasibility study are developing alternative solutions and deciding on a recommended course of action. This study is, however, different. The

Fig. 15.7: *A statement of scope and objectives for the prototype system.*

STATEMENT OF SCOPE AND OBJECTIVES (VERSION III): *August 18, 1984*

PROJECT: *GAMES AND RECREATION SYSTEM (PROTOTYPE)*

THE PROBLEM: *To provide marketing with the information needed to accurately estimate the revenue potential of the full system.*

THE OBJECTIVES: *To design a prototype on-line system to support:*

1. *Playing video games.*
2. *Writing original video games.*
    a. *Initially, only the BASIC language will be supported.*
3. *Providing system information to the customer.*
4. *Providing a communication link with the customer.*
    a. *The system will be limited to dial-up access.*
    b. *Access will be limited to ASCII, asynchronous terminals.*
    c. *The existing information services customer base will be tapped to the extent possible.*
5. *Accounting for customer usage.*
6. *Accounting for author royalties.*
    a. *Initially, all royalties will be at a constant rate.*
7. *Generating appropriate management reports.*
    a. *Management reports will be limited. Priority will be given to supplying marketing information.*

The prototype system will serve as a framework for eventually creating the full games and recreation system. The main purpose of the prototype is to provide marketing with sufficient information to estimate the revenue potential of the full system.

PROJECT SCOPE: *The cost of the prototype system is not to exceed $250,000. A maximum of three analysts and five programmers will be assigned.*

point of a feasibility study is not to complete a predetermined set of steps, but to answer certain questions. THINK, INC. is satisfied that the proposed system is technically and operationally feasible; the only remaining question is economic. To answer this question, management has authorized a prototype—in effect, an in-depth feasibility study in its own right. Since the questions have been answered, the preliminary study is over, and analysis can begin.

One task does remain, however. How long will it take to develop the prototype system? To answer this question, a rough implementation schedule is prepared (Fig. 15.8) calling for three analysts, five programmers, and a total of 18 months of elapsed time. Does management want it faster? If so, perhaps additional personnel can be committed. Can three analysts and (after six months) five programmers be spared? If not, perhaps the schedule can be extended. We'll assume that management considers the analysts' plan acceptable, and move on.

Fig. 15.8: *A rough implementation schedule.*

STEP	TIME (months)	ELAPSED TIME	PERSONNEL	
			ANALYSTS	PROGRAMMERS
Analysis	1	1	3	0
System design	2	3	3	0
Detailed design	3	6	3	0
Implementation	12	18	3	5

SUMMARY

This chapter discussed a preliminary feasibility study for the proposed games and recreation system. The initial statement of scope and objectives lacked an estimate of the system cost; thus the intent of the preliminary feasibility study was to clarify the system's scope. After verifying the objectives, the analysts prepared a high-level, data flow diagram of the system. Rough cost estimates of the individual modules were then prepared and accumulated; the resulting estimated system cost was incorporated into the statement of scope and objectives and presented to management.

Marketing was unable to provide reasonable estimates of the benefits that might be generated by the system, so it was seen as a gamble. Management, intrigued by the proposal, was willing to risk as much as $250,000 to determine economic feasibility, and thus the analysts were asked to investigate a prototype system given this financial limitation. In response, the analysts, with marketing input, prepared a priority list of system functions. By accumulating costs from the top, down, they compiled a list of prototype system functions, prepared a new statement of scope and objectives, and once again, presented it to management. The proposal was approved, and a rough implementation schedule prepared.

One final point must be stressed. Developing a prototype system means, in effect, conducting an in-depth, long-term feasibility study. The objective of the prototype will be to demonstrate economic feasibility.

EXERCISES

1. Why is the statement of scope and objectives from Chapter 14 incomplete?

2. Verbal descriptions often are inadequate as communication tools. Why? Why are graphic techniques often so much better?

3. In developing a high-level data flow diagram in the chapter, numerous processes were grouped by their timing requirements. Why? Does this technique seem sensible to you? Why (or why not)?

4. How does developing a high-level model of a system help an analyst estimate its cost?

5. Why was estimating system cost so important in this example?

6. What is a prototype?

7. In the text, the intent of the prototype is to demonstrate economic feasibility. A prototype might also be used to demonstrate technical and/or operational feasibility. Explain.

8. As the analysts started work on the new prototype system, a new statement of scope and objectives was referenced but not presented (page 190). Based on the earlier versions and the new information, write it.

9. Once the analysts were given a system scope of $250,000, the objectives from the initial problem definition became invalid. Why?

10. Why is it so important that marketing be involved in redefining the system objectives for the prototype?

11. Explain the purpose of the rough development schedule prepared by the analysts.

16

Case C Analysis

OVERVIEW

In this chapter, we consider the analysis step in the games and recreation system project. We begin with the data flow diagram from the feasibility study, and explode the processes, highlighting the group approach to analysis. Next, algorithms are defined and a data dictionary prepared. With the data dictionary and the algorithm definitions as references, the flow of data is traced through the logical model, allowing the analysts to clarify the exit criteria further. The act of tracing data flows raises new questions which must be anwered; the cyclic nature of analysis is stressed. Finally, the finished documentation is prepared, inspected, and reviewed by management.

ANALYSIS: OBJECTIVES AND EXIT CRITERIA

Generally, the objective of analysis is to develop a high-level, logical model of a system; in other words, to determine what the system must do. Does this objective change on a prototype? No. Although the prototype's intent is providing marketing with necessary information, it will also serve as a framework for constructing the complete system. Thus it is *essential* that the analysts plan a *complete* system. Perhaps, only selected portions of that system will be designed and implemented, but it is crucial that detailed design and implementation take place in the *context* of a system plan. As before, during analysis a logical system will be designed, documented by a data flow diagram (or diagrams), a data dictionary, and preliminary algorithm descriptions, and subjected to both inspection and management review.

THE DATA FLOW DIAGRAMS

A feasibility study is a compressed, capsule version of the entire analysis and design process; thus it is not surprising that some preliminary analysis has already been done. In particular, a high-level data flow diagram was prepared (reproduced as Fig. 16.1). While its level was certainly appropriate to the feasibility study, analysis demands more detail; the next step is to explode the processes.

Fig. 16.1: *The high-level data flow diagram from the feasibility study.*

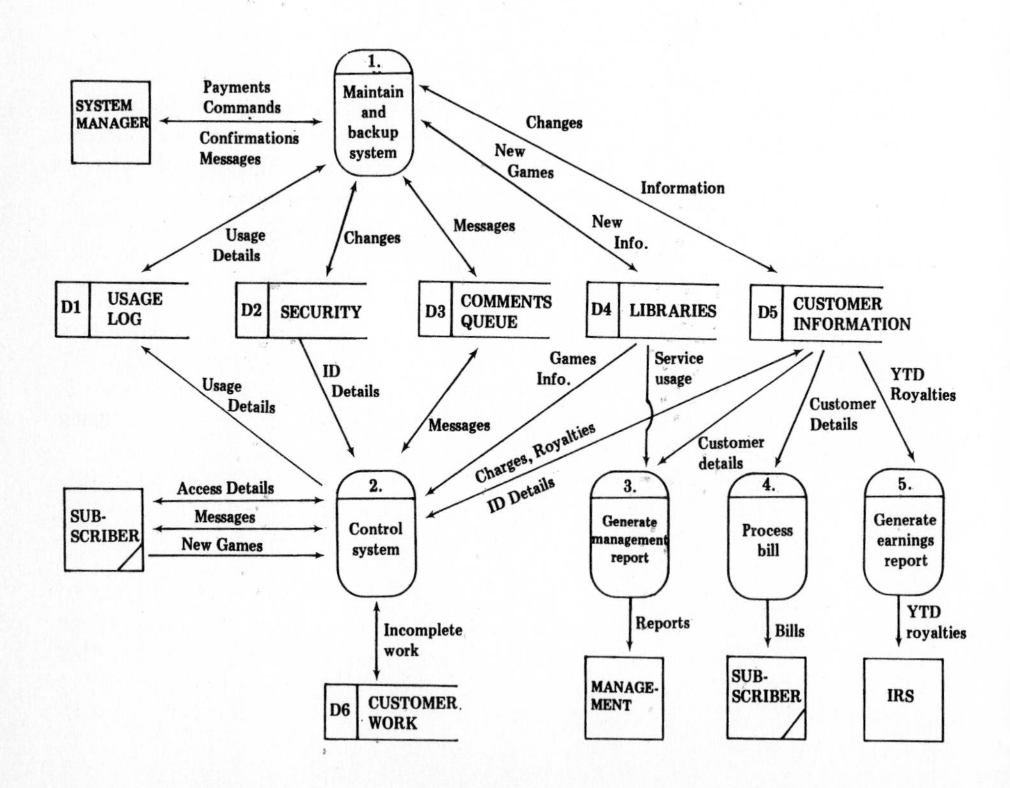

Look carefully at Fig. 16.1. Which of the five processes would you consider most crucial? Process 1 is concerned with maintenance and backup. While certainly important, these are support functions. Similar arguments might be advanced for processes 3, 4, and 5; all involve support functions that essentially summarize or report the results of the primary process, *Control system* (process 2). It makes sense to study the most important process first.

What functions are performed by *Control system*? Earlier, when the high-level data flow diagram was prepared, a list of functions was compiled and grouped by timing requirements (Fig. 15.3). Now the task is to decompose each process. Certainly, the basic functions listed earlier must be included. Additionally, it is reasonable to expect the analysts, who know something about on-line systems, to identify other crucial functions. Let's assume that they are able to develop the following list:

1. Log all customer transactions.
2. Retrieve system information.
3. Retrieve ski/entertainment/sports information.
4. Retrieve a game.
5. Control terminal access.
6. Retrieve/save subscriber work space.
7. Add a message to the comments queue.
8. Retrieve a message from the comments queue.
9. Verify customer identity.
10. Accumulate charges and/or royalties.

How would such a system work? On most systems, it is possible to identify a *driving mechanism*, and to base the system design around it. Carefully consider the list of functions above. Can you see that the driving mechanism must be number 5, *Control terminal access*? Why? Transactions arrive through the terminal. Logging transactions is a reaction to the primary event. System information is retrieved in response to one transaction, while ski information is retrieved in response to another. Read down the list. Convince yourself that every other function is a response to a transaction that enters the system through *Control terminal access*.

Given this observation, we can develop a logical model of the subprocesses involved in controlling the system (Fig. 16.2). Subscriber or customer information enters through *Control terminal access*; this module, in turn, controls access to the other functions. Follow the flow carefully. Identify each of the functions from the above list (note that charges and royalties are accumulated to *Process bill*, which is process 4).

Fig. 16.2: *An explosion of the system control process.*

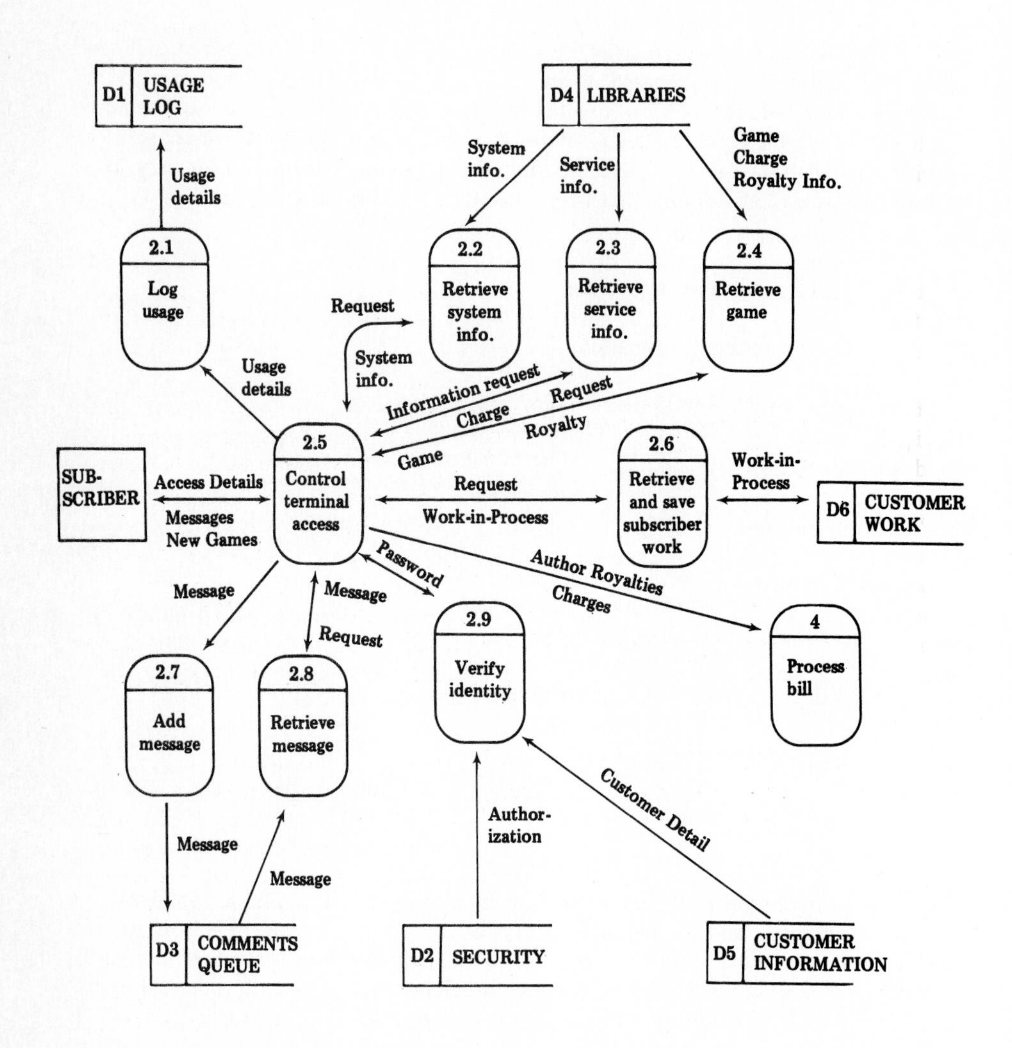

Take the time to convince yourself that the logical model is sensible. Only two key design assumptions have been made:

1. A system is controlled by its driving mechanism.

2. Modules that access data stores should be specialized, dealing with only a single data store.

Note that the design assumptions do not imply a physical implementation. The libraries, usage log, and security information are shown as separate stores because the data are logically different, not because they must be stored on physically separate files or devices. LIBRARIES (D4) could eventually be implemented as a single data base, or as three (or even more) independent data bases. The USAGE LOG (D1) could be stored on disk, tape, or even the printer, or it could be integrated, along with all the other data stores, into a single, massive, games and recreation system data base. The data flow diagram implies nothing about eventual physical implementation.

Briefly refer back to the general, high-level data flow diagram (Fig. 16.1). Figure 16.2 is an explosion of one function from that general model; note that the processes are identified as 2.1, 2.2, and so on. Should the symbols from Fig. 16.2 be incorporated into Fig. 16.1 to form a single, detailed data flow diagram? Probably not. Why? What is the purpose of a data flow diagram? In addition to documenting a logical model, a data flow diagram is a communication tool. The objective is not simply to document a model, but to document that model *clearly*. Mentally combine all the symbols from Figs. 16.1 and 16.2 to form a single data flow diagram. Would such a model be clear? Would it be easy to follow? Certainly not; it would be chaotic. A single, high-level summary diagram accompanied by separate models of the detailed processes conveys just as much information and is far easier to read.

Many analysts use a guideline to decide if symbols should be incorporated into an existing data flow diagram. The magic number is 7±2 processes; in other words, an ideal target is between five and nine process symbols on a single diagram. With fewer than five, the diagram is almost not worth drawing; with more than nine, it can become difficult to follow. The target is only a guideline, of course; depending on such factors as the number of other symbols and the complexity of the data flow lines, diagrams with fewer than five or more than nine process symbols might be appropriate. Use your own judgement, but remember the guideline.

One function from the high-level data flow diagram, *Process bill* (number 4) appears unchanged on the exploded diagram of Fig. 16.2. What happens within *Process bill*? Basically, data must be collected on-line; later, perhaps monthly, the data can be compiled and bills prepared (Fig. 16.3). Three new process symbols are shown. Should these three symbols replace process 4 in Fig. 16.2? Perhaps, but then Fig. 16.2 would show a dozen process blocks, and too many symbols can lead to visual pollution. Why not incorporate the new explosion into the high-level data flow diagram (Fig. 16.1)? We certainly could, but the high-level overview of the entire logical system is a very useful document, so why clutter it with details? Despite the fact that the billing explosion shows only four process symbols, the best choice is probably to keep it separate.

Fig. 16.3: *An explosion of Process bills.*

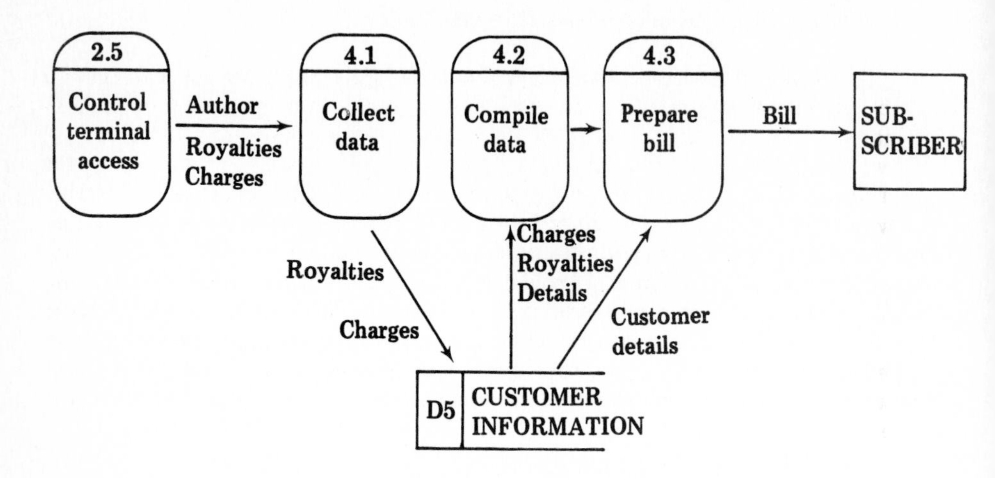

What about the other processes? For example, should we decompose system maintenance and backup? Perhaps. What functions are performed by this module? It allows the system manager to copy, modify, or access any of the data stores. The exploded data flow diagram will not be developed here, but will be left for an end of chapter exercise.

Consider *Generate management report* next. What management reports are required? Do we know yet? In part. During problem definition, some sense of need for customer, author, and game usage information was mentioned. Additionally, we might begin to think about the kind of information marketing needs. (Anticipate yet another end of chapter exercise on this topic.)

Finally, we have *Generate earnings report*. The function of this process is annually to copy royalty information from the CUSTOMER INFORMATION data store to a medium the internal revenue service accepts. Is there any need to draw a data flow diagram for such a simple process? Is there any value to be gained from decomposing it? Not really. Decomposing this module would be a waste of time.

GROUP DYNAMICS

Throughout our discussion of the data flow diagram, we concentrated on the results. Let's consider the process. This project involves an analysis team. How can a group of people develop a data flow diagram together?

One technique is a committee approach. One member of the group, probably the leader, starts the process with a preliminary diagram that serves as an agenda for a meeting. The team members work through it, block by block, questioning the leader's assumptions and recommending changes. At the close of the meeting, an updated version is prepared, disseminated, and studied before a follow-up meeting, with the new version as an agenda, is called. This approach is relatively efficient, but it does put a disproportionate workload on one team member.

There are other dangers, too. Imagine that you have just spent hours preparing a set of documentation. During the meeting, an error is pointed out, or another viewpoint suggested; there is a chance that the comments of someone who shared none of the work might force you to redo it all. How might you respond? Would you see the error or the better alternative clearly, and willingly redo it in a quest for perfection? Or would you attempt to defend your work? Perhaps you would feel that only "significant" problems justify changing the documentation, and insist on the right to define "significant." You might even try to change system requirements to fit your documentation, rather than change documentation to fit the system. A dominant team leader who aggressively definds his or her work creates conflict, not cooperation. A weak leader might be tempted to do sloppy work, knowing that it will probably be rejected anyway. In either case, the committee meeting will be seen as a waste of time, and the benefits that might be achieved from open interaction, lost.

Group design is an option. The process begins with the team members working individually to extract the data flow diagram elements from existing problem documentation. Then, they meet. The team leader starts by sketching a few key symbols on a chalkboard; through discussion and interaction, symbols are added, deleted, moved, and linked to form a data flow diagram. The chalkboard is easy to erase, so changes inconvenience no one. Symbols are drawn and lettered freehand; clearly such documentation could not be considered finished. Since all team members contribute to the design, no one can claim individual ownership. Finally, the burden is evenly distributed; no single member is disproportionately affected by a proposed change. Once an acceptable design is generated, one team member might be assigned to document it formally; if the design is still preliminary, a photograph of the chalkboard might be enough to define a starting point for the next meeting. The time for pen, ink, and template is after a stable design is produced; using "permanent" documentation techniques before that point can be counterproductive.

Why is effective group interaction so valuable? Consider that the analysts are currently in the analysis phase of a system life cycle. The truly creative idea, the concept of a games and recreation system, was generated some time ago. Such creativity generally comes from individuals, not groups, but the analysts are beyond that point now. They are grappling with the problem of bringing someone's creative idea to life; they are concerned with designing and implementing a games and recreation system. While some creativity is still required, close attention to detail is probably more important. A group approach is excellent for dealing with detail and complexity, as points missed by one member might well be caught by another. Good management, a good project leader, and a design methodology make the group approach even more valuable.

Developing a logical design is cyclic. An initial attempt to prepare a data flow diagram will probably raise more questions than it answers. By dividing interview responsibility among the analysts, answers can be found and a subsequent meeting scheduled. This time, perhaps, a better logical design is generated, but new questions will arise. Thus, added interviews and another meeting are scheduled. Eventually, the analysts will be satisifed with their design.

THE ALGORITHMS AND THE DATA DICTIONARY

Given a complete data flow diagram, the analysts now have a reference document for the entire system, and can begin to consider such details as the data elements and the algorithms. Initially, the work is divided, and each analyst works independently. Beginning with the data destinations, they define the elements composing the output data flows. Working backwards, they trace each data element through the data flow diagram, following it to its source. In the process, needed algorithms are identified. Considering the algorithms highlights additional data elements that must come from somewhere. These elements are traced to their source. Eventually, each analyst has a good picture of his or her piece of the system.

The next step is to meet and assemble the pieces. With the data flow diagrams as a reference, the process of tracing the data flows is repeated, but this time with other analysts offering insight and suggestions; the sense is almost that of a structured walkthrough. Why is the input of other analysts so valuable? It is easy to overlook details, and attention to details is essential if a project as complex as the proposed games and recreation system is to be designed successfully. There are, of course, dangers. A group meeting can easily turn into a battleground, with analysts competing with, attacking, and defending, rather than helping each other. A good project leader will keep this from happening.

Following the meeting (or meetings), the results are formally documented. This is the time to enter data elements into a data dictionary and to formally define the algorithms. For example, Fig. 16.4 shows the list of data elements identified as part of the CUSTOMER INFORMATION data flow; to the right are preliminary data dictionary entries for the first few. Figure 16.5 shows algorithm descriptions for the customer usage charge and several of the lower-level parameters that are summed to compute it.

Why is a data dictionary so important? The games and recreation system is complex. To implement it, subdivision will be necessary. The data are the glue that links together the various pieces. If modules are to be successfully linked, all participants must share common data definitions. The time to agree on data definitions is now.

Why are the algorithms so important? The objective of analysis is to define what must be done. In very general terms, every computer system converts input data into output information, and the algorithms define the rules for these conversions. Without a sense of the algorithms, we cannot say that we have a good idea of what must be done.

Both the data dictionary entries and the algorithm descriptions raise additional questions. For example, consider the algorithm description for the "overhead charge" from Fig. 16.5. It indicates that the charge is to be $2.00 per hour of acual use, which seems simple enough, but how is the data element "hour of actual use" defined? It is the difference between the sign on and sign off times, but what is its precision? One programmer might interpret the algorithm as meaning "per hour or fraction thereof."

Fig. 16.4: *CUSTOMER INFORMATION data elements.*

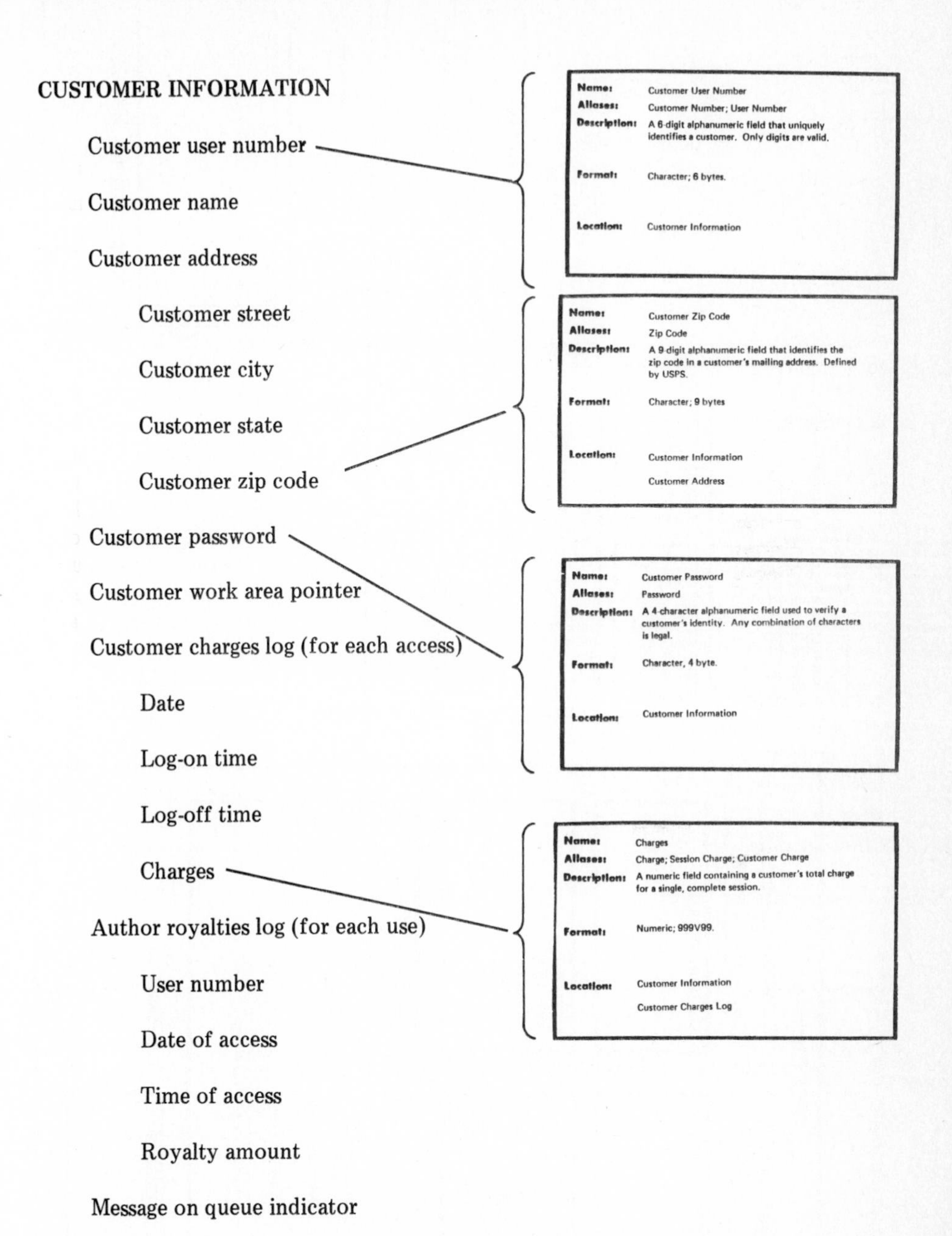

CUSTOMER INFORMATION
Customer user number
Customer name
Customer address
Customer street
Customer city
Customer state
Customer zip code
Customer password
Customer work area pointer
Customer charges log (for each access)
Date
Log-on time
Log-off time
Charges
Author royalties log (for each use)
User number
Date of access
Time of access
Royalty amount
Message on queue indicator
Name: Customer User Number
Aliases: Customer Number; User Number
Description: A 6-digit alphanumeric field that uniquely identifies a customer. Only digits are valid.
Format: Character; 6 bytes.
Location: Customer Information
Name: Customer Zip Code
Aliases: Zip Code
Description: A 9-digit alphanumeric field that identifies the zip code in a customer's mailing address. Defined by USPS.
Format: Character; 9 bytes
Location: Customer Information
Customer Address
Name: Customer Password
Aliases: Password
Description: A 4-character alphanumeric field used to verify a customer's identity. Any combination of characters is legal.
Format: Character, 4 byte.
Location: Customer Information
Name: Charges
Aliases: Charge; Session Charge; Customer Charge
Description: A numeric field containing a customer's total charge for a single, complete session.
Format: Numeric; 999V99.
Location: Customer Information
Customer Charges Log

Fig. 16.5: *Some algorithm descriptions.*

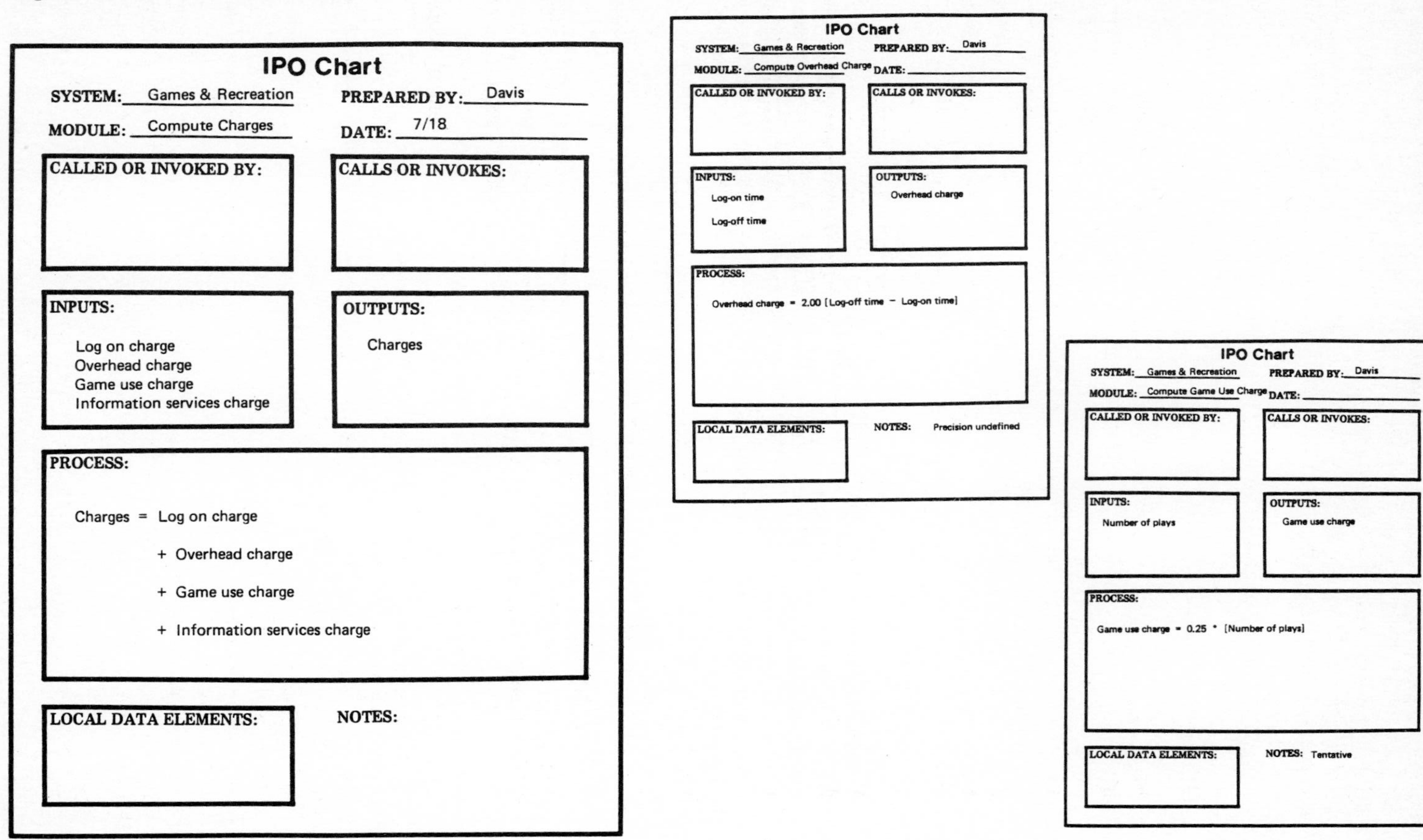

Using such a rule, a customer who used the system for 1.25 hours would be charged $4.00. Another programmer might assume that the rule calls for rounding; 1.25 hours rounds to 1.0 integer hours, and generates a charge of $2.00. Yet another programmer might use fractional time without rounding or truncating: 1.25 hours times $2.00 per hour is $2.50. Which number is correct? With the same calculations performed on thousands of customers twenty-four hours per day, can you see that the answer is significant! Assume that fractional hours are to be used? Should the precision be one decimal place? Two? Ten? It does make a difference. This might be a question for accounting or marketing.

Preparing a data flow diagram led to the generation of data dictionary entries and algorithm descriptions. The data dictionary entries and algorithm entries, in turn, led to new questions. Possibly, the answers may require a change to the data flow diagrams. Sense the cyclic nature of this process. A preliminary logical design gives people a basis for discussion. Answers to the resulting questions lead to a revised logical design. The new model becomes a new target, generating new questions, and leading to yet another revision. Eventually, the logical model settles down, invoking agreement rather than objections. When this happens, the analysts can confidently preceed to the next step.

Analysis is now complete. A logical model of the system, documented by data flow diagrams, a data dictionary, and algorithm descriptions, has been prepared. The users like it, agreeing that it represents what must be done to solve their problem. Now, the finished documentation can be prepared. Perhaps a professional documentor will be called in to draw the data flow diagrams; perhaps the analysts will complete the work. A complete data dictionary and a set of algorithm descriptions can be prepared. The exit criteria are now ready for inspection.

THE INSPECTION

An inspection (see Module A) is a formal technical review of the exit criteria, performed by technical personnel (not by management). The intent is to determine the technical accuracy of the exit criteria. Do the exit criteria meet the requirements defined in prior steps; in other words, following analysis, are the data flow diagram, data dictionary, and algorithm descriptions compatible with the statement of scope and objectives? Can the next step in the analysis and design process be built on the exit criteria being inspected? We are about to move to system design. Can a system be designed to implement the logical functions by using the stated data elements and algorithms?

The inspection team consists of four members; a moderator, the author, and two inspectors. One inspector comes from the marketing department (the user); her responsibility is determining if the logical system meets marketing's needs. The other is a systems analyst who has been working on another project; his task is to view the exit criteria technically—can this system be designed? The moderator is another project leader. The author is, of course, the leader of the games and recreation system project.

One of the inspectors is asked to serve as the reader. Starting with the destinations, this individual begins to trace the data flows verbally. The data dictionary provides details on individual data elements. Tracing these data elements from their destination to their source highlights algorithms, and the algorithms, in turn, suggest other data dictionary elements. The inspection cannot possibly consider every detail of the design, of course; by its very nature, the process must focus on a sample. Given a reasonable sample, however, technically competent people can verify (or challenge) the accuracy and completeness of the exit criteria. Questions are asked; the author responds. Answers lead to followup questions. The informed outsider's perspective can be very valuable. A few minor oversights are noted, but generally the inspection is a success. Following some brief rework to correct those minor oversights, the moderator signs a form indicating that the exit criteria have passed inspection. Now, it's time for a management review.

THE MANAGEMENT REVIEW

Since the system passed its inspection, technical accuracy can be assumed, and management can concentrate on return on investment, cash flow, and personnel allocation. Does marketing think that the proposed logical system will give them the information they need to estimate the revenue potential of the games and recreation system? Yes. Can the manager of systems analysis and design spare the members of the team, or are they needed to help on another, more crucial job? They are still available. Does Mary still believe in the games and recreation system? More than ever; the analysis team has done an excellent job and she is excited at the prospects. Can the company continue spending money on a prototype. The money is still available, says the financial manager.

The management review culminates in a decision, and the next step in the process, system design, is authorized. The three analysts will be assigned to the games and recreation system project for the next two months, and will have sufficient funds to do the job. Thus, its on to system design.

SUMMARY

In this chapter, we considered the analysis step for the games and recreation system project. Starting with the data flow diagram generated during the feasibility study, the analysts exploded each process to a lower level of detail (functional decomposition). The text stressed the cyclic nature of this process, pointing out that a data flow diagram is typically sketched, erased, resketched, and revised again and again until an acceptable version evolves. Once the data flow diagram was complete, a data dictionary and a set of preliminary algorithm descriptions were generated. The data flow diagram served as a basic reference document, and starting with the destinations, data elements were traced back through the data flows; this, in turn, highlighted the needed algorithms which identified still more data elements. Finally, an acceptable logical design was completed and documented using a data flow diagram, data dictionary, and algorithm descriptions. These exit criteria were then subject to a formal technical inspection and a management review.

EXERCISES

1. "It is crucial that detailed design and implementation take place in the context of a system plan." Why?

2. Why is it a good idea to begin analysis by focusing on the most important process?

3. In the text, *Control terminal access* was identified as the driving mechanism of the on-line portion of the system. Briefly explain how this driving mechanism becomes the basis for a system design. To be more specific, explain the thought process that might have taken the analysts from process 2 of Fig. 16.1 to Fig. 16.2.

4. Another assumption made in preparing a logical system design was that modules should be specialized, directly accessing only a single data store. Does this standard seem sensible to you? Why, or why not? How does it relate to structured or modular programming?

5. In the text, a guideline of 7±2 process symbols was suggested for a data flow diagram. Why?

6. In Fig. 16.2, the analysts used process 4 from the higher-level data flow diagram rather than create a new process (perhaps 2.10) to handle billing information. Why?

7. Decompose *Maintain and backup system*, process 1 on Fig. 16.1.

8. Decompose *Generate management report*, process 3 on Fig. 16.1. You may want to refer back to the problem definition and/or the feasibility study to find necessary information.

9. What advantages or benefits might be realized from a group or committee approach to systems analysis and design?

10. What disadvantages or dangers might be associated with a group or committee approach to systems analysis and design?

11. The development of a logical design is a cyclic process. Explain.

12. Complete the data dictionary entries for the data elements referenced in Fig. 16.4.

13. Briefly explain why a data dictionary is so important in a project as complex as the games and recreation system.

14. Why is the precision of the numbers used in a calculation so important?

15. What is the purpose of an inspection? Following analysis, the inspectors were selected from a user group and from systems and programming. Why?

16. Refer To Fig. 16.1. The data elements flowing to process 5, *Generate earnings report*, are an author's social security number and accumulated royalty earnings for the year. Trace these elements to their source, and identify any needed algorithms.

17. Refer to Fig. 16.1. The data elements flowing to process 4, *Process bill*, include a customer's identification number, name, and address, total charges, total royalties, net charge, hours of use, and available workspace. Trace each of these data elements to its source. (Don't forget the detailed data flow diagrams—Figs. 16.2 and 16.3.) Identify any needed algorithms.

17

Case C System Design

OVERVIEW

Analysis is now complete, so we turn to system design. The objective is to plan a prototype within the context of a complete system. We'll begin by generating several alternative system structures. One design will then be selected, and decomposed into functional blocks. The cost of each block will be estimated, and the functions prioritized; then the specific feastures to be included in the prototype will be identified using this information. Finally, a detailed design and implementation schedule will be prepared, and the exit criteria subjected to inspection and management review.

SYSTEM DESIGN FOR A PROTOTYPE

The basic objective of system design is to develop a blueprint for the physical system; at this point, the analyst's focus turns from *what* to *how*. Our three analysts are about to design a prototype. Does this fact change the objective? No. A good prototype serves as a model or framework for the final system, which implies *complete*, not partial, system design. The prototype must be developed *in the context* of a complete system plan.

The first step is to identify several broad alternative design strategies; one of these options will be selected. The physical elements of the system—specific programs, files, data bases, and operating procedures—will then be decomposed, and detailed cost estimates prepared. Next, a prioritized list will be generated, with each system element designated as essential, highly desirable, or postponable. Working with marketing, the analysts will then study the prioritized list, and select the functions to be included in the prototype.

The objective is answering two key questions. Given the $250,000 limit, can a workable system be created? Will this prototype allow marketing to access the games and recreation system's revenue potential accurately? Weren't these questions answered during the feasibility study? Only partially. Don't they relate to technical and economic feasibility? Yes, of course. A prototype is an extended feasibility study; its intent is to *demonstrate* feasibility. Three months ago, the system was just an idea. Now, the analysts know a great deal about what the system must do and how much it will cost. Can the objectives be met within the established scope? If the answer is no, work on the games and recreation system should stop now, before more money is wasted.

GENERATING ALTERNATIVE PHYSICAL DESIGNS

How can alternative physical design strategies be developed? One approach is to start with data flow diagrams, and to hypothesize several different sets of automation boundaries. Another technique is to use a model; for example, THINK, INC.'s existing sports and ski information services are commercially successful, and might suggest a proven design for the new system. Let's concentrate on the data flow diagrams (see Figs. 16.1, 16.2, and 16.3).

To simplify documentation, several related data flow diagrams were prepared during analysis. While most useful for documenting the logical model, there is a danger that the decomposed subdiagrams will restrict the physical design unnecessarily. For example, Fig. 16.2 showed a decomposition of the system control process, while Fig. 16.3 showed a breakdown of billing. Isn't it possible that at least one of the processes from the billing data flow should be on-line with the system control modules? Certainly; how else could charges and royalties be accumulated on-line? The problem with multiple, linked data flow diagrams, is that it is difficult to see the legitimate physical relationships between processes recorded on separate pieces of paper. Thus, for working purposes, it might be useful to prepare a single, all-inclusive data flow diagram. Such diagrams can be quite large, and very difficult to follow; they are not documentation tools, but working tools. Consider using a large bulletin board. Tack up color-coded index cards to represent the processes; use other cards or slips of paper for the data stores, sources, and destinations. Because the cards are so easy to move, such a model makes an excellent focus for the discussion and experimentation marking the early stages of system design.

The key to grouping processes within an automation boundary is the timing requirements of those processes. For example, each time a subscriber signs onto the system, his or her identity must be verified, and this task must be performed on-line, with a response time of fractions of a second. At the end of the month, each subscriber must be billed for services used; billing is a batch activity demanding only monthly response. It is possible to design a system on which bills are compiled continuously and merely dumped at the end of the month. In other words, billing and identity verification could be grouped within a common "immediate response" automation boundary. The converse is not, however, true; verifying identity could not be grouped with billing in a *monthly* automation boundary. The timing requirement of an automation boundary is determined by the member process demanding the shortest response time.

Assume that the analysts have used the index card technique suggested above. Processes demanding an immediate response (a second or less) are identified by red cards; those needing an hour or less are yellow; weekly processes are green, and so on. One alternative set of automation boundaries should be immediately obvious: group processes by color. Thus, we might have an on-line program to implement the red functions. A terminal driven module with a batch option might be written to implement the yellow functions. Traditional batch programs might be fine for the green functions. Once the data stores have been recast as appropriate files and/or data bases, we have one alternative strategy.

Next, the analysts might consider moving some less timely functions into the on-line module. For example, the "yellow" or hourly processes from above might include generating certain management reports. Could management use an on-line report generation or query feature? Perhaps. Maybe some of the hourly functions might be shifted on-line, while others remain in the short-term batch category. Moving functions on-line makes the system easier or more convenient to use. The disadvantage is cost. Adding on-line functions generally makes the on-line program bigger, more complex, and more difficult to maintain. Thus we face a trade-off; is the added convenience worth the added cost?

Perhaps the ultimate system would have all functions on-line. With such a system, customer bills might be compiled and printed on request, rather than just at the end of each month. Another option might be mailing bills throughout the month, thus avoiding the problem of staffing a clerical function for a peak workload. Even generating earnings reports for the internal revenue service might be improved, with customers having the option to receive end-of-year reports that correspond with a fiscal rather than calendar year (school employees, for example, might prefer a year that begins in September and ends in August.) Would the added cost be worth the benefit? At this stage, that's not the point; if the alternative is not identified, the question will not even be asked. The intent, early in system design, is to identify, not to evaluate, alternatives.

EVALUATING ALTERNATIVES

Once alternatives have been identified, evaluation can begin. Ideally, at least three—inexpensive, intermediate, and expensive—should be seriously considered. Space does not permit an in-depth study of three options within the text; thus we will select one and concentrate on it. You should keep in mind, however, that the work described in the balance of this chapter will be repeated at least three times.

We'll study a design that consists of five programs (Fig. 17.1). An on-line program, system control, contains all the *immediate* functions; in other words, all functions that must be performed in direct support of a customer interactively using the system through a terminal. The tasks of maintaining and backing up the system are treated as a special set of routines within the system control module; these functions are accessed through a maintenance terminal. Three batch programs, the management report generator, the earnings report generator, and the billing routine are also incuded. Look back at the general data flow diagram of this system (Fig. 16.1); you should have little difficulty relating it to the system flowchart. The final program, log analysis, is a new routine that reads log tapes and allows a system manager to study them. Initially, it was seen as a part of the backup and maintenance process. However, the decision to use tape for the system usage log makes log analysis a poor candidate for on-line execution.

DEFINING THE PROTOTYPE

Given a broad system design, how might a prototype be implemented within its framework? The basic limitation on the prototype is financial: no more than $250,000 can be spent. How much of the proposed system design (Fig. 17.1) can be implemented for $250,000? To answer this question, it is first necessary to know how much the proposed system will cost. Unfortunately, it is extremely difficult to estimate the cost of a complex module such as system control. Perhaps if the system were decomposed into single-function modules, an intelligent analyst might be able to estimate cost accurately.

Decomposing the Programs

What functions must be performed by the system control program? First, every data flow diagram process inside the automation boundary defining the module must be included. The algorithms provide another clue; each will eventually be implemented as one or more modules in a structured program. Another technique for identifying functions is to "walk through" a typical user session. What happens when a user establishes contact with the system? First, that user must be identified. What is involved in identifying a customer? The user number and password must be checked; this, in turn, requires reading the customer information file. Next, a physical device identification number is checked; what functions must be performed to implement this security precaution? Once a customer is accepted by the system, what happens next? The customer requests a service. Which one? The system must contain a module to respond to customer commands. This, in turn, might require accessing the games, sports or ski information, and/or system information libraries. Of course, the user will need a work space. With multiple users, an internal security module to keep one user from interfering with another must be implemented. Additionally, services must

Fig. 17.1: *A flowchart of one alternative system design.*

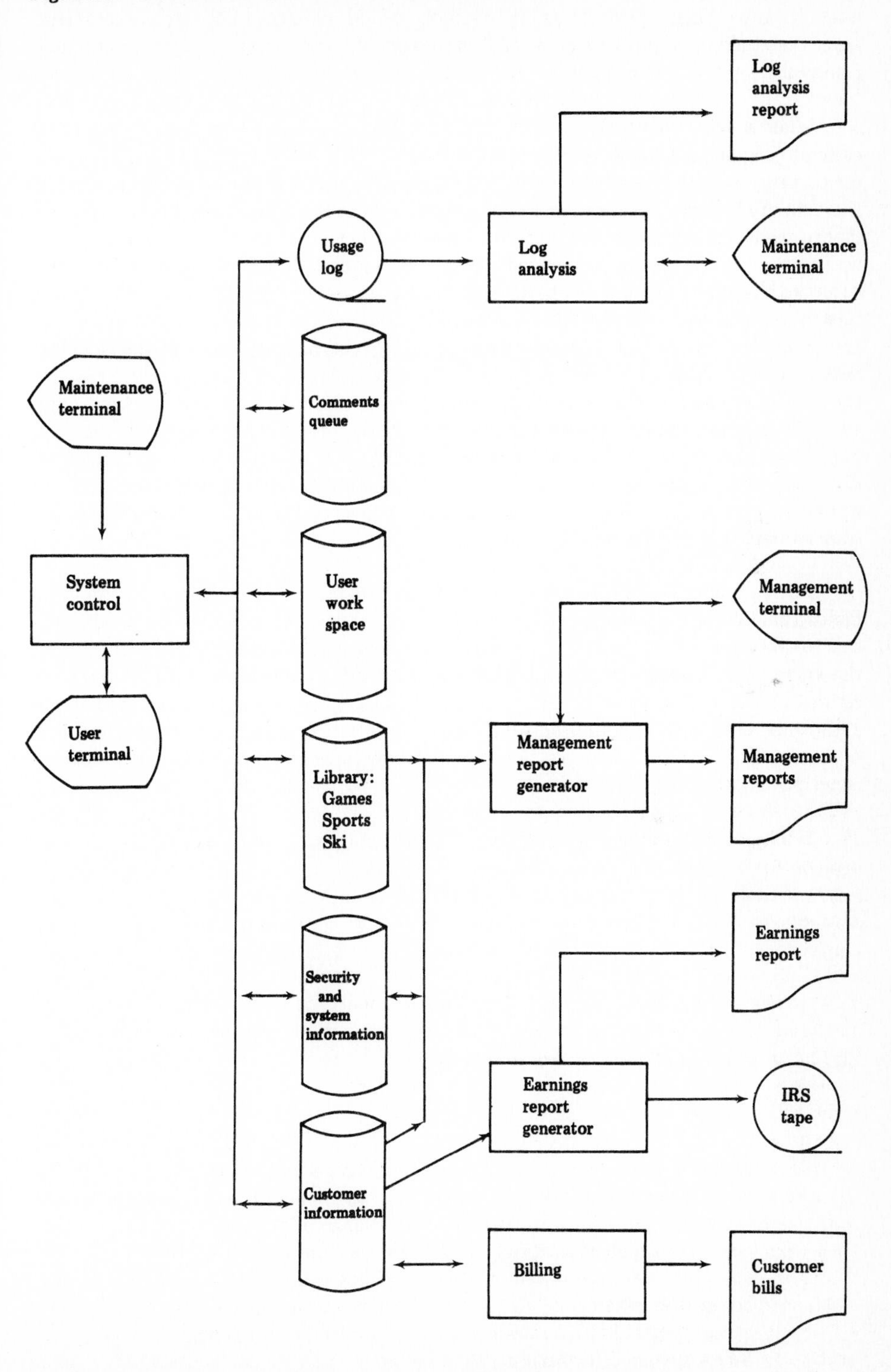

be charged, and royalties accumulated, and each transaction must be recorded on the usage log tape. Finally, customers might choose to add messages to or retrieve messages from the comments queue. Figure 17.2 summarizes these basic system control module functions.

What is the advantage of compiling such a list? The cost of the complete system control program is difficult to estimate accurately. The costs of the individual elements making up that module are, however, much easier to estimate. For example, consider the task of checking a user number and password. What is involved. What might you do if you were a programmer assigned this task? Clearly you would read a user number, use it to access the customer file for a password, request the password from the user, compare the user's input with the official version as stored on the customer information file, and accept or reject the request for access to the system. While the crucial nature of this module will certainly demand significant, additional planning, the point is that you can visualize it. If someone asked you to estimate how long it might take to plan and write this module, you could provide a realistic estimate based on your own experience with similar programming problems. By breaking each process into its constituent elements, estimating the time requirements of each, and then summing these times, it is possible to develop an accurate estimate of the personnel time required to plan and implement the system. Since people are paid a salary, estimates of personnel time can be used to compute cost.

For example, Fig. 17.3 shows the estimated personnel time to plan and code each major function within each program in the proposed system; to do the entire job will call for 83 analyst months and 121 programmer months. Let's use $2500 per month as the value of an analyst's time, and $2000 per month for the programmers'. With these figures, systems analysis will cost an estimated $207,500, while programming is an additional $242,000, bringing the total to $449,500. Add computer time, testing, and management, and the system cost is probably three times the prototype limit. Clearly, functions must be cut.

Look at the problem another way. Three analysts have been assigned to this project for a total of eighteen months. How much money is already committed? Assume that the lead analyst earns $3000 per month; over eighteen months, that amounts to $54,000. Two junior analysts, at $2500 per month for eighteen months comes to $90,000. The total is $144,000; the limit is $250,000; the balance remaining

Fig. 17.2: *A list of system control functions.*

check user number	access work space
check password	implement internal security
access customer information	charge for service
check device identification	accumulate royalty
identify service requested	log usage
access games library	store comments
access sports/ski information	retrieve comments
access system information	

Fig. 17.3: *Time estimates (in personnel months) by function.*

FUNCTION	PLANNING	PROGRAMMING	TOTAL	CUMULATIVE
Overhead	24	0	24	24
System Control Module				
Check user number	1	1	2	26
Check password	1	1	2	28
Access customer information	1	1	2	30
Check device identification	3	3	6	36
Identify service requested	6	6	12	48
Access games library	1	3	4	52
Develop new game	1	1	2	54
Access sports/ski information	1	1	2	56
Access system information	1	1	2	58
Access work space	1	1	2	60
Implement internal security	3	3	6	66
Charge for service	6	8	14	80
Accumulate royalty	6	18	24	104
Log usage	1	3	4	108
Store comments	2	4	6	114
Retrieve comments	2	4	6	120
Analyze log contents	6	18	24	144
Report generator				
Library usage	2	6	8	152
Analyze comments	2	6	8	160
Analyze customer information	2	6	8	168
Billing	4	8	12	180
Earnings report	4	8	12	192
Maintenance and backup	2	10	12	204
TOTALS	**83**	**121**	**204**	

is $106,000. Programmers earn, on average, $2000 per month. Divide the remaining balance by $2000, and you get 53 programmer months. We need 83 analyst months to design the complete system; we have 54 (three times eighteen). We need 121 programmer months to implement the system; we have 53. Once again the message is clear; the prototype will contain only *some* of the system functions. It might be useful to list the personnel limits in the form of a table (Fig. 17.4); using these limits, the analysts can now begin to plan the prototype.

PLANNING THE PROTOTYPE

Clearly, some functions must be cut from the list of Fig. 17.3. Which ones? Which functions absolutely must be included in the prototype? Which functions might be postponed? Let's assume that, after extensive discussions with marketing, the system's functions are grouped into three categories. The first list (Fig. 17.5) is essential; without these tasks, the system is not technically feasible. The second (Fig. 17.6) contains functions that marketing considers necessary, but not *absolutely* essential; they could, reluctantly, live without them. The third list (Fig. 17.7) might be categorized as "bells and whistles." These are the nice features that will help to market a full system, but that are not essential on the prototype. Note that some functions appear on more than one list; perhaps a partial implementation of a given function is essential, but full implementation can wait.

Look carefully at the first list (Fig. 17.5). Note that it calls for 36 analyst months and 50 programmer months, for a total of 86 personnel months. Are these requirements within the limits set for the prototype system (see Fig. 17.4)? Yes, they are; in fact, the basic system demands only 86 of the available 106 personnel months. Can all of the functions on the second list (Fig. 17.6) be included as well? No. The "like to have" list comprises 34 personnel months; adding it to the "essential" list's 86 produces a total of 120 personnel months. Only 106 are available. Some of the functions from the second group can be included; others cannot. Which ones should be included? Once again, the problem is prioritizing needs, only this time with a smaller list. Let's assume that marketing chooses to incorporate fixed royalty payments and the associated basic earnings report, and is willing to wait to see if actual costs are lower than expected before suggesting other functions. We now have a good grasp of the functions to be performed by the proposed system (Fig. 17.8), and can move on to detailed design.

Fig. 17.4: *Games and recreation system prototype personnel limits.*

TYPE OF PERSONNEL	ELAPSED TIME	PERSONNEL MONTHS	NUMBER PEOPLE
Project leader	18	18	1
Programmer/analyst	18	36	2
Programmer	12	48	4
Reserve	12	4	
		106	

Fig. 17.5: *Prototype system—essential functions. Note that some functions are only partially implemented.*

FUNCTION	PLANNING	PROGRAMMING	TOTAL	CUMULATIVE
Overhead	12	0	12	12
System control				
Check user number	1	1	2	14
Check password	1	1	2	16
Identify service requested	3	3	6	22
Charge for service	2	4	6	28
Log usage	1	1	2	30
Access games library	1	3	4	34
Access sports/ski information	1	1	2	36
Access system information	1	1	2	38
Develop new game	1	1	2	40
Access work space	1	1	2	42
Store user comments	2	4	6	48
Access customer information	1	1	2	50
Analyze log contents (basic)	2	6	8	58
Report generator (partial)				
Library usage	1	3	4	62
Analyze customer information	1	3	4	66
Billing	2	6	8	74
Maintenance and backup	2	10	12	86
TOTALS	36	50	86	

Fig. 17.6: *Prototype system—desirable functions. Once again, look for partial functions.*

FUNCTION	PLANNING	PROGRAMMING	TOTAL	CUMULATIVE
Overhead	6	0	6	6
System control				
Identify service requested	3	3	6	12
Charge for service	2	2	4	16
Accumulate (fixed) royalty	2	4	6	22
Report generator				
Library usage (full)	1	3	4	26
Analyze customer information	1	3	4	30
Earnings report (basic)	2	2	4	34
TOTALS	17	17	34	

Fig. 17.7: *Prototype system—postponable functions. In many cases, these functions represent the completion of tasks already partially implemented.*

FUNCTION	PLANNING	PROGRAMMING	TOTAL	CUMULATIVE
Overhead	6	0	6	6
System control				
Check device identification	3	3	6	12
Charge for service	2	2	4	16
Accumulate royalty	4	14	18	34
Log usage (full)	0	2	2	36
Implement internal security	3	3	6	42
Retrieve comments	2	4	6	48
Analyze log contents	4	12	16	64
Report generator				
Analyze comments	2	6	8	72
Billing (full services)	2	2	4	76
Earnings report (full)	2	6	8	84
TOTALS	30	54	84	

Fig. 17.8: *Functions that will be included in the prototype.*

System control
- Check user number
- Check password
- Identify service requested
- Charge for service
- Log usage
- Access games library
- Develop new game
- Access work space
- Access sports/ski information
- Access system information
- Store comments
- Access customer information
- Accumulate (fixed) royalty

Analyze log contents (partial)

Report generator (partial)
- Library usage
- Analyze customer information

Billing (partial)

Maintenance and backup

Earnings report (basic)

THE EXIT CRITERIA

What has system design demonstrated? First, it is indeed possible to construct a prototype system that meets marketing's needs within the limitations set by management. This conclusion is important. Up to this point, management, marketing, and the analysts were really guessing. Now, there is a concrete plan. Now, cost estimates are based on concrete modules, rather than on logical concepts. Before, we thought we could do it; now, we *know*.

What are the exit criteria from system design? Ideally, several alternatives should be suggested. (Although this chapter concentrated largely on one, imagine a similar study with similar documentation for at least two others.) We have a system flowchart and a list of functions to be performed by each program on that flowchart. A detailed cost/benefit analysis is probably not called for, since this is a prototype system that is not expected to yield direct benefits by itself; the only real cost/benefit consideration is: How much can be done for $250,000? The analysts should, however, prepare a complete design and implementation schedule, perhaps in the form of a project network or PERT diagram (see Module L). All the necessary information is available in the chapter, so we'll leave this task to the student; in other words, expect an end of chapter exercise.

THE INSPECTION AND THE MANAGEMENT REVIEW

On such a large project, a technical inspection and a management review will almost certainly mark the end of system design. The analysts have essentially said: Yes, we can do it! Can they? Have they overlooked anything? Management is about to commit the bulk of the $250,000 allocated to the prototype. To this point, the work has involved a few analysts for a few months; following the next step, detailed design, additional personnel will be assigned, and computer time utilized. If the job cannot be done, it can be scrapped now with minimum loss. Waiting involves more time and more personnel, and the project becomes increasingly difficult to stop. Thus, the end of system design is a crucial check point, and the technical inspection and management review take on added significance.

We'll assume that the games and recreation system passes both the inspection and the review. The system looks good, so it's on to detailed design.

SUMMARY

In this chapter, we considered the system design step in the games and recreation system project. First, a set of alternative implementation strategies was developed and studied, and at least three reasonable options identified; one alternative was selected for study in the text. Isolating the programs from a system flowchart, the analysts used functional decomposition to prepare a list of detailed system functions. The cost of each function was then estimated; summing these costs showed that not all system features could be implemented in the prototype. Thus, working with marketing, a prioritized list was prepared. By following this list from top to bottom, the functions to be included in the prototype were identified, and an implementation schedule was prepared. Following a successful inspection and management review, the analysts were ready to move on to detailed design.

EXERCISES

1. The text suggested that a prototype should be designed "in the context" of a complete system design. Why?

2. As we approach the end of system design, management faces an important decision: Should the system be continued or dropped? Why is it so crucial to make this decision now? Why might it have been difficult to decide earlier?

3. In developing alternative system designs, we first combined the several data flow diagrams developed during analysis into one. Why?

4. "The key to grouping processes within an automation boundary is the timing requirements of those processes." Explain.

5. Generally, the timing requirement of an automation boundary is determined by the member process with the shortest response time requirement. Why?

6. Explain how functional decomposition can help simplify estimating costs.

7. Imagine that you have been asked to design a system to track library circulation. "Walk through" the process of circulating a book, and list the functions that might be involved. In other words, what happens between the time you decide to check out the book, and when that book is returned to the stacks by library personnel?

8. Explain the process of compiling a priority list of system functions. Why was this such an important step in designing a prototype system?

9. What are the exit criteria from system design?

10. Why didn't the analysts prepare a complete cost/benefit analysis including an estimate of the internal rate of return?

11. Using the estimated planning and coding times from the chapter, prepare a complete project network for the proposed prototype system.

12. Assume that management has decided to skip the prototype and develop the complete system using the design described in the chapter. Prepare a project network showing how the job might be scheduled. Assume that you are limited to five analysts and ten programmers.

13. Assume that management wants to accelerate the prototype system. The initial estimate was eighteen months. The feasibility study and analysis have accounted for three, leaving fifteen. Management wants it in nine. Can it be done? Prepare a project network. How many analysts and programmers will be needed?

14. Why are the technical inspection and the management review that follow the system design step so important?

18

Case C
Detailed Design

OVERVIEW

This chapter discusses the detailed design phase of the proposed games and recreation system. In it, we will develop implementation specifications, a test plan, and an implementation schedule for one program—system control. The HIPO technique will be used to document the program design, and a project network will be developed for the implementation plan. Following detailed design, the exit criteria will be inspected.

THE OBJECTIVES OF DETAILED DESIGN

During analysis, the functional requirements of the logical system were defined. Later, as an exit criterion from system design, a system flowchart (Fig. 17.1, reproduced as Fig. 18.1) was drawn to document the high-level physical design. Since this is to be a prototype system, the analysts then carefully defined the functional modules to be included in the prototype (Fig. 17.8), reproduced as Fig. 18.2). The programs, the basic functional elements of those programs, and the necessary files are now known. It is time to plan the details.

The objective of detailed design is a complete implementation plan. Since data link the various components of the system, the first task will be defining the file contents. Next, HIPO specifications will be written for each of the programs. Given the data and program definitions, a test plan will then be developed. Finally, a detailed implementation schedule will be prepared and documented with a Gantt chart and/or a project network. Of course, these exit criteria will be inspected and reviewed.

The system flowchart (Fig. 18.1) identifies five different programs, six files or libraries, and a number of reports. To describe, in detail, the planning for each component would require hundreds of pages of text and untold figures and diagrams. We won't do that. Instead, we will select one program, system control, concentrate on it, and highlight only the key points in the detailed design process. Look for several end-of-chapter exercises asking you to complete portions of the work.

Fig. 18.1: *The system flowchart.*

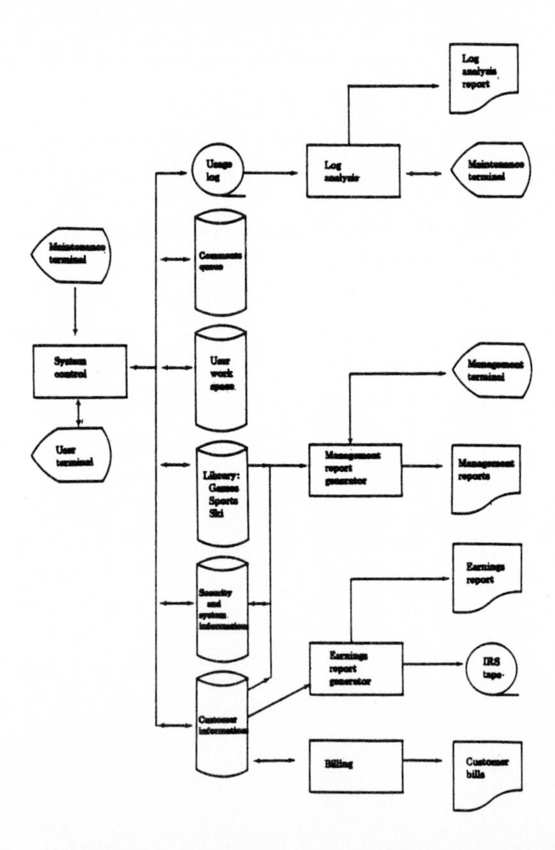

Fig. 18.2: *Functions that will be included in the prototype.*

System control
- Check user number
- Check password
- Identify service requested
- Charge for service
- Log usage
- Access games library
- Develop new game
- Access work space
- Access sports/ski information
- Access system information
- Store comments
- Access customer information
- Accumulate (fixed) royalty

Analyze log contents (partial)

Report generator (partial)
- Library usage
- Analyze customer information

Billing (partial)

Maintenance and backup

Earnings report (basic)

DEFINING THE DATA

Let's begin by defining the data. A key source will be the data dictionary, which identifies each data element that must be stored or manipulated by the games and recreation system. Each element must occupy one or more physical files or libraries, or at least appear on a CRT screen or on a report. Which data elements go with which files? To answer this question, the analysts must group data elements to form structures. Throughout the text, we have used index cards to simulate a data dictionary; let's assume that the analysts have done this. Going through the cards, one by one, they note the file or files where each element is stored. The cards are then sorted by file name—imagine creating a separate pile of data elements for each system file. By rearranging the index cards for a given file, a data structure is developed. Finally, using the structure as a reference, source code (for example, COBOL DATA DIVISION entries) can be written for each structure. Perhaps these source statements might be placed on a COPY library, so that programmers can access a common set of data descriptions.

The analysts' objective is to define the data completely. Often, commands or job control language statements might be coded to describe each file, and this task should not be left to the programmers. It's not that the programmers are incapable. The job control language statements represent formal definitions of the file linkages, showing *exactly* how the various components of the system will fit together. It is essential that these linkages be consistent. The best way to ensure consistency is to have the analysts code all the job control statements before implementation begins.

DESIGNING THE SYSTEM CONTROL PROGRAM

Selecting a Model

Now that the data have been defined, let's turn to the system control program. It will control on-line user access to the system and to the various files and libraries. How might the analysts go about designing it?

We'll begin with a user's perspective. When a customer signs onto the system, he or she expects to encounter a consistent set of rules for accessing any basic service. If each service requires a different set of access rules, customers will find the system difficult to use, and may decide not to subscribe. We cannot allow individual programmers to design the various services independently. We do not want a loosely linked collection of independent modules; we want an integrated system.

Fortunately, there is a model—the sports and ski information service. Figure 18.3 shows the basic structure of the program that controls access to this service. It begins with the display of a menu that lists available options. The customer selects, by number, one item from this menu, types the number, and presses the enter or return key; the system then reads the response and identifies the user's choice. This cycle—write a menu, read a response, interpret the response—may be repeated several times, as the customer's needs are defined in more and more detail. Finally, the system has enough information to access the appropriate library or libraries, compose an answer, and display it on the terminal. To continue with a subsequent request, the user depresses any key on the terminal, and, in response, the system displays the original menu. The process is repeated until the user selects a sign-off option.

Fig. 18.3: *The sports/ski information routine serves as a model for key elements of the new system.*

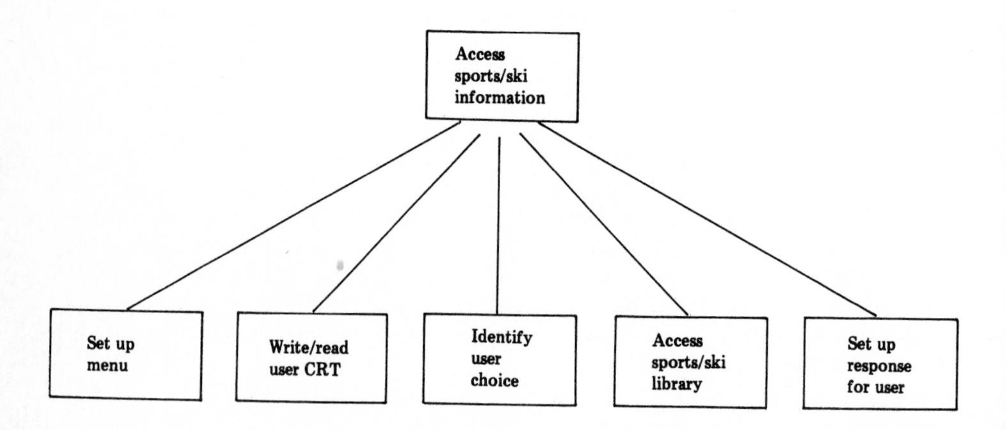

Other services will be similarly designed. For example, consider the game playing service. It will begin (Fig. 18.4) with a menu listing the available games. The customer will select one, type its number, and hit the return key. The program will then identify the user's choice, load the appropriate game into the user's work area, and transfer control to the game. Following the game, the program will return to the original menu and await the user's next choice; this cycle will continue until the sign-off option is selected. The basic control structure of each of the customer services can be consistently defined using this model.

Designing the Control Structure

The unique feature of this games and recreation system is that it allows a customer to access a variety of such services. How can all these services be linked and supported within a common control structure? How might the analysts design this control structure?

Several documents prepared in earlier steps might help. One is the system flowchart of Fig. 18.1, which defines the various files and I/O devices that the program must access. The system flowchart is a physical representation of one set of automation boundaries from the exploded data flow diagrams. The diagram showing the functions contained within this boundary (Fig. 16.2, reproduced as Fig. 18.5) could prove useful. A series of brief algorithm descriptions was prepared during analysis, and each algorithm must be incorporated. Finally, we have the list of functions to be included in the prototype system (Fig. 18.2).

Fig. 18.4: *A similar model will be used to access games.*

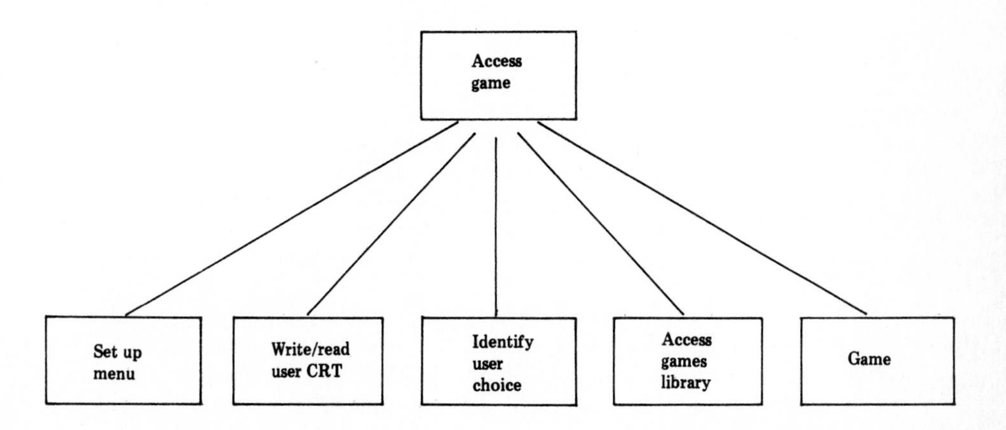

Fig. 18.5: *A functional view of the system control program.*

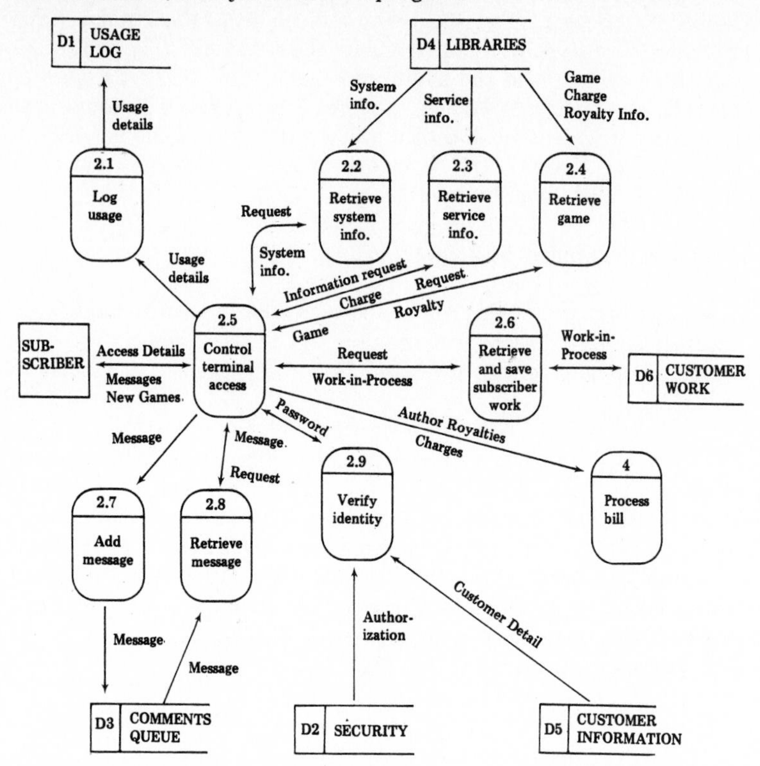

The key to linking the functions is to identify the system's driving mechanism. The games and recreation system will offer customers a choice of several services; how will the proper service be selected? A customer will access the system through a terminal and request a service by selecting a code from a menu; the structure of this basic service routine is shown in Fig. 18.6. Note that certain overhead functions must be performed; for example, a subscriber's work area must be set up and later saved, and billing data must be collected. Otherwise, the module that services customer requests resembles the sports/ski information model, with a menu displayed, the user's choice read, and an action taken in response.

Next, we might decompose the module labeled *Identify service* (Fig. 18.7). At this point the system first learns exactly what the customer intends to do. Remember the requirement that all transactions be logged? Can you see why this is an ideal time to log the transaction? Once this overhead function has been performed, the user's choice is identified, and the appropriate low-level service given control.

Do we need a level of control above the basic service routine (Fig. 18.6)? Yes. For example, it would be unwise to attempt to service a customer request before identifying that customer. Also, consider the problem of controlling access to the maintenance and backup procedures. The analysts decided to treat maintenance and backup as part of the system access program; imagine the danger of an unauthorized user playing with maintenance routines. (As a parallel, imagine a student gaining access to the routines that modify grade records.) A suggested high-level control structure is shown as Fig. 18.8.

Fig. 18.6: *The structure of Service customer request.*

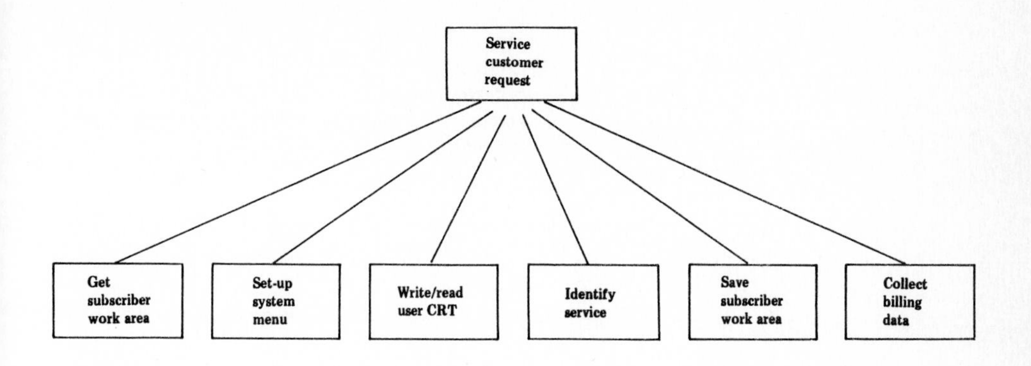

Fig. 18.7: *The Identify service structure.*

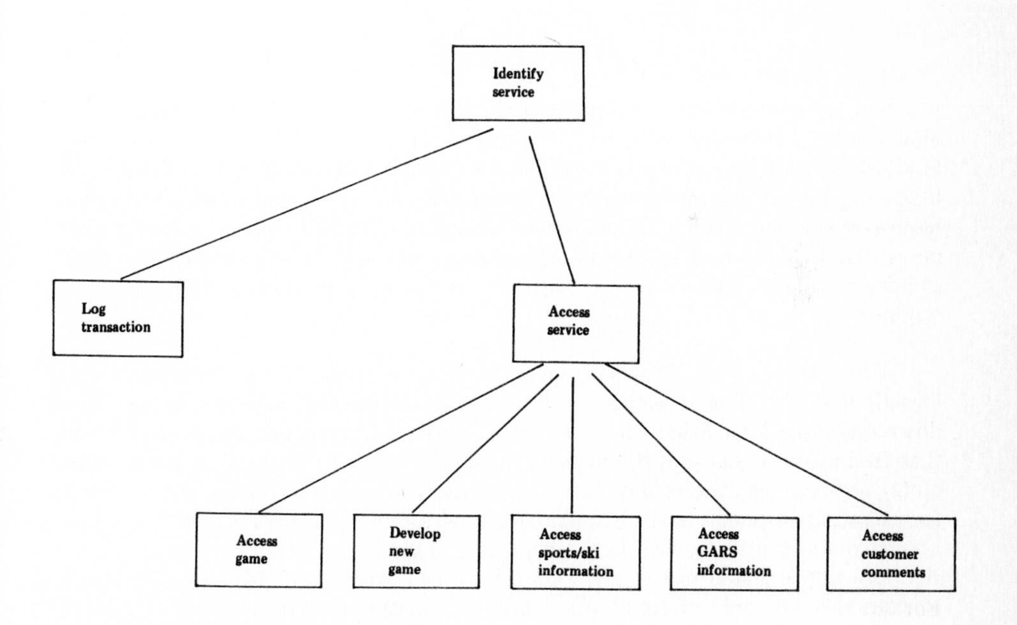

Fig. 18.8: *The high-level control structure of the games and recreation system's on-line program.*

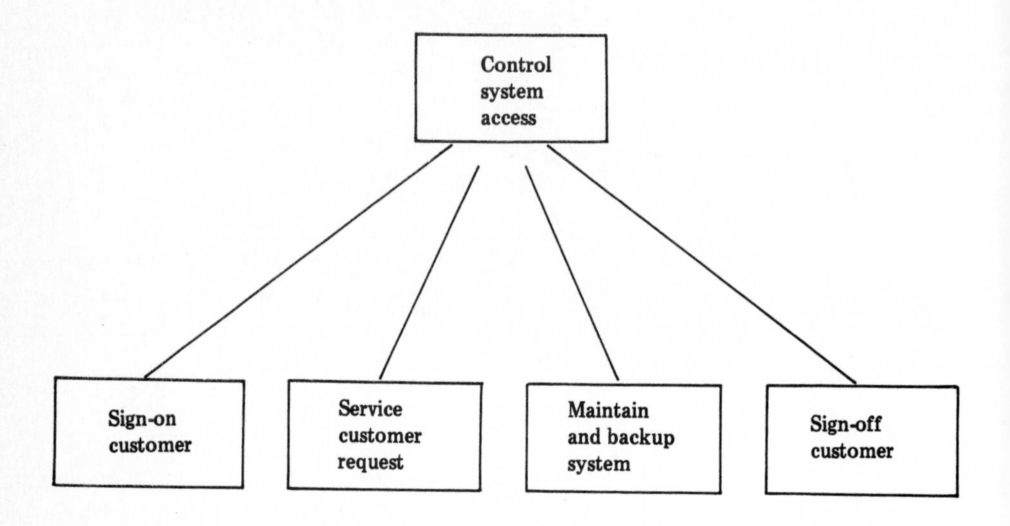

Let's pause briefly, and link all these partial hierarchy charts (Fig. 18.9). Note how the top module controls access to the system. A customer or manager is first identified, and then allowed to access either the customer modules or the maintenance modules; following a session, the user is signed off. Once a customer gains access, what happens? A work area is assigned and, using the menu/response approach, a service request is obtained. Next, the requested service is identified. After a service is complete, the work area is saved, and billing data collected. Now, it's back to *Service customer request*, where the user has the option of requesting another service or signing off.

Drop down a level. What happens under *Identify service*? First, the transaction is logged; then the desired service (a game, sports/ski information) is accessed. Move down one more level. What happens when a customer accesses the games service? Note that below *Access games* is the detailed hierarchy chart of Fig. 18.4. Study the complete hierarchy chart carefully. Do you see how system control functions are performed at the top and detailed functions are performed near the bottom? Do you see how common functions, such as logging each transaction or compiling billing information, are implemented nearer the top? Note how control flows from the top, down. Imagine that you are a customer who wishes to play the latest video game—MADMAN. Picture the control structure asking you the following questions:

1. Who are you?

2. What service do you want?

Fig. 18.9: *A program hierarchy chart.*

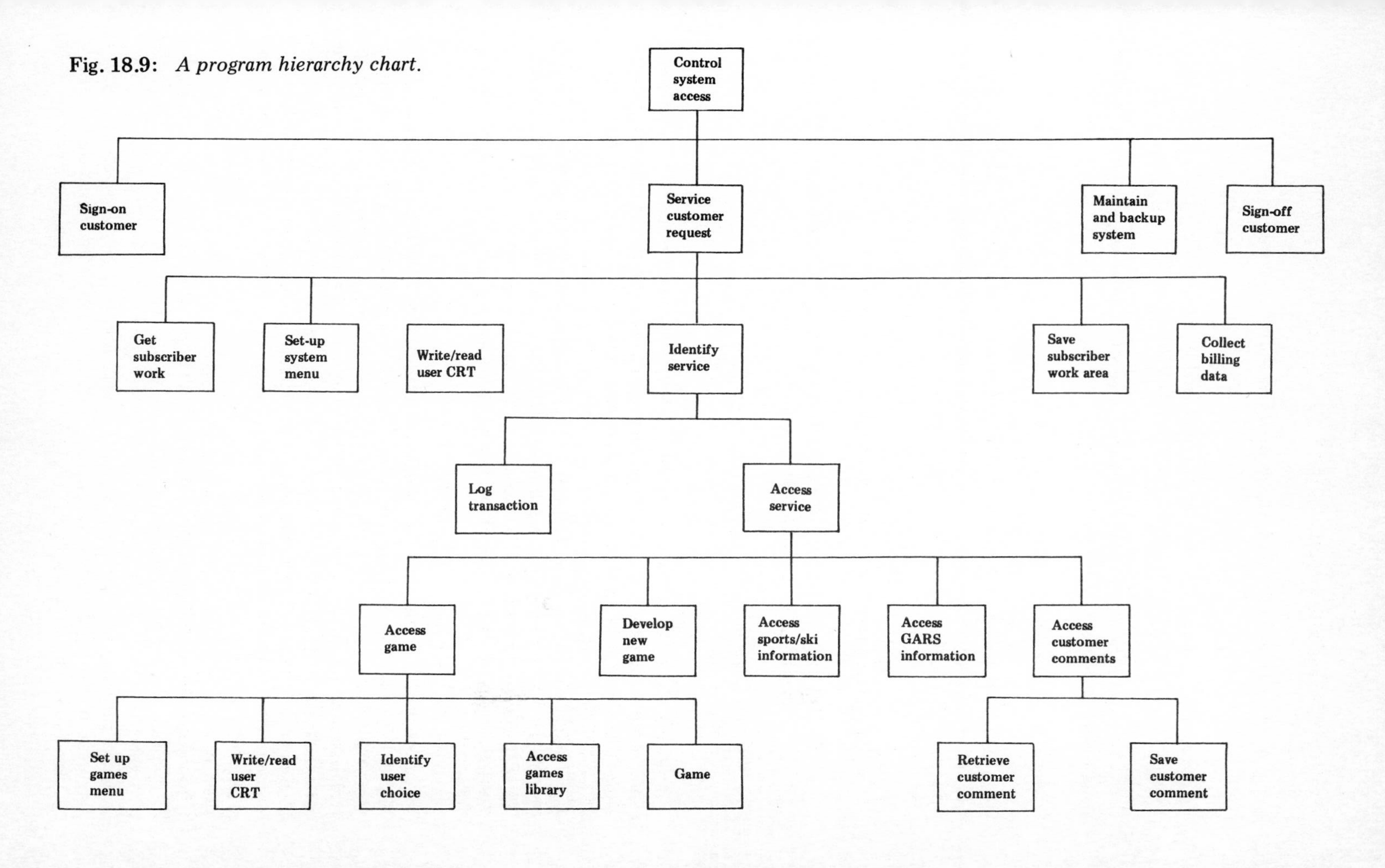

3. What game do you want to play?

4. How many cream pies do you want the madman to have?

Notice how the questions move from the general to the specific. At which level in the control structure would you expect each question to be asked?

We now have a good idea of the basic structure of the system control program. Let's consider our model in a bit more detail.

Decomposing the Control Structure

Can any of the modules on Fig. 18.9 be further decomposed? Certainly. Consider, for example, *Sign-on customer*. What functions must be performed in order to sign-on and identify a customer? We might identify the following three:

1. Collect a user number and password from the customer.

2. With the user number as a reference, read the customer's approved password from the customer information and/or security files.

3. Compare the data as entered to the standard.

If the customer passes the test, access to *Service customer request* is authorized; if not, access is denied. Note that these same steps might also serve to identify a system manager who wants to maintain or backup the system.

Might any of these modules be further decomposed? Consider data collection; to collect a user number and a password, the system must set up a message requesting a user number, write it to a terminal, read the user's response, and then repeat these steps for the password. Accessing the system files might be decomposed, too; see Fig. 18.10.

What of the other modules? Clearly, several detailed functions must be performed in order to maintain and backup the system. What is involved in signing-off a customer? What functions are controlled by *Collect billing data*? What is involved in logging a transaction? At the bottom, note that *Develop new game*, *Access sports/ski information*, *Access GARS* (for Games And Recreation System) *information*, *Retrieve customer comment*, and *Save customer comment* will each have a structure similar to *Access games*. While we will not decompose each of these modules in the text, look for several end of chapter exercises.

Once the hierarchy chart has been fully decomposed, the next step is to check it thoroughly. A good starting point is the list of prototype functions prepared during system design (Fig. 18.2). Can all the listed functions be found on the hierarchy chart? If not, the analysts may have missed something. Are there modules not identified on the list of prototype functions? If so, are these "extra" modules really necessary? After the function list has been checked, the system flowchart might be referenced.

Fig. 18.10: *A decomposition of the "sign-on customer" routine.*

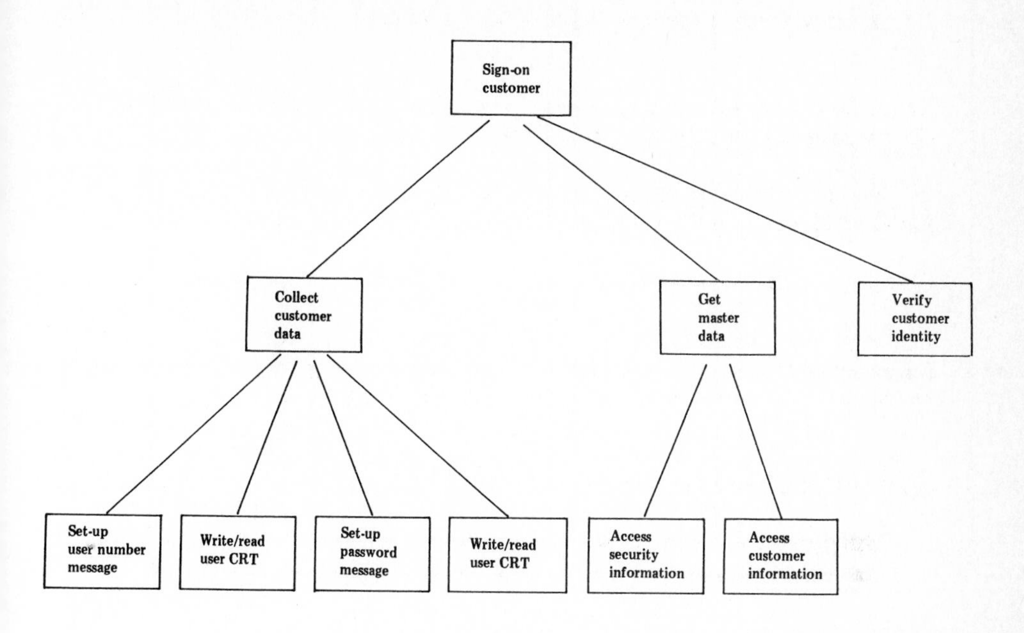

There should be one module to access each of the files, the user terminal, and the maintenance terminal. Is there? If not, is further decomposition necessary? Are there multiple modules that access the same I/O device? If so, these modules should be identified as common routines, and called from a variety of locations; generally, all I/O to a given device should be done by a common low-level module. The data flow diagram (Fig. 18.5) is another source; all functions should appear on the hierarchy chart. Finally, the analysts should be able to identify the module that will eventually hold each of the algorithms defined (at a high level) during analysis. These checks allow the analysts to test for completeness.

Assuming all the required elements are present, does the hierarchy chart represent a good program design? How might the analysts answer this crucial question? Two general quidelines are suggested. The first is *cohesion*: Each module should perform a *single*, *complete* function. The second is *coupling*: Each module should be *independent* of the rest of the program. Both criteria call for more detail than is presently available. Thus the analysts turn to developing input/process/output charts for each of the modules in the hierarchy chart.

Developing the IPO Charts

HIPO documentation consists of two parts: a hierarchy chart and an associated set of IPO (input/process/output) charts. The IPO charts define the inputs to, outputs from, and process performed by each module on the hierarchy chart. For example, Fig. 18.11 shows two IPO charts from the decomposition of *Sign-on customer*

Fig. 18.11: *Two IPO charts.*

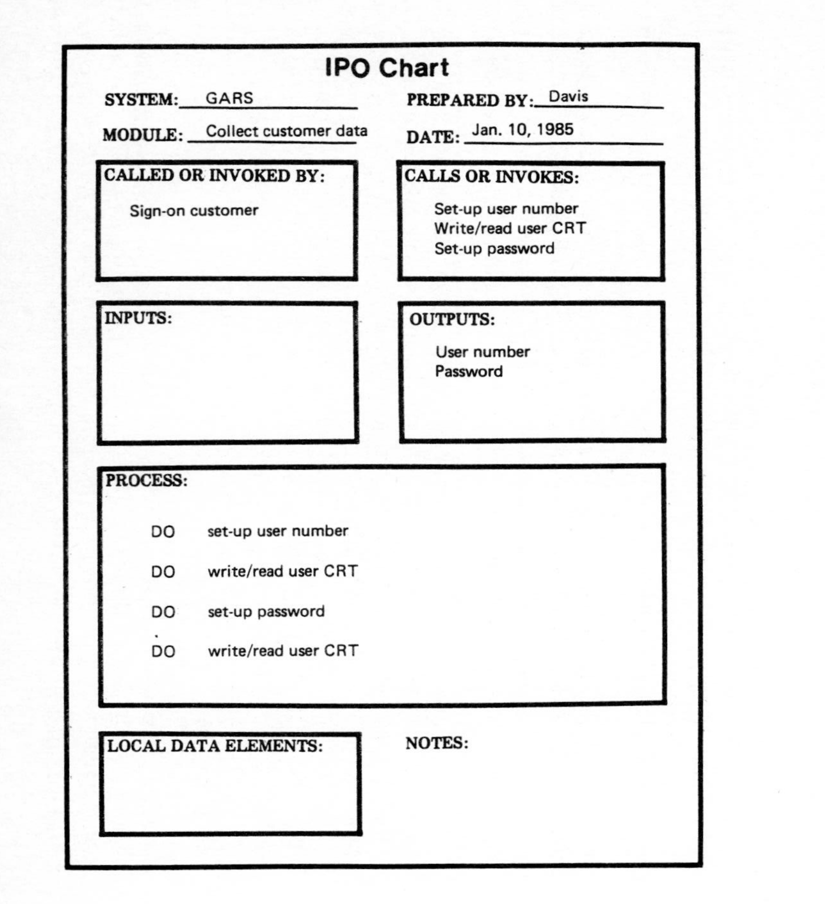

IPO Chart

SYSTEM: GARS

PREPARED BY: Davis

MODULE: Write/read user CRT

DATE: Jan. 10, 1985

CALLED OR INVOKED BY:

Collect customer data
Access sports/ski info.
Access game
etc.

CALLS OR INVOKES:

INPUTS:

User screen

OUTPUTS:

User screen

PROCESS:

WRITE user screen
PAUSE
READ user screen

LOCAL DATA ELEMENTS:

NOTES:

(Fig. 18.10). Note *Collect customer data* first. It is called by *Sign-on customer*; it, in turn, calls three lower level modules; the "called by" and "calls" blocks document the module's position on the hierarchy chart. There are no inputs; the outputs are a user number and a password. Four process steps are cited. Is this a reasonable module? It seems to be. It performs a single, complete function, collecting all the data needed to identify the customer attempting to sign on the system. Granted, this function is subdivided into four lower-level, more detailed steps, but these steps are logically related; *Collect customer data* seems cohesive. Is it independent? Only two data elements are returned by this module, and both are essential; it appears loosely coupled.

Turn now to the second IPO chart (Fig. 18.11): *Write/read user CRT*. This is the common user-terminal access routine called by any module that requires terminal access (not all are listed). Its function is to display a message on a terminal, read a response, and pass the response back to the calling program; note that it calls no lower-level routines. Read the inputs, the outputs, and the procedure; the function of the module should be obvious. Is it cohesive? Does it perform a single, complete function? At first glance, it seems to perform two functions—write and read a terminal. However, what is the purpose of these steps? A menu is written so a user's response can be read. The complete function requires both steps. Can you see why this module is cohesive? Is it independent? Can you imagine performing this function with less input and output? This module is loosely coupled to the rest of the system.

A single IPO chart should be prepared for each module on the hierarchy chart; we won't present the complete set in the text, but expect several end of chapter exercises. As you list the inputs to and outputs from each module, think about the module's independence. Are there too many variables being passed? Too many variables suggests excessive coupling, implying a lack of module independence; perhaps a redesign of the control structure should be considered. Is each variable essential to the function performed by the module? Unnecessary variables create unnecessary coupling, which creates an unnecessary risk.

Excessive variables can also be a sign of poor cohesion. If, for example, a module returns two or more unrelated data elements to the calling module, it must be performing more than one function. As you document the process, ask yourself a few questions. Are all these steps associated with a single process? Does each step contribute to that process? Is the process complete? A cohesive module performs a single, complete function.

A key objective of structured programming is producing programs that are easy to understand, and hence easy to debug and maintain. Large modules work against this objective. A common rule of thumb is that no module should exceed a single page of a program listing. In IPO chart terms, a structured English, pseudo code, or logic flowchart version of a process should fit within a single process block; if it doesn't, further decomposition may be necessary. This informal standard represents a common-sense limit that an analyst might use in conjunction with the more theoretical concepts of cohesion and coupling.

The Structure Chart

Preparing IPO charts is time consuming. It is not a good group activity; the work should be divided among the analysts. Whenever portions of a job are performed independently, there is a danger of inconsistencies and misinterpretations. Also, checking for cohesion and coupling as each IPO chart is prepared is not, by itself, enough, since the relationships between modules can cause problems, too. Before accepting the HIPO documentation as valid, it is essential for the analysts to check each others' work and verify their control structure. Preparing a structure chart (see Module H) is an excellent way to achieve both objectives.

A structure chart begins with a hierarchy chart (Fig. 18.9, plus the additional decompositions). Begin at the top. What data elements are passed from *Control system access* to *Sign-on customer*? The IPO charts are the source of this information. Write the data element names on the line linking these two modules; use a small arrow to indicate the direction of flow. Is every data element essential to the function performed by the lower-level module? If not, unnecessary data coupling may be present. Are all the data elements related to a single logical function? If not, cohesion is suspect.

Next, the analysts might consider the link between *Control system access* and *Service customer request*. Again, data elements (from the lower-level IPO chart) are written on the line connecting these two modules, and coupling and cohesion are evaluated. Step by step, they work through the entire hierarchy chart until each module has been checked. If the linkage between modules cannot be easily explained on a structure chart, imagine the problems the programmers will have. The time to correct design deficiencies is now, not during programming.

A structure chart is not particularly useful as documentation; it is a working tool. The analysts should begin with a hierarchy chart, and add the data flows as they go; the value is in developing the chart, not referencing it. We will not include a structure chart as a textbook figure, but you will be asked to develop one in the exercises.

DESIGNING THE SCREENS, THE REPORTS, AND THE FILES

We now have complete data structures, a hierarchy chart, and a set of IPO charts. Only a few details remain before implementation can begin. For example, consider screen layouts (see Module N). The system is designed around a menu concept; what should be displayed for each menu? This is an important question, as the screen is the user's window on the system, and a poorly designed, ambiguous, or unclear menu will make potential users wonder about the quality of the *entire* system. Screen design cannot be left to the individual programmers; careful planning is essential. Often, the starting point is a simple sheet of graph paper. Look at a CRT terminal. Any terminal. How many characters does it display across a line? How many lines can be displayed? Some terminals display twenty lines of forty (relatively large) characters each. Others display twenty or twenty-five eighty-character lines. It is reasonable to view a screen as a rectangular grid, with one character in each square; for example, Fig. 18.12 shows how the menu for the *Service customer request* module might be displayed on a forty by twenty screen.

What will he programmer do with a screen layout? Simply, a screen represents a set of program constants. In COBOL, for example, the layout might be coded as a data structure in the DATA DIVISION's WORKING-STORAGE SECTION, with a single

Fig. 18.12: *A screen layout for Service customer request.*

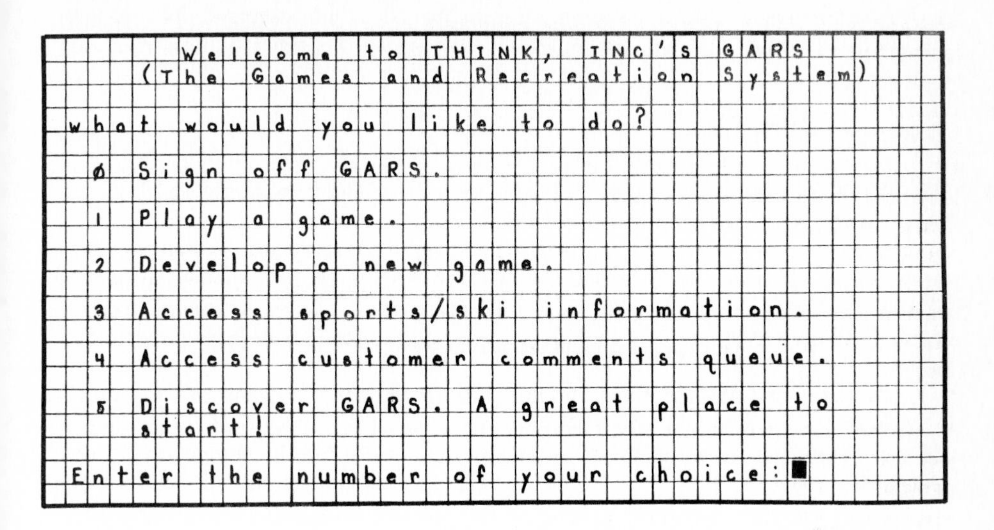

source constant defining each line. To display a screen, the program writes the complete structure; to read a screen, it reads the complete structure.

The idea of the menu approach used in this system is to display a list of options, and to allow the user to choose one. Clearly, when a screen is read, the user's response must appear somewhere on that screen. Where? How can the programmer find it? Look near the lower right of Fig. 18.12. One square on the grid is blackened; it shows the location of the *cursor*. The cursor indicates the position on the screen where the next character typed by the user will be displayed. On output, the programmer can specify the cursor's location; on input, the user's response will be found here.

An organization as large as THINK, INC. will almost certainly have screen design software. With such software, the analyst might lay out a screen such as the one in Fig. 18.12 by typing it on a CRT terminal, shifting and retyping material until it looks right. At this point, a command is issued to the software package, and the information on the screen is automatically converted to a set of constants in a source language—a COBOL DATA DIVISION structure, for example. Refer to Module N for more detail on screen design.

Report design is similar. The idea is to lay out a rough version of a report on paper, show the draft to the user, make necessary corrections, and generate the source code (constants, again) to define the structure of the report. Once again, see Module N.

Space estimates (Module M) are also necessary. Earlier, we developed a series of data structures, one for each file or library. A structure defines the number of characters per file entry (or per record). How many entries will there be? The analysts should be able to estimate these values, and use them to compute space requirements. These space requirements, in turn, determine the number of on-line storage units that must be allocated to this system, an important element of the expected hardware cost.

DEFINING A TEST PLAN

The detailed design of the program is now essentially complete; how will we test it? A well-designed test plan is essential to the successful implementation of the system. Testing will be performed at two levels. First, each module must be independently tested. When all have passed their individual tests, they must be combined so that the system can be tested. Appropriate data and procedures must be established for each phase in the plan.

Consider first the individual modules. The programmer assigned to work on the *Access games* logic should not have to worry about the status of any other module in the system until the subsystem is finished and ready for the system test. An excellent approach is to write a dummy *Access service* module to call *Access games* (or any of the other low-level services) so that normal system flow can be simulated. Additionally, the subsystem should be given realistic test data, including extreme conditions, to process; perhaps a few common games might be placed in a library, and other programmers invited to play them during free time.

The system test might begin with the test data for all the individual low-level functions; essentially, the low-level tests are repeated in the context of the complete control structure. Next, a selected group of ten or twenty, nontechnical "testers" might be given a brief overview of the system and turned loose for a few days. Finally, a select group of technical experts might be given full details on the system's design and told to try and "break" it. With such exhaustive testing, most of the serious bugs will be found before the product is released to the customers.

DEVELOPING AN IMPLEMENTATION PLAN

The final step in detailed design is preparing an implementation plan. The various programs and files are defined. Who will write those programs? In what order will they be written? When will they be complete? Is the preliminary implementation schedule still valid? Management will insist that such questions be answered before authorizing the funds and personnel needed to implement the system.

During the feasibility study, preliminary time estimates were made in years. The time estimates of system design were stated in months. Why the change? Simply, as system development progresses, the analysts know more about the system, and are thus able to estimate more precisely. We are now in detailed design. We have a hierarchy chart that defines individual program modules. We should be able to estimate programming time with great precision—programmer *days* might be reasonable. Personnel days can, of course, be converted to personnel months and years, so the new estimate can be compared with old ones. Is the expected completion date identified at the end of system design still valid? If not, it should be changed. If these more precise estimates suggest that the job won't take as long as the analysts thought, perhaps more features can be added to the prototype.

How might the analysts estimate programming time for each module? Often, they begin with another, easier to visualize parameter, lines of code. Each module on the hierarchy chart is one program routine. If structured design has been conducted properly, each will represent no more than one page on a compiler listing—that's between fifty and sixty lines per module. Let's assume that a study of past programmer performance suggests that a typical programmer generates, on average, about ten lines of tested, debugged, documented code per day. Using this standard, lines of code can be converted directly to programmer days. Figure 18.13 shows module by module time estimates for the system control program.

Four programmers will work on this system; perhaps two on this program. How should the program functions be distributed? While in theory a random assignment of these independent modules might work, there is a better way. For example, the control structure of each of the lower-level services is similar. If one programmer is given all these modules, he or she might learn from the first one, and thus do a better job on the others. Another argument is consistency; if one programmer does all the low-level routines, it is likely that all will follow a common pattern. The same arguments might be advanced for the modules concerned with backing up and maintaining the system. Finally, since the high level control structure is essential to the proper functioning of the entire system, it might be assigned to the more experienced programmer. Figure 18.14 shows one possible set of programming assignments.

With only two programmers, how long will it take to finish the program? A Gantt chart (Fig. 18.15) is an excellent tool for planning the work of a limited number of people. It shows an estimated total of 187 working days. Management will probably want a concrete schedule with firm targeted completion dates. Once a starting date is established, programmer days can be defined relative to this date, simply by counting. (Remember that most programmers work a five-day week, not a seven-day week; for example, seventy programmer days represents not ten, but fourteen calendar weeks.)

Can the system be completed more quickly? Certainly. Consider, for example, the project network of Fig. 18.16. It shows that, although certain functions must be implemented in a fixed order, others can be done concurrently. Of course, multiple concurrent functions will require more than two programmers, and the cost of the program may increase; we'll return to this idea in Chapter 19. For now, let's assume that the Gantt chart of Fig. 18.15 represents the planned schedule.

PREPARING A SYSTEM IMPLEMENTATION PLAN

The discussion of this chapter has focused on one program. The system flowchart (Fig. 18.1) identifies four programs: a log analysis routine, a management report generator, an earnings report generator, and a billing program. The data must be defined, HIPO documentation developed and checked, report and screen formats generated, and an implementation plan designed for each program. The last step in detailed design is to combine the component implementation plans into an integrated system implementation plan.

Fig. 18.13: *A list of system functions with time estimates.*

MODULE	LINES OF CODE	PROGRAMMER DAYS	WORK UNIT
Control system access	100	20	20
Sign-on customer	40	4	
Collect customer data	30	3	
Set-up user number message	40	4	
Write/read user CRT	20	2	
Set-up password message	20	2	
Get master data	40	4	
Access security information	30	3	
Access customer information	30	3	
Verify customer identity	100	10	35
Service customer request	100	10	
Get subscriber work area	40	4	
Set up system menu	20	2	
Identify service	60	6	
Log transaction	30	3	
Access service	40	4	
Save subscriber work area	50	5	34
Collect billing data	50	5	
Compute current charge	50	5	
Access customer information	20	2	
Update charges	50	5	
Rewrite customer information	20	2	19
Access game	40	4	
Set-up games menu	20	2	
Identify user choice	40	4	
Access games library	60	6	
Accumulate royalty	100	10	
Access customer information	30	3	
Rewrite customer information	40	4	33
Develop new game	40	4	
Set-up new games menu	20	2	
Identify user choice	40	4	
Implement security	100	10	
Load BASIC compiler	30	3	
Access customer storage	50	5	28
Access sports/ski information	100	10	
Modify existing service	100	10	20

Fig. 18.13: *Continued.*

MODULE	LINES OF CODE	PROGRAMMER DAYS	WORK UNIT
Access GARS information	50	5	
Set-up display area	20	2	
Access GARS library	40	4	
Display information	40	4	
Scroll display	50	5	
Identify user choice	30	3	23
Access customer comments	60	6	
Set-up comments menu	20	2	
Identify user choice	20	2	
Retrieve customer comment	40	4	
Identify customer	30	3	
Set-up comment entry	50	5	
Verify comment validity	50	5	
Store customer comment	30	3	30
Maintain and back-up system	80	8	
Set-up maintenance menu	30	3	
Write/read maintenance terminal	20	2	
Identify system manager	60	6	
Access security file	40	4	
Identify maintenance service	50	5	28
Back-up files	40	4	
Back-up security file	40	4	
Back-up user work space	40	4	
Back-up library	50	5	
Back-up games library	50	5	
Back-up sports library	30	3	
Back-up ski library	30	3	
Back-up GARS library	20	2	
Back-up comments queue	50	5	
Back-up customer information	50	5	36
Maintain a file	40	4	
Maintain security file	100	10	
Maintain user work space	40	4	
Maintain games library	100	10	
Maintain sports library	50	5	
Maintain ski library	50	5	
Maintain GARS library	50	5	
Maintain comments queue	100	10	
Maintain customer information	100	10	63
Signoff customer	50	5	5

Fig. 18.14: *One possible work distribution.*

Programmer A		Programmer B	
Control system access	20	Service customer request	34
Sign-on customer	35	Collect billing data	19
Maintain and back-up system	28	Access game	33
Back-up files	36	Develop new game	28
Maintain files	63	Access sports/ski information	20
Sign-off customer	5	Access GARS information	23
		Access customer comments	30
	187		187

Fig. 18.15: *A Gantt chart can be used to graphically illustrate the implementation schedule.*

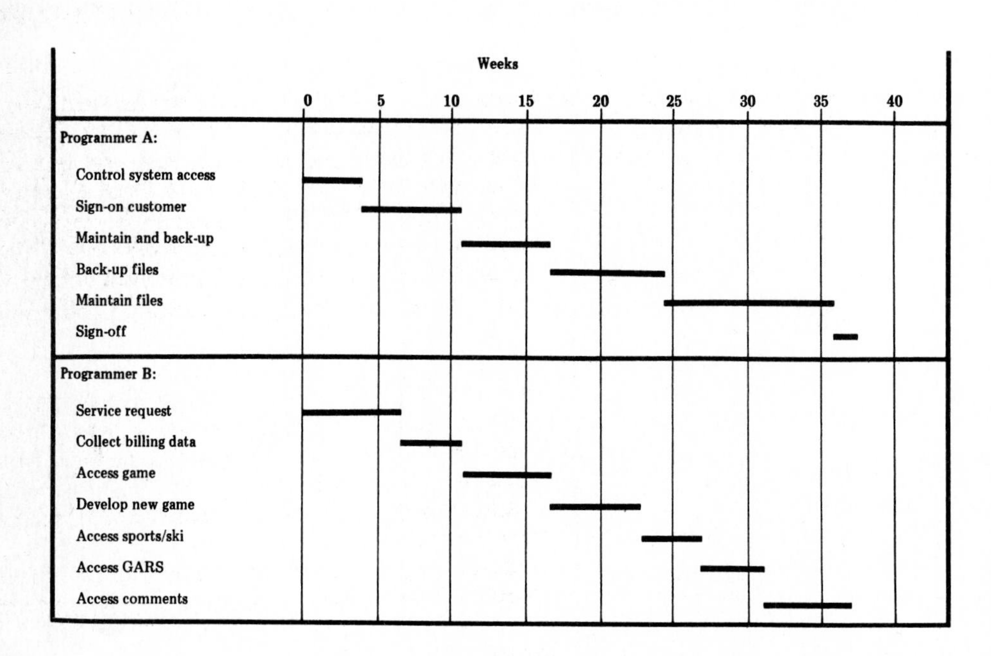

Fig. 18.16: *A project network showing how nine programmers might code the system control program.*

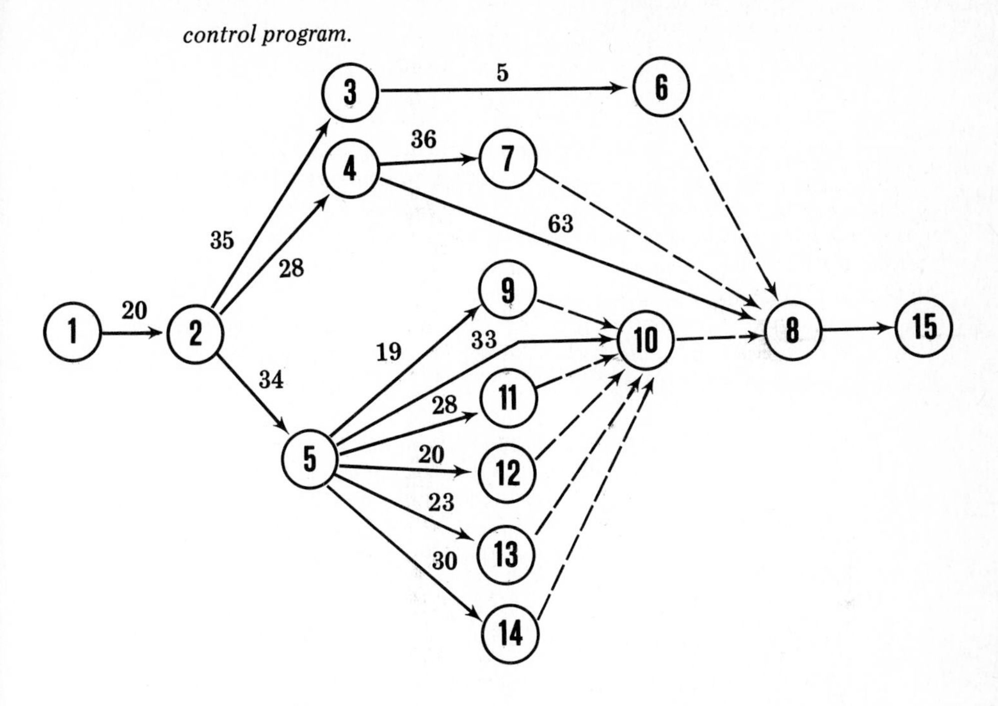

1-2 Control system access
2-3 Sign-on customer
2-4 Maintain and back-up system
2-5 Service customer request
3-6 Sign-off customer
4-7 Back-up files
4-8 Maintain files
5-9 Collect billing data
5-10 Access game
5-11 Develop new game
5-12 Access sports/ski information
5-13 Access GARS information
5-14 Access customer comments
8-15 System test

THE INSPECTION AND THE MANAGEMENT REVIEW

The exit criteria are now complete, and a formal technical inspection can begin. Each program should be inspected separately; in other words, since the games and recreation system is composed of five programs, five inspections should be scheduled. The inspectors should be skilled programmers. The key question is: Can valid code be written to these specifications?

The management review is just as crucial. What if, for example, the new cost and time estimates indicate that the job should no longer be done? Management should have the option to cut its losses. Three analysts have been working in the project to this point. If implementation is authorized, four programmers will be assigned, more than doubling the firm's financial committment. If errors have been made, now is the time to find out.

Our analysts have done a good job. The program specifications are approved; thus it's on to implementation.

SUMMARY

This chapter covered the detailed design stage of the games and recreation system project. The high-level system design from the prior step was our starting point. One program, system control, was selected for detailed study. The data formats of all files accessed by this program were defined first. We then developed a hierarchy chart for the functions performed by the program, and used the data flow diagram, system flowchart, algorithms, and a list of prototype functions to check it for completeness. The next step was preparing an IPO chart for each module on the hierarchy chart. Finally, the design was carefully checked by using the hierarchy chart and the IPOs to develop a structure chart.

Screen and report formats were designed next, and the amount of space required by each system file was estimated. The analysts then prepared an implementation plan including a complete schedule; a Gantt chart and a project network were used to illustrate the plan. Finally, the exit criteria were inspected and reviewed by management.

EXERCISES

1. What is the objective of detailed design?

2. Why should detailed design begin with the data? Explain how the data dictionary is used to help develop data structures.

3. In designing the system control program for the games and recreation system, using a consistent structure for all the low-level service functions was considered important. Why?

4. The system's driving mechanism was described as the process of serving customer requests. What is a driving mechanism? Why is it such an important element in developing a system design?

5. Part of the system design called for transactions to be logged under control of *Identify service* (Fig. 18.7). Why here? Why not at some other spot in the hierarchy chart?

6. If servicing a customer request is the system's driving mechanism, why was the high-level control structure of Fig. 18.8 needed?

7. A control structure for *Access sports/ski information* was suggested in Fig. 18.2. Add it to the hierarchy chart of Fig. 18.9.

8. On page 230, the text cited four questions that the system might pose in supporting a user who wants to play a specific game. Refer to the hierarchy chart of Fig. 18.9, and identify the module responsible for each question.

9. Refer to Fig. 18.9. Decompose the module *Maintain and backup system.* Remember that each file must be backed up and maintained.

10. Decompose *Develop new game* from Fig. 18.9.

11. What steps might be required to sign a customer off the system. Decompose this high-level module from Fig. 18.9. Don't forget to log the transaction.

12. Prepare Warnier-Orr diagrams (Module K) to parallel the hierarchy charts of Figs. 18.3, 18.4, 18.6, 18.7, 18.8, and/or 18.10.

13. Prepare a Warnier-Orr diagram (Module K) to parallel the system hierarchy chart of Fig. 18.9.

14. Prepare Warnier-Orr diagrams (Module K) to parallel the partial hierarchy charts prepared for exercises 9, 10, and/or 11.

15. Information from the data flow diagram, the system flowchart, and preliminary algorithm descriptions was used to check the hierarchy chart. Explain how these documents are related.

16. Get together with three or four other students from your class, and prepare a set of IPO charts to accompany the hierarchy chart of Fig. 18.9.

17. Figure 18.11 shows two IPO charts from the partial hierarchy chart of Fig. 18.10. Complete the other IPO charts.

18. Prepare IPO charts to accompany the partial hierarchy charts you prepared in exercises 9, 10, and/or 11.

19. What is cohesion? What is coupling? Why are they important?

20. Working in a group, prepare a structure chart for the hierarchy chart of Fig. 18.9.

21. Prepare a structure chart (or charts) for the partial hierarchy charts you prepared in exercises 9, 10, and/or 11.

22. Compile a list of ten popular video games. Imagine that these games are available on our system, and prepare a screen layout for a game selection menu. Assume an 80 by 25 grid; then use a 40 by 20 grid.

23. In the language of your choice, code a set of constants to represent the screen layout of Fig. 18.12.

24. Assume that customer comments queue entries are limited to a maximum of 256 characters each. In the prototype, we anticipate roughly 1000 customers,

and expect the average customer to have 3.5 messages on the queue at any time. How much space is required on a 256-byte sectored disk? On an IBM 3330? See Module M.

25. Estimate the size of the customer information file. Assume (for the prototype system) 1000 customers. Use the data dictionary prepared earlier to estimate the logical record length; if you did not create a data dictionary, assume a length of 120 characters. Refer to Module M, and base your estimate on a 3330 disk system. Begin with unblocked data, and then repeat your calculations with a blocking factor of ten.

26. Assume that management wants the system control program in six months rather than twelve. How many programmers will be needed? Develop a project network, and use it to prepare an implementation plan.

19

Case C Implementation And Maintenance

OVERVIEW

This chapter completes the games and recreation system case study. In it we will consider some problems associated with implementing a large system. The value of walkthroughs and inspections will be illustrated, and documentation standards described. We'll consider the system test in some detail. Finally, the operation of the prototype will be briefly discussed, and the evolution of the full system previewed.

IMPLEMENTING THE PROTOTYPE

What is unique about the games and recreation system? Relative to the first two case studies, it is a big project. Implementing it is a group activity, involving several programmers. What problems does such a group approach create? The first concern is assigning the work. As implementation begins, the HIPO program specifications define an integrated set of independent modules which suggest how the work might be divided. After the work is assigned, it must be coordinated, integrated, and managed. The HIPO documentation shows how the pieces fit together; these interfaces, in turn, suggest how the work of the programmers can be linked and tested. The schedule (a Gantt chart or a project network) shows the estimated completion time for each module, and thus supports management by exception. In short, the plan is the key to controlling group program implementation.

What if the plan changes, however? What if a programmer discovers a better way to implement the logic of a module? What if an analyst's error or oversight is not discovered until after coding begins? What if a perfectly good module just happens to be difficult to code? What if a user changes an objective at this late date? No plan is perfect—it will change. The problem is managing change. What does this mean?

Programmers tend to focus on their own code. Unfortunately, a change in one module can affect others; the impact of the change on the entire system must be carefully evaluated. Once again, the plan provides a basis for evaluating change. The hierarchy chart shows the control structure that links the modules; the associated IPO charts define the input and output data. The data dictionary lists where each data element is used. If the modules are independent, the data should represent the only source of intermodule interference; in other words, by tracing the data elements entering and leaving the to-be-changed module, the potential impact of the change can be studied.

For example, imagine that a module accepts as input the variable X, and produces Y as output. Assume that X is precise to four decimal places, while Y is precise to two. A programmer, looking only at this module, might be tempted to question the precision of X, and suggest that it be truncated to two places. Another programmer might arbitrarily decide to change Y to match X. In the context of this module, either change seems reasonable. An impact analysis of the proposed change might, however, identify another module, written by another programmer, that requires X to four places and/or Y to two. Changing the precision of a global variable can affect other modules. If the change is a good one, the plan must be changed; if not, the programmer responsible for this module may be forced to write suboptimum code for the good of the system.

Evaluating the impact of proposed changes is the analyst's responsibility. Generally, each module should be independent, linked to the system only through its global inputs and outputs. Changes within a module that affect only the implementation of an algorithm or the value or precision of a local variable are probably limited to that module; thus only the appropriate IPO diagram need be changed. Modifications involving global variables are potentially much more severe; changing a global variable can impact every other module using that variable. Programmers must be aware of the potential impact of a deviation from the plan, and clear all such changes with the analysts.

Dealing with Scheduling Pressures

Scheduling pressures are almost traditional during implementation; if a system looks good, management usually wants it as quickly as possible. Our analysts have devised a plan to complete the prototype system in one year. If more programmers are assigned, can the completion date be accelerated? This is where the project network comes into play. For example, contrast Figs. 18.15 and 18.16 from the last chapter. The first is a Gantt chart showing how two programmers might code and debug the on-line system control program in 187 days. The second is a project network describing the logical relationships between various work units; it shows that much of the work can be done in parallel. As the project nears completion, Fig. 18.16 shows nine concurrent tasks in process. What if nine programmers were assigned; how long would it take? The critical path represents a total of 111 days; two programmers will need 187 days; the difference, 76 days, is significant. What would these 76 days cost, however? Assume that a programmer earns $100 per day. One programmer for 187 days would cost $18,700; two, $37,400. With the accelerated plan, each programmer would be needed for only 111 days ($11,100), but nine would be required, giving a total cost of $99,900. The difference, $62,500, is *also* significant. Is accelerating the program 76 days worth $62,500? Management will have to decide.

Part of the plan is a set of checkpoints or target dates for completing major segments of the program. What happens when these target dates slip? They will; no plan, no matter how good it is, can forsee every contingency. A module might be more difficult than anticipated. A programmer might be less skilled than imagined. Personnel problems will occur; people do have accidents, become ill, are reassigned, or leave the company. Equipment will fail. If anything can go wrong, it will. When it happens, what can be done about it? The plan provides a structure for evaluating the impact of the change. New target dates can be reestimated, based on the new data. Perhaps programmers might be added in an effort to keep the project roughly on schedule.

New people should be added with extreme care, however. For example, assume that, 100 days into the project, one of the two programmers breaks a leg, and thus must miss three weeks of work. Looking strictly at the numbers—two programmers, 187 days—management might be tempted to replace the injured programmer with another one. Is it realistic to assume that the new person will be able to finish the job in 87 days? No. He or she will need time to understand the system before starting to write code. Who will provide this training? Probably, the uninjured programmer. How much coding will this other programmer do while the new person is being trained? Relatively little. In other words, adding a new programmer negatively impacts the entire system, so the schedule will almost certainly slip. In fact, it might be better to wait two or three weeks until the original programmer is able to return to work; at least then the impact on the schedule will be predictable.

What if the schedule does slip, and management wants to get it back on track? The tendency is to throw more programmers into the project. It seems reasonable to argue that if one person can do a job in one hundred days, that two should need fifty, and four should need only twenty-five. It rarely happens that way. For example, picture the problem of cutting the grass on a football field. Imagine that two people are assigned to the job, and that each is given half the field to mow. Some time later, it becomes obvious that work is not progressing as rapidly as it should, so ten more people are assigned. What happens? First, there is a management problem—how should the remaining work be divided? The original two workers must stop what they are doing while a new plan is developed, which wastes time. More time is lost as the workers take their positions and start their mowers. The problems don't stop as the work begins, however. Suddenly twelve different people must coordinate their work, and additional management time is needed to answer such questions as who cuts the grass that lies on the dividing lines between work areas? Also, some workers will be slower than others, and, since the job isn't done until all the grass is cut, more idle time will result. Clearly, adding people will increase the cost of the job, but that's not the point. People are added in an attempt to accelerate the schedule. Because of the need for more training, planning, management, inspection, and testing, adding people after a project has begun may have exactly the wrong effect, *delaying* rather than accelerating the expected completion date. Management must be aware of this potential problem; an interesting and informative discussion of this subject can be found in *The Mythical Man Month* by Frederick P. Brooks, Jr.

Walkthroughs and Inspections

Writing code demands precision, concentration, and attention to detail; it is essentially an individual task. The specifications provide a framework within which the programmer works. Is it reasonable to assume that a programmer's code properly implements the specifications? No; that's one reason why programs and individual modules are subject to tests. Unfortunately, a test is performed after coding and debugging are complete. What if the wrong code were written and successfully debugged? Time and money would be wasted.

While writing code is an individual task, evaluating that code is not; often, the perspective of other technical professionals can be most valuable. Structured walkthroughs are an excellent tool for evaluating code. A structured walkthrough is much like an informal inspection. Following the first clean compilation, the author meets with one or more other analysts or programmers to review the code; in effect, the group attempts to "play computer," suggesting input data conditions, and then stepping through the logic to see if the module properly handles those conditions. Often the group is able to spot subtle logical errors, and suggest corrections. Note that the participants should not simply evaluate the programmer's code; they should also relate that code to the specifications. A "well coded" module that fails to perform its function is useless.

You may have had an experience that illustrates the value of a walkthrough. Your program isn't working properly. You have stared at the code for hours, and can't find the error. Finally, in frustration, you ask a friend for help, and within minutes an obvious bug is spotted. In effect, your friend is conducting a walkthrough. When people work with details, they tend to see what they *think* they wrote. Consequently, it is extremely valuable to have someone else go through the code.

In many organizations, the walkthrough is formalized, and conducted as an inspection. The inspection process is described in Module A; it is rigidly structured, and involves formal reporting. A walkthrough is less formal; the atmosphere is generally more relaxed, with no notes taken, and no reports forwarded to management. Often, informal walkthroughs are used at the module level, and formal inspections at the program level.

Documentation Standards

Documentation is an important part of implementation. Merely writing the documentation is not enough, however; it must be maintained. Why? The idea is to provide a maintenance programmer with accurate information about a program. Inaccurate, incomplete, or out-of-date documentation is, at best, useless, and at worst, misleading. Programmers will generally maintain their source code, simply because most program changes are implemented by revising and recompiling the source code. All too often, however, the programmer hasn't time to dig out the documentation manual and revise the associated IPO chart. Consequently, the HIPO documentation quickly becomes out-of-date and useless. Now, the questionable validity of the HIPO documentation becomes an excuse to avoid future maintenance—it's out-of-date anyway, so why bother?

Since programmers do update source code, why not bury the documentation in the source code? A typical module in a structured program consists of one source-listing page. Why not preceed each module with a one-page IPO chart written as source language comments? Would programmers maintain such comments? Although some management control will be required, the proximity of the documentation and the code does tend to create a subtle peer pressure to maintain both together. Coupled with the system flowchart and the program hierarchy charts, which are maintained by the project leader, the source module IPO charts form an outstanding documentation package.

THE SYSTEM TEST

At the end of the implementation process comes the system test. Ideally, the test should proceed module by module. For example, a preliminary system test might be conducted on the control structure, with dummy low-level modules in place; the purpose of this first test is to ensure that the control structure properly transfers control. Next, the detailed modules are added, one at a time; this approach helps to isolate problems when they occur, simplifying debug. This first series of tests exercises the test data, with the application programmers verifying that the results match those obtained earlier when the modules were tested independently.

The second phase might be termed *benevolent* system testing. Selected employees are given an overview of the system's functions and told how to access and use it. Over a two-week period they are encouraged to access GARS on their lunch hours, during their free time, and even from their homes after work and on weekends. In exchange, they are required to keep a log in which problems, useful features, and weeknesses are noted. A hot line is available to report serious problems. At the end of the two-week period, this group attends a debriefing session, where they comment on what they liked and didn't like about the system. The idea is to identify problems so that they can be corrected before the system is released to the customers.

The final phase might be termed *malevolent* system testing. A team of technically expert "commandos" is selected, given an in-depth overview of the system, and told to try to "break" it. A time limit, perhaps two full days, is set, and the commandos are turned loose. Using terminals similar to those that might be available to a user, they try to force system failure. For example, one might find a way to drop the system into an endless loop, while another might discover how to access the control module and modify his or her authorization codes, thus gaining access to a maintenance module and cancelling all charges. Serving on a commando team is fun, and THINK, INC.'s programmers and analysts willingly volunteer. The objective is simple: identify serious system defects before the customers do.

OPERATING THE PROTOTYPE SYSTEM

Following a successful test, the games and recreation system will be released to selected customers. They will be charged a reduced rate in exchange for keeping a log; heavy use of the customer comments queue to report problems and successes will be encouraged, too.

What happens if problems are discovered? There will be bugs. All should be carefully documented, but should they be corrected? The top priority, obviously, is to keep the system running. Next comes direct marketing support—the programmers or analysts will be expected to fine-tune portions of the data collection logic. Finally, changes that might improve the system can be considered, but should be implemented only with marketing approval. Why? Some will be of direct concern; for example, marketing may want to know how correcting a particular bug might affect customer use. Other changes could, however, invalidate all the marketing data collected before the change. Remember that a prototype system represents a controlled experiment, and the parameters of the experiment should not be changed without good reason.

By studying the prototype, marketing expects to obtain the information it needs to estimate the revenue potential of the games and recreation system. Let's assume that marketing's reaction is positive—the system promises to be a major new revenue producer. Economic feasibility has now been established at a cost of roughly $250,000. What happens next? In effect, a feasibility study has just been completed. Analysis normally follows the feasibility study. Why go all the way back to analysis? Why not simply begin adding functions to the prototype to produce a finished system? Because the prototype was just that: a prototype or model. In building and using it, a great deal was learned about how the system should be constructed. Weaknesses, inefficiencies, and problems in the initial design should be corrected, not propogated in the real system, and such problems can occur in the basic system design, as well as in a detailed computational module. Of course, adding to the prototype should be viewed as one alternative during analysis and system design; the point is that the prototype should not be seen as the only alternative.

Throughout this text, we have used a methodology in analyzing and designing systems. We have seen how the methodology can be applied to a variety of problems; it is a most effective tool. Do not, however, forget that the methodology is just a tool; by itself, it does not guarantee success anymore than a quality typewriter guarantees a good term paper. The structured approach to systems analysis and design generates good systems only when it is used by a good analyst; systems analysis still demands skill, experience, creativity, and hard work. On the other hand, it is difficult for even a skilled, experienced, creative, hard working analyst to do a good job with inferior tools, and that's why a good methodology is so important.

Note that the methodology presented in this text is not the only one available; in fact, the author's methodology has been generalized from a number of sources, particularly, those referenced throughout Part IV. It is likely that you will encounter other approaches. You should, however, discover that the principles and objectives of those other approaches are similar, and that the real differences are in details, not philosophy. Soon, you will be designing your own systems. It is interesting and challenging work. Good luck!

SUMMARY

This chapter discussed the implementation of the games and recreation system prototype. First, we covered several problems associated with a group approach to programming. The implementation plan was seen as crucial to coordinating the activities of the programmers, and provided a means for evaluating the impact of proposed changes. Next, we briefly considered how a schedule might be accelerated, and pointed out how belated efforts to "catch up" can be counterproductive. Structured walkthroughs and the importance of good documentation were then discussed, and the system test overviewed. Finally, the chapter turned to the prototype system itself, and outlined a number of operating considerations.

REFERENCES

Brooks, Frederick P. Jr. (1978). *The Mythical Man Month.* Reading, Mass.: Addison-Wesley.

EXERCISES

1. Discuss some of the problems associated with a group approach to implementation. Explain how the implementation plan helps solve many of these problems.

2. "Programmers tend to focus on their own code." Do you agree? Why, or why not? Why is this a potential problem?

3. Explain how the impact of a proposed change in one module can be traced through an entire program by using the HIPO documentation and the data dictionary as references.

4. Explain the difference between global and local variables. Why is a change to a global variable potentially more significant than a change to a local variable?

5. Accelerating system implementation generally adds to system cost. Why?

6. Adding people to a project after that project has begun can be counterproductive. Explain.

7. A programmer's code should be evaluated early in the implementation process. Why?

8. What is a structured walkthrough? Why are walkthroughs valuable? Contrast a walkthrough with an inspection.

9. Why is it such a good idea to bury documentation in the source code?

10. The first phase in the system test included adding detailed modules one at a time. What is the point of this step-by-step approach?

11. Why did THINK, INC. have two groups of its own personnel use the games and recreation system prototype before it was released to the customers?

12. As the prototype is running, bugs will be found. They must be carefully documented, but not necessarily fixed. Why? What maintenance task must be performed on a prototype?

PART IV

The Analyst's Tools

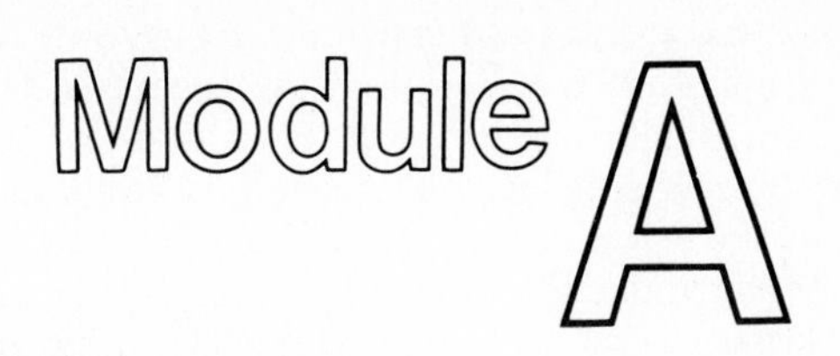

Inspections and Walkthroughs

OUTLINE

INSPECTIONS AND WALKTHROUGHS

If the systems analysis and design process is to be controlled, we must have a defined event or milestone that occurs at the end of each step in the process. Management review is certainly required, but tends to focus on such factors as cost, return on investment, and schedule, and often lacks technical depth. An *inspection* is a formal review of the exit criteria conducted by technical personnel. The intent of the inspection is to determine the technical accuracy of the exit documentation. When a project passes an inspection, it is assumed that the work up to this point is both technically acceptable and consistent with the objectives of the system. Often, an inspection is used as a prerequisite to a management review.

A *walkthrough* can be viewed as an informal inspection. Before presenting documentation to an inspection team or to management, the analyst is strongly advised to preview the material to several colleagues; such "dry runs" help to identify rough spots, and can save the analyst considerable embarassment. Although quite valuable at any stage in the systems analysis and design process, walkthroughs are particularly important during the implementation stage as means of checking the accuracy of the code.

THE INSPECTION TEAM

An inspection team normally consists of four individuals: the moderator, the author, and two inspectors. The key member is the *moderator*. Ideally, this individual should be respected technically, and should be unbiased, with no direct involvement in the project. The moderator runs the inspection, scheduling all meetings, distributing all necessary documentation, conducting all sessions, and making certain that the inspection is both thorough and fair. This is a very difficult job, with a significant amount of responsibility.

The *author* is usually the person who wrote the documentation or the code being inspected; if an analysis or programming team is involved, the project leader normally performs this role. The author is expected to answer technical questions, but should refrain from defending the work. It is very difficult for anyone to evaluate his or her own efforts; thus the outside perspectives of the moderator and the inspectors are crucial to the inspection process.

The *inspectors* should be technical professionals who, while not directly involved in preparing the documentation, have a stake in the outcome. The individual who was responsible for the prior step would be a good choice; a member of the group that will subsequently use the output of this step would be another. Normally, two inspectors are assigned; under unusual circumstances, the team can be larger or smaller.

THE INSPECTION PROCESS

The inspection process consists of six steps:

1. planning,

2. overview,

3. preparation,

4. the inspection session,

5. rework,

6. follow-up.

Planning

As soon as the documentation for a given step is completed, the author contacts the moderator and asks that the inspection process begin. The first task is to select an inspection team. In many organizations, the moderator selects the team; in others, management assumes this responsibility. Once the team has been named, the moderator must distribute all materials and schedule the inspection meeting or meetings.

The steps in any well-run project will, of course, be carefully scheduled in advance. The inspection is a key step in this process; if the inspection is late, the project will almost certainly fall behind schedule. Ideally, the inspection should not begin until the author has completed all the documentation; realistically, management will pressure both the author and the moderator to begin on time.

Overview

The overview step is the only optional step in the inspection process. If a project is particularly extensive or involves a number of concepts or techniques that are not apparent to the inspectors, it may be valuable for the author to present a brief technical overview of the project and the documentation. The moderator must control the overview. It should not be allowed to degenerate into a question and answer session, a sales pitch, or a preliminary inspection. The objective is to save the moderator and the inspectors some time. The danger is that the author may bias the other members, making it easy for them to overlook errors. The author's presentation should stick to the facts, stressing what was done and how—not why it was done that way. Later, after the other members of the team have had an opportunity to review and understand the documentation, the reasons behind the technical decisions should be considered.

Preparation

The preparation step calls for individual work on the part of each of the participants. The moderator and the inspectors should read the documentation and note any questions or potential problems. Should the author be asked to answer questions about the project during preparation? Although asking for clarification can save some time, there is a danger that the author may explain away a possible error, and thus bias the inspection process. In some organizations, contact between the inspectors and the author during the preparation step is officially prohibited, but such rules are very difficult to enforce. At the very least the participants should be aware of the potential for bias, and should avoid nonessential contact.

The Inspection Session

The inspection session is conducted by the moderator. One of the inspectors (not the author) is asked to be the reader; this individual reads aloud or paraphrases the documentation. Since paraphrasing involves at least an element of interpretation, it is better than simply reading the material word for word; if the reader paraphrases incorrectly, the documentation is probably unclear, and unclear documentation often accompanies a technical error. During the inspection session, the author's primary responsibility is to answer technical questions.

The objective of the inspection session is to find errors. Note that the inspectors are not to correct these errors; their responsibility is to find them. All participants, including the moderator, the author, and the reader, should inspect the documentation; anyone, including the author, is allowed to identify an error. The moderator maintains an error log (Fig. A.1), noting each error and estimating its severity: trivial, moderate, significant, severe, or fatal. The inspection session should be limited to perhaps 90 minutes, and all participants should be aware of this time limit. If excessive errors are encountered, the moderator has the authority to terminate and reschedule the inspection session; inspecting incomplete or sloppy documentation is a waste of time. Finally, the moderator has the right to schedule a reinspection after rework has been completed.

The moderator must control the inspection process. One danger is that the author will become a proponent or defender of the work, and attempt to discredit any suggestion that an error might be present. At the other extreme, one or more inspectors might become excessively negative, and conduct a "witch hunt" rather than an inspection. It is always possible that one individual may try to dominate the inspection by force of personality. The moderator's job is to see that the inspection is conducted fairly and impartially, and that everyone has an opportunity to participate.

A common point of disagreement involves assigning a measure of severity to each error. The author may see an error as trivial, while an inspector may consider it severe; the result could well be a protracted argument. After a reasonable discussion, the moderator must break in, arbitrarily assign a severity level to the error, and move on. The estimate of severity is, after all, merely a label that is attached to the error. The important thing is that the error be detected; its classification is secondary.

Rework

Following the inspection, the moderator and the author meet to discuss the results; the focus of this meeting is the error list compiled during the inspection session. Each error should be discussed, and the rework time estimated. The responsibility for actually doing the rework belongs to the author. As each error is corrected, the author should note the actual rework time. A key management concern is that rework time be estimated accurately. An inspection data base containing a history of estimated and actual rework tmes can be used to help improve the estimation process.

Follow-up

When the rework is completed, the author and the moderator meet once again to review the results. If the moderator is satisfied with the rework, the inspection process

Fig. A.1: *A typical error log.*

INSPECTION ERROR LOG

PROJECT: ______________________

MODULE OR COMPONENT: ______________________

INSPECTION LEVEL: ______________________

SESSION DATE: ______________________
TARGET DATE: ______________________
MODERATOR: ______________________
AUTHOR: ______________________
INSPECTOR: ______________________

	ERROR DESCRIPTION	SEVERITY	EST. TIME	ACT. TIME	DATE COMP.	CHECK
1						
2						
3						
4						
5						
6						
7						
8						
9						
10						
		TOTALS				

INSPECTION TIME

MODERATOR	
AUTHOR	
INSPECTOR	
INSPECTOR	
INSPECTOR	
TOTAL	

We have inspected the unit of work described above and have found it technically acceptable.

MODERATOR ______________________

AUTHOR ______________________

INSPECTORS

____

DATE INSPECTION COMPLETED: ______________________

ends. If not, the moderator may request additional rework and another follow-up session, or perhaps schedule a reinspection. If this is necessary, the inspection team is reconvened, and the inspection session, rework, and follow-up are repeated.

THE INSPECTION AND THE MANAGEMENT REVIEW

Following the successful completion of an inspection, the moderator must "sign-off." Sometimes, a simple memo will suffice. In other organizations, a standard form is completed and signed. Often, the moderator, the author, and the inspectors sign the error list (complete with rework notations) at the end of the process. This formal documentation notifies management that the project has been technically reviewed and found acceptable. In the subsequent management review, technical aspects of the system can be assumed valid, and management can concentrate on costs, benefits, and schedule.

INSPECTION POINTS

In this text, we recommend that formal inspections be conducted at the end of the analysis, system design, detailed design, and implementation steps (Fig. A.2). Problem definition and the feasibility study do not require a formal inspection, as they tend to be very limited in a technical sense; the systems analyst may, however, choose to conduct a walkthrough of the feasibility study before presenting it to management and the user. The implementation step ends with a formal system test, and many organizations consider this test a sufficient check of the technical accuracy of the code. Given this viewpoint, a walkthrough might be enough during implementation, although the case studies in this text will call for an inspection.

An inspection marks the end of one phase of the system life cycle. It is usually followed by a management review. The inspection and the review represent the essential defined events or milestones that mark the transition from one stage to another.

The first formal inspection follows completion of the analysis stage (Fig. A.2). The key question to be answered in this first inspection is: Does the analyst really understand the problem? Key documentation includes a data flow diagram, an elementary data dictionary, and brief descriptions of the important algorithms. This documentation should be aimed at the user, and not at the programmer. At this stage, it is important that user representatives be on the inspection team.

The end of the system design phase marks the second inspection point (Fig. A.2). The intent of system design is to set the technical direction for the system; thus both users and programmers should be on the inspection team. In addition to the documentation from the analysis stage, the analyst should identify at least three alternative solutions, and provide system flow diagrams and cost estimates for each; often the analyst will recommend one of these options. The inspection team should review the alternatives as presented, and should make certain that the analyst has not simply bypassed this step by presenting only one serious alternative. Comments on the recommended alternative are certainly in order. The final decision, however, will be made by management.

Fig. A.2: *A summary of key inspection points.*

Analysis: What must be done to solve the problem?

Inspection criteria:
- data flow diagram
- data dictionary
- algorithms

Inspectors:
- users

Key inspection concerns:
1. Does the analyst *really* understand the problem?
2. Has the analyst defined what must be done to solve the problem?

Objective: To ensure that the analyst is on the right track.

System Design: How, in general, should the problem be solved?

Inspection criteria:
- alternative solutions
  - system flow diagrams
  - cost estimates
- recommendations

Inspectors:
- users
- programmers
- technical personnel
- author - analysis

Key inspection concerns:
1. Are the alternative solutions technically realistic?
2. Do the alternatives represent a reasonable choice?
3. Will the alternatives solve the user's problem?

Objective: To set a reasonable technical direction for the system.

Detailed Design: How, specifically, should the system be implemented?

Inspection criteria:
- Hierarchy charts
- IPO charts
- complete data dictionary
- file specifications
- pseudo code algorithms
- cost estimates
- implementation schedule
- hardware specifications
- rough test plan

Inspectors:
- programmers
- author - system design
- perhaps a user

Key inspection concerns:
1. Can the code be written from these specifications?
2. Are the cost estimates and schedule reasonable?
3. Are implementation specifications consistent with the system objectives?

Objective: To ensure that the system will solve the user's problem.

Implementation: Write the code and install the system.

Inspection criteria:
- source listing
- procedures

Inspectors:
- programmers
- author - detailed design
- users of procedures
- system users

Key inspection concerns:
1. Does the code meet the specifications?
2. Are the procedures reasonable?

Objective: To ensure that the system as implemented matches the system as planned.

The third inspection follows detailed design (Fig. A.2). The key question at this stage is: Can programming write code based on these specifications? The analyst is expected to provide literally everything the programmers need to write the code—as a minimum, hierarchy charts, input/process/output charts, a complete data dictionary, file specifications, pseudo code versions of the algorithms, cost estimates, and implementation schedules. The inspection team should contain at least one programmer, and probably more. If someone other than the author prepared the system design specifications, that individual should be on the team as well. Following this step, management will be asked to commit sufficient funds to implement the system; thus an in-depth management review can be expected. Note: although our emphasis has been on software development, other exit criteria for the detailed design stage, hardware specifications for example, might be subjected to an inspection as well; it is even possible that different teams of experts might inspect different elements of the documentation.

Several inspections might occur during implementation (Fig. A.2 again). Many organizations use code inspections, with a reader paraphrasing a source listing, while the inspectors look for logical errors. Code walkthroughs might be used in addition to (or in place of) an inspection. A code walkthrough is less formal than an inspection, with teams consisting exclusively of other programmers, and no standard error reporting procedures. Because a system test is normally scheduled at the end of the implementation step, a formal inspection may not be needed.

During implementation, the analyst normally prepares operating procedures, security procedures, auditing procedures, and similar documents. It is important that these be checked. Often, a formal inspection involving the affected people is scheduled for each set of procedures; for example, computer operators may be asked to inspect the operating procedures.

SOME POLITICAL CONSIDERATIONS

Excessive management involvement can destroy the inspection process. A manager's comments tend to take on added significance simply because they come from a manager. If this happens, what should be a technical review process can easily be dominated by non-technical management issues. Given a reasonable level of independence, the inspection process will generate a solid technical review of the project. The fact that the project passed an inspection can then be accepted as proof of its technical soundness, allowing management to concentrate on such issues as cost, benefit, personnel, equipment commitments, and the schedule.

The error reports generated during the actual inspection session represent another point of concern. People naturally fear that an error report will in some way be used against them—that error rates will eventually creep into personnel evaluations. This puts unnecessary pressure on the inspection team. Management must avoid misusing these data.

Another danger is that an analyst or programmer, fearing criticism or the misuse of an error report, wlll simply postpone the inspection until everything is perfect. A project schedule is essential, and the moderator must have the authority to insist that the schedule be followed (within reason), or that the schedule be officially changed. A change in the schedule is, of course, a legitimate reason for management to become involved.

The inspection process puts a great deal of pressure on the moderator. Without management's authority, this individual must perform several management-like functions, including scheduling meetings, conducting meetings and limiting their scope, and, perhaps most significantly, ordering and evaluating rework. The moderator is in a particularly uncomfortable position when a reinspection is required, because the need for reinspection implies that the author did not do a very good job the first time through. Some personal friction is inevitable. In many organizations, a reinspection is made a standard part of the inspection process, and, if the first inspection goes well, the moderator is given the authority to cancel it. As a result, the moderator can make a positive decision rather than a negative one. In effect, the moderator is asked to say "good job" or "no comment" rather than "bad job" or "no comment", a much more confortable choice.

REFERENCES

1. Freedman and Weinberg (1982), *Handbook of Walkthroughs, Inspections, and Technical Reviews*. Boston: Little, Brown and Company, Inc.

2. IBM Corporation (1977), *Inspections in Application Development — Introduction and Implemention Guidelines*. White Plains, New York: IBM Corporation. (IBM publication number GC20-2000.)

Module B

Interviewing

OUTLINE

INTERVIEWING

An effective systems analyst must be able to conduct interviews. An interview can be an extremely valuable source of information, particularly during the early feasibility study and analysis stages of the system life cycle. This module is written to give you some general hints on how to conduct an interview.

From the analyst's perspective, the basic reason for conducting an interview is to collect information. As the system life cycle begins, the analyst usually has numerous bits and pieces of information to work with: existing documentation, procedures, and so on. A good analyst can often form a surprisingly complete picture of the existing or proposed system from these sources, but invariably questions and ambiguities will arise, and key pieces will be missing. Only people directly involved in the system can answer these questions, clear up the ambiguities, or supply the missing pieces. Thus, the systems analyst must conduct interviews.

PREPARING FOR THE INTERVIEW

Effective interviewing is the result of careful preparation. Don't "wing it." Know why you want the interview and what you hope to accomplish before you schedule an interview.

Begin by defining the purpose of the interview. Go through the formal documentation, and develop a picture of the existing or proposed system. Identify questions, missing pieces, and ambiguities. These unknown factors or components represent an initial outline of the interview objectives. Note that you may have to interview several people to meet these initial objectives.

Next, *select the person or group to be interviewed.* Obviously, you want to find the individual who can best answer the questions. How do you find this person? Several clues might be used, including the formal organization chart, a work flow analysis, or a report distribution list. Often, the best thing to do is to begin with the organization chart and interview the manager who seems most likely to be responsible for a given objective. Although the manager may not be able to answer specific questions with the required level of detail, he or she should be able to tell you who can. If nothing else, starting with the manager makes political sense; people are less hesitant to give you their time if the boss knows about the interview and has approved it.

Before actually conducting the interview, *do your homework.* Know the topic. Read the relevant documentation. If you are about to interview a manager, know that manager's position on the organization chart, and know the basic functions of that manager's department or group. If you are about to interview a clerical employee, be familiar with the relevant documents or procedures that employee uses. An unprepared interviewer is resented; people do not like to have their time wasted.

Prepare specific questions aimed at the individual (or group) you are planning to interview. Refer to the objectives outline and select all questions that this individual might be able to answer. Develop a written list of questions, and consider follow-up questions to use in case the interview begins straying from the key point. Remember, however, that you can't anticipate everything, so don't try. The list of questions is a guide to the interview, and not an absolute.

Schedule the interview. You need the information. You are asking another person to give up some time. You must be willing to arrange your schedule and to travel to the other person's office or workplace. If you expect cooperation, schedule the interview at the subject's convenience.

THE INTERVIEW ITSELF

A well-conducted interview consists of three distinct parts: an opening, a body, and a closing. Let's consider each of these phases one at a time:

The Opening

The key objective of the opening is to establish rapport. Begin by identifying yourself, the topic you plan to discuss, and the purpose of the interview. Be honest. If there is an established project, you might offer to share the statement of scope and objectives prepared during the problem definition step; even if the subject chooses not to read the statement, its very existance gives the interview an added touch of legitimacy. Tell the individual why he or she was chosen for the interview. Where appropriate, identify the managers who have authorized the interview.

In an attempt to establish a relaxed atmosphere, many good interviewers begin with a brief period of smalltalk by discussing the weather, the exploits of a local sports team, or similar trivia. While this technique can be effective, it can also backfire. If you have an established relationship with the person being interviewed, or if the subject is obviously nervous, a brief period of casual conversation might help. Avoid wasting time, however. When in doubt, get to the point.

The Body

If you are conducting the interview, you are responsible for actually getting things started. Have your first question prepared—an open question, for example:

> When I read the documentation for this system, I had some trouble with (mention the part or section). Can you explain it to me?

Or, consider asking the subject how his or her job relates to the project, or how a particular procedure or system works. Ask follow-up questions to help focus the interview. Many people tend to concentrate on how things work; ask why they work as they do. Another effective technique is to say something like: "Let's see if I understand what you're saying", and then offer a brief summary. If your understanding is wrong, the other person will probably correct you; if it's accurate, you establish that effective communication is taking place.

You should generally begin with a relatively broad, open question, and gradually, through increasingly specific follow-up questions, focus the interview on particular points of concern. Often, an individual will explain a detail while reacting to an open question. Also, the interviewer might learn something, getting an answer to an important question that had not been anticipated. This type of questioning is called a funnel sequence.

Listen to the answers. Don't concentrate so intently on your next question that you miss the answer to the current one. (This is a common beginner's mistake.) Be flexible. Try to stick to the subject, but allow a certain amount of spontaneous discussion; you might learn something. Your list of prepared questions (or your objective outline) should be used as a guide or a memory jog, and not as an absolute. Delete questions that seem unimportant, or that, based on earlier responses, you know cannot or will not be answered. Bypass questions that have already been answered. Make sure your questions are relevant. Avoid needlessly complex questions; ask one clear question at a time.

Your attitude toward the interview is important in determining its success or failure. An interview is not a contest. Avoid attacks; avoid excessive use of technical jargon; conduct an interview, not a "snowjob." Talk *to* people, not up to them, down to them, or at them. An interview is not a trial. Do ask probing questions, but don't cross-examine. Remember that the interviewee is the expert, and that you are the one looking for answers. Finally, whatever you do, avoid attacking the other person's credibility. Don't say, "So and so told me something different", or, "You don't know what you're talking about." You will sit through an occasional useless interview. An early closing might be in order, but always act professionally in spite of your disappointment.

Should you take notes during the interview? Unless you have an excellent memory, it is always a good idea to jot down key points, but don't overdo it. Be unobtrusive and selective; it is not necessary to record every word. One suggestion is to leave space for notes on your objective outline or interview outline; if nothing else, a prepared note-taking structure can eliminate the need to write the question during the interview.

Don't be a compulsive note taker. Don't concentrate so much on recording every word, that you miss the meaning. You must listen to the answers. You must be prepared to ask follow-up or probing questions, and you can't do that if your concentration is focused on a piece of note paper. Be honest with yourself. If you feel compelled to take extensive notes, consider taping the interview. A caution: if you plan to use a tape recorder, get the permission of the subject. Also, check out your equipment before the interview, and take along an extra tape or two, just in case.

The Closing

As the interview draws to an end, it is important to maintain a sense of rapport. Pay attention to the time. If the interview runs longer than anticipated, ask permission to continue, and offer to reschedule. When you have all the information you need, thank the subject for cooperating, and offer to make your written summary available for review. If you anticipate the need for a follow-up or subsequent interview with the same person, say so. Some interviewers like to "wind down" with a brief period of casual conversation. If you feel comfortable with this approach, use it; casual conversation is not, however, required—don't force it.

FOLLOW-UP

As soon as possible after the interview has ended, transcribe your notes. Ideally, the notes should concentrate on key ideas; use your memory to fill in the details. If you have recorded the interview, listen to the tape and compile a set of selective notes. In either case, have your summary typed, and be sure to identify the person, the date, the place, and the topic of the interview. Share your summary with the interviewee; it's good public relations, and provides an excellent opportunity for correcting misunderstandings.

One or more follow-up interviews may be necessary. If the follow-up involves a single question or two, consider using the telephone. If you anticipate an interview of more than a few minutes, say so, and offer to schedule an appointment.

REFERENCES

1. Stewart, Charles J. and Cash (1978). *Interviewing Principles and Practices, second edition*. Dubuque, Iowa: W.C. Brown.

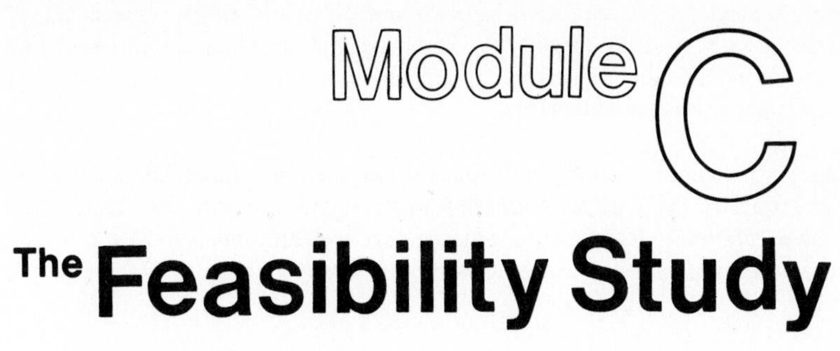

The Feasibility Study

OUTLINE

WHAT IS A FEASIBILITY STUDY?

A feasibility study is a compressed, capsule version of the entire systems analysis and design process. The study begins by clarifying the problem definition. The initial statement of scope and objectives is confirmed or corrected, and any constraints imposed on the system are identified.

Once an acceptable problem definition has been generated, the analyst develops a logical model of the system. A search for alternative solutions then begins, using this model as a reference. Next, the alternatives are carefully analyzed for feasibility. At least three different types of feasibility are considered:

1. *Technical:* Can the system be implemented using current technology?

2. *Economic:* Do benefits outweigh costs?

3. *Operational or organizational:* Can the system be implemented in this organization?

For each feasible solution, the analyst prepares a rough implementation schedule.

The results of the feasibility study are presented to both management and the user. A written report is almost always required, and oral presentations are common. "Drop the project" is one possible recommendation. Assuming that the analyst has found a feasible solution, the feasibility study should provide a broad sense of technical direction for the project; *i.e.*, a plan to proceed.

How much time should be spent on the feasibility study? The answer depends on the project's scope. For a relatively minor change in the format of an existing report, a brief telephone conversation might be adequate. On an estimated quarter million dollar project to develop a new accounting system, a feasibility study of a few weeks would probably be reasonable. For the software company developing a new, multi-million dollar package, it might make sense to spend six months or a year designing and testing a prototype system. The cost of the feasibility study should be approximately 5 to 10 percent of the estimated total project cost.

Over the next several pages, we will outline the steps involved in a typical feasibility study. Use the outline, as a general guideline, but remember that no two systems are exactly the same. The steps describe a feasibility study of several days' duration.

THE STEPS IN A TYPICAL FEASIBILITY STUDY

1. Define the scope and objectives of the system. During problem definition, a statement of scope and objectives was prepared. The analyst should confirm the problem definition, the anticipated scope of the project, and the system objectives. Any constraints should be clearly identified. Interviews with key personnel and a review of written material will certainly be required. Essentially, the analyst is attempting to answer a very simple question: Am I working on the right problem?

2. *Study the existing system (if there is one).* The existing system is an important source of information. Obviously, if a system is being used, it must be performing some useful work, and its basic functions must be incorporated into the new system. On the other hand, if the existing system were doing a perfect job, there would be no need for a new system; thus, any problems identified in the old system must be corrected. Finally, the cost of operating the existing system represents an economic target; if the new system does not provide additional benefits and/or reduce costs, the old system should be retained.

Carefully analyze written procedures and documentation. Learn the informal system, too. Track the work flow; a good place to start is with the distribution list for any reports generated by the system. Learn what the system does, and why it does it that way. Get cost data; know how much it costs to operate the present system. You will find it necessary to interview people (see Module B). Remember that the relationship between a systems analyst and a user resembles that of a doctor and a patient. The user will often describe symptoms rather than real problems, and the analyst must interpret the information.

A common error is spending too much time analyzing the existing system. The objective is not to document what is done, but to *understand* what is done. The analyst should not be concerned with *how* the existing system works, but should concentrate on *what* it does. Construct a data flow diagram (Module D) and a data dictionary (Module E) for the old system; a high-level system flow diagram (Module F) is another possibility. Do not, however, spend a great deal of time on the implementation details; for example, avoid drawing program logic flowcharts from the code unless you are trying to define a particularly crucial algorithm.

The analyst should get one final piece of information from the present system. Few systems exist in a vacuum; most interface with several others. Define these interfaces; they represent very important constraints on the design of a new system.

3. *Develop a high-level logical model of the proposed system.* At this point, the analyst should have a good sense of the functions and constraints of the new system. A logical model of the new system can be constructed using a data flow diagram and perhaps a data dictionary. Later, this logical model can be used in designing the new system (Fig. C.1).

4. *Redefine the problem in the light of new knowledge.* By developing a logical model of the proposed system, the analyst is essentially saying: "Here is what I think the system must do." Does the user agree? It's easy to find out. Ask. At this stage, the analyst should review the problem definition, scope, and objectives with key personnel, using the logical data flow diagram and the data dictionary as a basis for the discussion. If the analyst has misunderstood, or the user has overlooked something, now is the time to find out. Think of the first four steps of the feasibility study as a loop. The analyst defines the problem, analyzes it, develops a tentative solution, redefines the problem, reanalyzes it, revises the solution, and continues this cyclic process until the logical model meets the system objectives.

5. *Develop and evaluate alternative solutions.* Given a logical model of the proposed system, the analyst can begin to generate high-level, alternative physical solutions. How does an analyst generate these alternatives? Perhaps the easiest approach is to start

Fig. C.1: *Good design begins with the existing physical system, develops a logical model of that system, uses the model to construct a logical model of the proposed system, and then bases the new physical system on that logical model.*

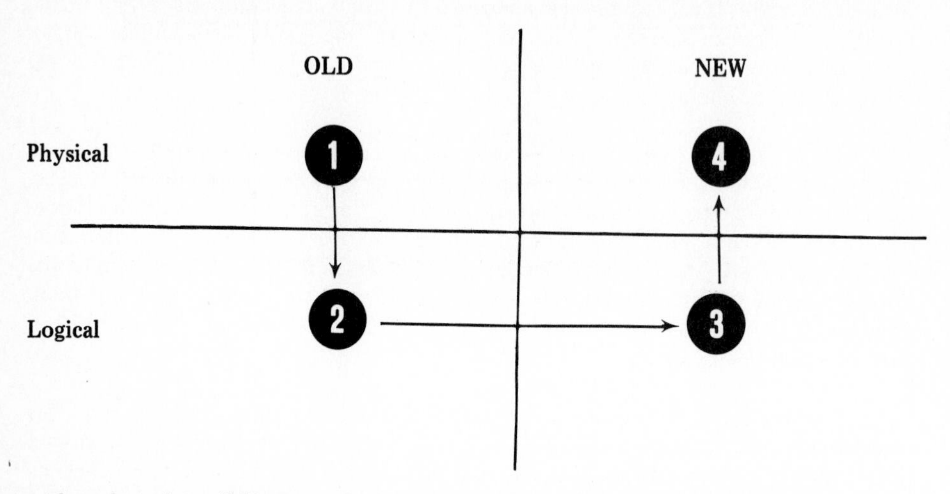

with technical feasibility as a driving mechanism, and to think of a variety of ways in which the problem could be solved. For example, Module D describes the process of identifying automation boundaries on a data flow diagram. Each of these automation boundaries represents one possible physical solution; the analyst might define several sets of automation boundaries, and then determine how the system might be implemented with each set.

Another option is to use a technique known as *brainstorming*. Given the logical model of the system and an hour or two of uninterrupted time, the analyst and a few technical colleagues meet to dream up possible solutions. The rules are simple. Everyone is expected to contribute possible solutions to the problem. No one is allowed to criticize, and no attempt is made to evaluate the suggestions. The point is simply to compile a list of ideas. A brainstorming session might end at a predetermined time, or when a predetermined number of alternatives has been generated. Following the session, the analyst is expected to evaluate the suggestions and select a few that seem reasonable.

Many analysts prefer to use a checklist. For example, several different types of systems might be developed on each of several different types of computers; a few representative examples are summarized in Fig. C.2. At each block in the matrix is a particular type of system—a batch system on a microcomputer, a batch system on a mainframe, an interactive system on a minicomputer, and so on. Using the matrix as a guide, the analyst can attempt to envision a system for each block. Two additional alternatives should always be considered—the existing system and a manual system.

Once a set of technical alternatives has been generated, the initial weeding can be done on the basis of technical feasibility. If, for example, a system requires 3 to 4 second response time, any batch-oriented alternative can be ignored. If the programmers are not trained in interactive programming, the interactive systems might be ruled out. At the end of this process, the analyst should have a set of technically feasible alternatives.

Fig. C.2: *A checklist of system types.*

SOURCE	TYPE	COMPUTER				
		MICRO-COMPUTER	MINI-COMPUTER	MAIN-FRAME	SERVICE BUREAU	TIME-SHARING SERVICE
INTERNAL	BATCH					
	INTERACTIVE					
	REAL-TIME					
EXTERNAL	BATCH					
	INTERACTIVE					
	REAL-TIME					
TURNKEY	BATCH					
	INTERACTIVE					
	REAL-TIME					

EXISTING SYSTEM___________

MANUAL SYSTEM___________

Operational or organizational feasibility might be considered next. For example, the user could be opposed on general principles to an outside service bureau or time-shared service. Another organization might have a bias against a particular vendor's hardware, or for a particular type of system. Company policy or a union agreement might dictate either for or against a particular option. Basically, the analyst should check each remaining alternative against the way the organization does business, and eliminate any that might be operationally unacceptable.

Consider economic feasibility next. Estimate both the development and operating cost of each remaining alternative. Estimate the cost savings and/or revenue increase relative to the existing system (if there is one). Perform a cost/benefit analysis for each alternative (see Module G). Only those alternatives that promise a positive return on investment should be considered further.

Finally, for each alternative that passes the technical, operational, and economic feasibility tests, develop an implementation schedule. This schedule does not have to be detailed; tie it to the system life cycle, and estimate the date of completion for each phase.

6. *Decide on a recommended course of action.* The key decision arising from the feasibility study is whether to continue or to drop the project. Indicate this essential "go/no go" recommendation clearly. Assuming that the recommendation is to continue with the project, the analyst should select the best alternative, and justify that choice. It is important to remember that management must constantly consider alternative sources for investment, and that developing a new system is a form of investment. If the best alternative for the proposed system offers a projected return on investment of 15 percent, and the prime interest rate is 16 percent, management may

Fig. C.3: *An outline of a typical feasibility study.*

A. **TITLE PAGE.** *Project name, report title, author(s), date.*

B. **CONTENTS.** *A list of report sections with page numbers.*

C. **PROBLEM DEFINITION.** *A clear, concise, one-page description of the problem.*

D. **EXECUTIVE SUMMARY.** *A clear, concise, one-page summary of the feasibility study, the results, and the recommendations. Include necessary authorizations, key sources of information, alternatives considered, and alternatives rejected. Highlight the costs, benefits, constraints, and time schedule associated with the recommended alternative.*

E. **METHOD OF STUDY.** *A reasonably detailed description of the approach and procedures used in conducting the feasibility study. Mention your sources and references, and identify key people. Briefly describe the existing system (if appropriate). Much of the detail belongs in the appendix (see item J): include only those facts directly relevant to the study or to your conclusions.*

F. **ANALYSIS.** *A high-level analysis of the proposed* logical *system. Include a statement of the system objectives, constraints, and scope; it should be more detailed than the one developed during problem definition. Include a logical data flow diagram and perhaps an elementary data dictionary for the proposed system. Identify key interrelationships with other systems.*

G. **ALTERNATIVES CONSIDERED.** *For each alternative seriously considered, include a statement of its technical feasibility, economic feasibility, operational feasibility, a rough implementation schedule, and a high-level system flow diagram or other system description. Be thorough, but don't overdo it—much of the detail belongs in the appendix.*

H. **RECOMMENDATIONS.** *Clearly state the recommended course of action. Provide material to support and justify your recommendation; in particular, provide a cost/benefit analysis.*

I. **DEVELOPMENT PLAN.** *Include a projected schedule and projected costs for each step in the system life cycle, assuming that the recommended course of action is followed. Provide detailed time and cost estimates for the next step in the process—analysis.*

J. **APPENDIX.** *Charts, graphs, statistics, interview lists, selected interview summaries, diagrams, memos, notes, references, key contacts, acknowledgements, and so on; in short, the details that support the study. Consider making the appendix available on a demand or need basis.*

well choose to put its money into Treasury bills rather than into this project. Occasionally, a project is justified on non-economic grounds; for example, if new laws or a new union contract make the payroll system obsolete, a new system must be developed. Generally, however, investment decisions are based on the expected returns, so always include a cost/benefit analysis (Module G).

7. *Rough out a development plan.* Assuming that management will accept the recommended course of action, develop an implementation schedule. Estimate personnel requirements, and indicate when analysts, programmers, technical writers, and others will be needed. Estimate the cost of each stage in the system life cycle. Finally, provide a clear, detailed schedule and a set of cost estimates for the next stage—analysis.

8. *Write the feasibility study.* An outline of a typical feasibility study is illustrated in Fig. C.3. Remember that all feasibility studies are different; this outline is intended to be used as a guideline only.

9. *Present the results to management and the user.* The decision to commit funds to the project must be made by management, and not by the analyst.

REFERENCES

The sources listed below take a somewhat different view of feasibility studies, but might prove interesting:

1. Clifton, David S. and Fyffe (1977). *Project Feasibility Analysis: A Guide to Profitable New Ventures.* New York: John Wiley and Sons, Inc.

2. Fitzgerald, J., Fitzgerald, and Stallings (1981). *Fundamentals of Systems Analysis, second edition.* New York: John Wiley and Sons, Inc.

Data Flow Diagrams

OUTLINE

DATA FLOW DIAGRAMS

During the early stages of the systems analysis and design process, the analyst collects a great deal of relatively unstructured data from such sources as interviews, written memos, documentation manuals, notes, and even casual conversation. It is important that all this data be summarized. Ideally, this summary should serve a variety of functions. It should simplify communication with the user, and should be useful in supporting the future development of the system. Also, the summary should not force the analyst into premature physical design decisions. The analyst needs something analogous to the architect's preliminary sketches.

A data flow diagram is a logical model of a system. The model does not depend on hardware, software, data structure, or file organization: there are no physical implications in a data flow diagram. Because the diagram is a graphic picture of the logical system, it tends to be easy for even nontechnical users to understand, and thus serves as an excellent communication tool. Finally, a data flow diagram is a good starting point for system design.

CONSTRUCTING A DATA FLOW DIAGRAM

A data flow diagram uses four basic symbols to form a picture of a logical system (Fig. D.1). A square defines a *source* or *destination* of data. A rectangle with rounded corners (some experts use a circle) represents a *process* that transforms data. An open-ended rectangle is a *data store*. An arrow is used to identify a *data flow*.

Note that a process is not necessarily a program. A single process might represent a series of programs, a single program, or a module in a program; it might even represent a manual process, such as keypunching or the visual verification of data. Note also that a data store is not the same as a file. A data store might represent a file, a piece of a file, elements on a data base, or even a portion of a record. A data store might reside on disk, drum, magnetic tape, main memory, microfiche, punched card, or any other medium (including a human brain).

What is the difference between a data store and a data flow? A data flow is data in motion; a data store is data at rest. In other words, they are simply two different states of the same thing. We'll return to this idea later.

Typically, a number of simplifying assumptions are made on a data flow diagram. For example, error processing or handling unusual conditions is ignored, as are housekeeping functions such as opening and closing files. No attention is paid to how the data are processed, or how the data flow from process to process. The point is to describe what happens, without worrying about how it happens.

With traditional logic flowcharts, the direction of flow is from top to bottom and from left to right. A good data flow diagram tends to follow a similar convention, with data moving from its source (at the upper left) to its destination (at the lower right), but the rules are much less rigid. For example, data sometimes flow back to a source. One way to indicate this is to draw a long flowline from one side of the diagram to the other. As an alternative, the symbol for the data source might simply be repeated (Fig. D.2)—in other words, the same symbol representing the same data

Fig. D.1: *Data flow diagram symbols.*

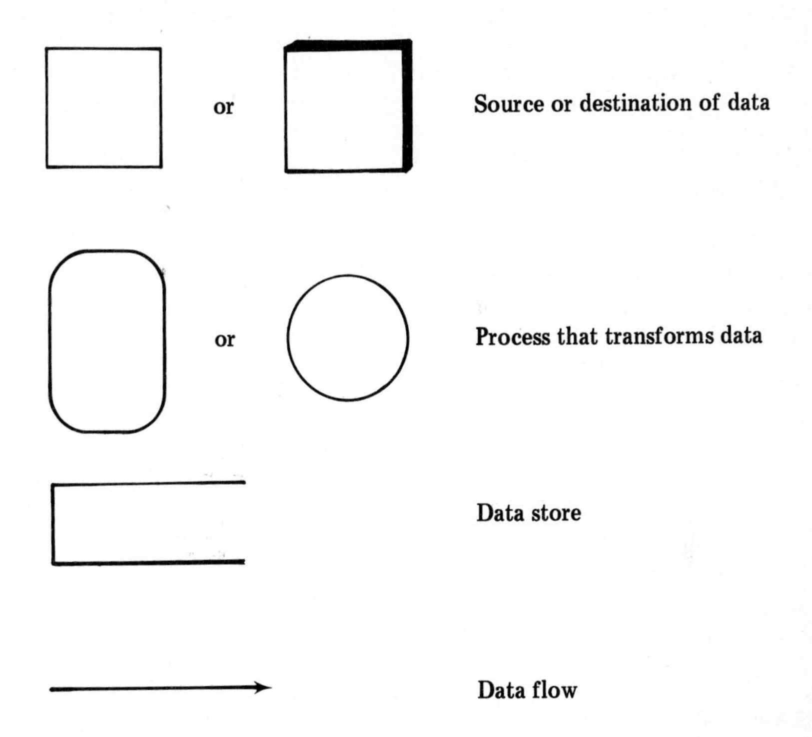

source can appear more than once on a data flow diagram. On a traditional flowchart, repeating a block of logic is considered poor form; on a data flow diagram, if repeating a symbol improves the clarity of the diagram, fine. To avoid possible misunderstanding, a symbol that is used more than once is normally marked with a short diagonal line in one corner; note, for example, the squares labeled *User* in Fig. D.2. Data stores are sometimes repeated, too.

AN EXAMPLE

Perhaps the best way to introduce data flow diagrams is in the context of a simple example. Imagine that the purchasing department needs a daily inventory exception report listing, by part number, all items to be reordered to replenish stock. For each reorder item, purchasing requires: the part number, part description, reorder quantity, current price, primary supplier, and secondary supplier. An item is reordered when the stock on hand drops below a critical level. Inventory transactions (additions and deletions) are reported to the system as they happen through a CRT terminal located in the warehouse. Purchasing wants its list once a day, at the start of business. We'll assume that this simple description of the system requirements has been distilled by the systems analyst from a number of memos, interviews, telephone conversations, and documentation manuals.

Fig. D.2: *A symbol can be repeated on a data flow diagram. When a symbol is used more than once, it is normally marked with a diagonal slash.*

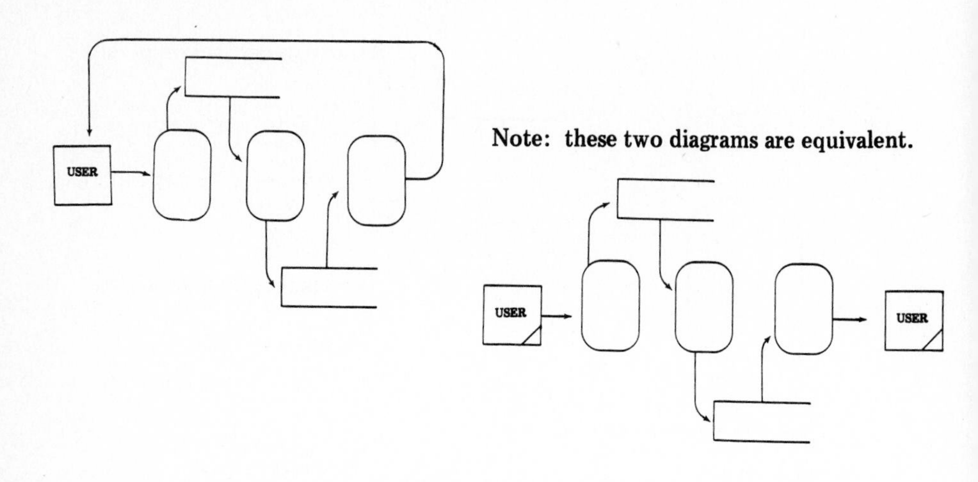

How does the analyst begin developing a data flow diagram? What are the components of a data flow diagram? There are four: sources or destinations, processes, data stores, and data flows. Thus, the first step is to extract the components from the description. Let's start with data sources and destinations (Fig. D.3). Reread the problem description. Clearly, purchasing needs the report; thus *purchasing* is a data destination. Keep reading. Inventory transactions are reported . . . through a CRT terminal located in the warehouse. That terminal is a source of data; let's identify the source as *warehouse*.

There are no more sources or destinations in the problem description, so we move on to processes. Read the problem description once again. Purchasing needs a report. Clearly, they do not yet have that report, so it must be generated: *generate report* is one process (Fig. D.3).

We must also process inventory transactions. Right now, there might be plenty of stock on hand for part number 123. A few minutes from now, however, a transaction might cause several units of this part to be shipped, and thus removed from inventory; this changes the stock on hand. A process is anything that changes or transforms data. Since the inventory transaction changes the stock on hand, processing that transaction belongs on the data flow diagram. Note that the problem description did not explicitly mention the need to process inventory transactions; the systems analyst must frequently interpret the specifications.

Now, let's consider data stores and data flows. Reread the problem description. The *exception report* is certainly a data flow; Fig. D.3 summarizes the data elements

Fig. D. 3: *The elements that make up a data flow diagram can be extracted from descriptive information.*

Source/Destination	**Process**
Purchasing	Generate Report
Warehouse clerk	Process Transaction*

Data Flow	**Data Store**
Exception report part number part description reorder quantity current price primary supplier secondary supplier	REORDER see exception report INVENTORY* part number* stock-on-hand reorder level
Transaction part number* transaction type transaction quantity*	*by implication

making up the report. *Inventory transactions* come from the warehouse; thus we have another data flow. There is an obvious timing mismatch between processing the inventory transactions and generating the exception report—note that transactions are processed as they occur, and the report is generated only once a day. The data making up an exception report must be held for a time, giving us a data store.

Not all data stores and data flows can be extracted directly from the problem description; some are merely implied. For example, the fact that "an item must be reordered when the stock on hand drops below a critical level" implies that the stock on hand and the critical reorder level must exist somewhere. Since these data elements would seem to exist for a period of time longer than a single transaction, it is reasonable to assume that there must be a data store holding inventory data. In Fig. D.3, data elements that are implied by the system description are identified by an asterisk.

Getting Started

Once the component parts have been isolated, the analyst can begin drawing the data flow diagram. It is best to start at a very high level (Fig. D.4), showing the entire system as a single logical process and clearly identifying the sources and destinations of

Fig. D.4: *A high-level data flow diagram, highlighting data sources and destinations.*

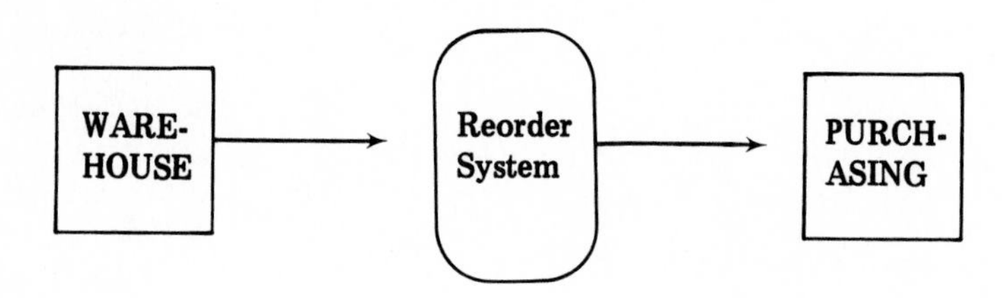

data. Although none are shown here, most systems will interface with other systems, and those other systems should be shown as data sources or destinations.

Even at this very high level, the data flow diagram is most useful as a communication tool. Has the analyst correctly identified all the data sources and destinations? Based only on written and oral documentation, this is not always an easy question to answer; all too often, details can be overlooked. With a high-level data flow diagram, however, the presence or absence of a given data source or destination can be verified (by the user or management) at a glance.

Exploding to a Major Function Level

Fig. D.4 represents perhaps the highest possible view of the system. Except for highlighting data sources and destinations, however, this data flow diagram is not particularly useful. The next step is to "explode" the process into its functional parts. To do this, we must refer back to the list of elements derived from the problem statement (Fig. D.3). Two processes were identified: *generate report* and *process transaction*. These two processes represent the basic functions that must be performed by the system; they will replace *reorder system* in Fig. D.4. Exploding a data flow diagram means replacing a high-level process with its lower-level components (Fig. D.5).

Note that two data stores have been added to the new data flow diagram. *Process transactions* needs inventory data; thus the data store known as *inventory*. Remember the difference in timing between processing the inventory transactions and generating the reorder report? Because of this difference, reorder information must be stored; Thus the *reorder* data store.

Fig. D.5 illustrates several conventions used in drawing a data flow diagram. Processes are numbered for easy reference. Data stores are labeled with "D" followed by a

Fig. D.5: *A functional level data flow diagram.*

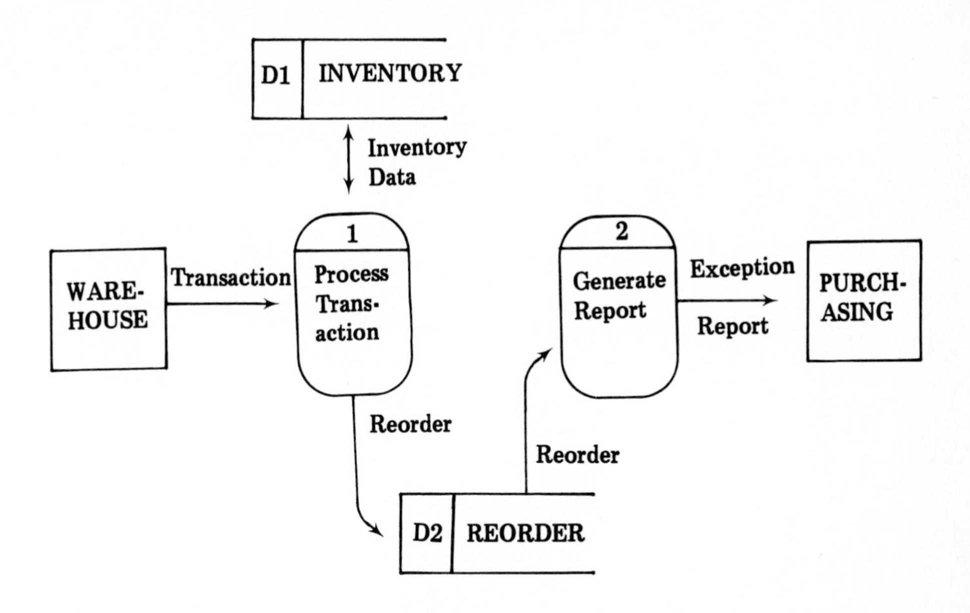

number; again, these labels are for reference only. The names of the data stores, sources, and destinations are written in all capital letters, while the process and data flow names are, except for the first letter, written in lower case. These conventions help to make the data flow diagram easier to follow.

The names of the various data flows are written on the data flow diagram (Fig. D.5). Like the processes, sources, destinations, and data stores, they are derived from the problem description (see Fig. D.3 again). Consider one of those data flows—*reorder*. Is *reorder* a data flow or a data store? At first glance, distinguishing between stores and flows may seem confusing. Don't worry about it. A data flow is data in motion; a data store is data at rest. What elements are found in a data store? Only those elements that entered the data store through a data flow. In other words, a data store and a data flow are just two different versions of the same thing.

Exploding the Major Functions

Once the major functions have been identified and incorporated into the data flow diagram, the analyst can begin to explode each of these functions to a lower level of detail. For example, consider the *process transactions* function of Fig. D.5. Logically, it might be reasonable to break this process into three steps (Fig. D.6): *accept transaction*, *update inventory*, and *process reorder*. Why these three steps? Think about the logical flow of data through the system. First, the transaction must occur and be accepted. Next the transaction can be processed. Finally, once the processing step has determined that a reorder is necessary, the reorder data can be processed. The three steps must occur in the prescribed order. They are relatively independent, linked only

Fig. D.6: The data flow diagram with the "process transactions" process exploded to a lower level.

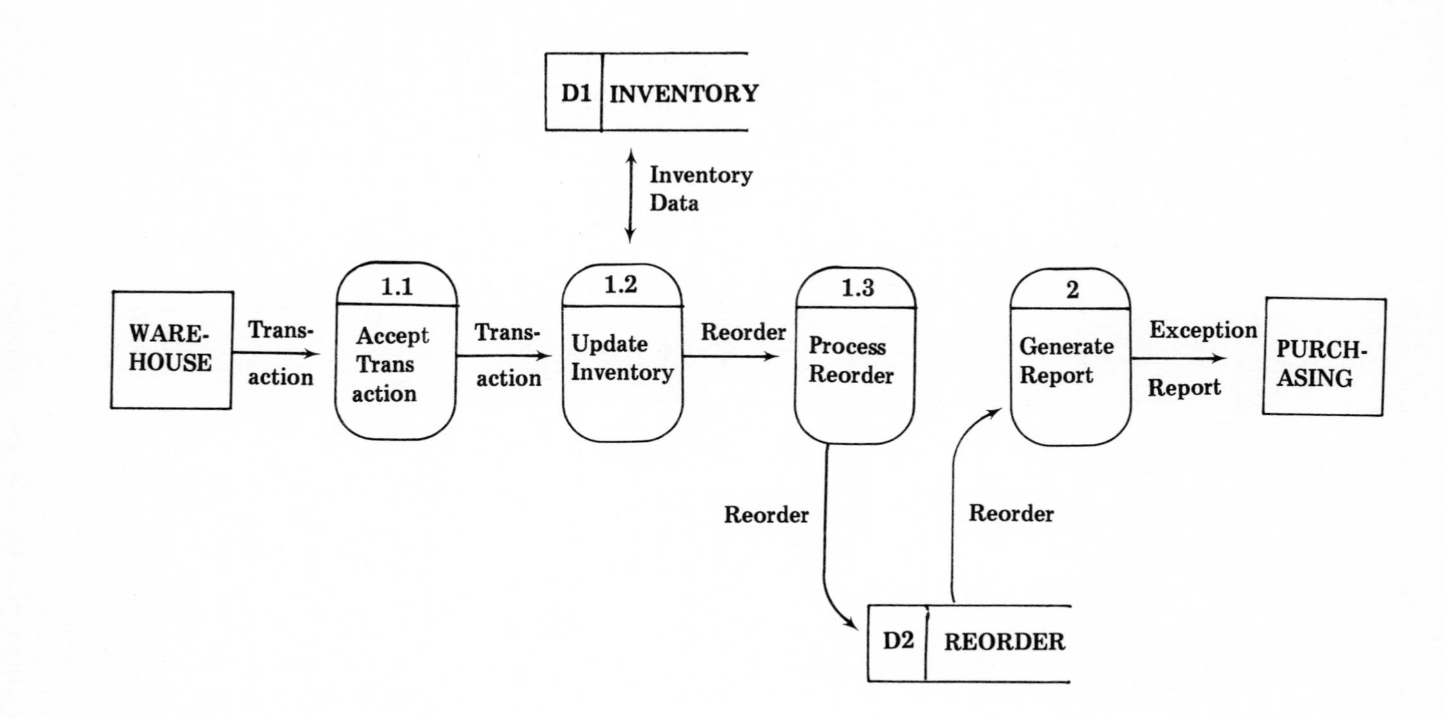

by a data flow. A similar breakdown can be imagined for many functional-level processes.

Note how the processes have been numbered on the exploded data flow diagram. Processes 1.1, 1.2, and 1.3 are component parts of what was process 1. If process 2 were to be exploded, its components would be numbered 2.1, 2.2, and so on.

Should we explode the second process, *generate report*? No. It is relatively easy to visualize what a report generation process will do. Although you might imagine a number of different ways to generate a report using a sort or some other sequencing mechanism, such details involve physical specifications, and are inappropriate on a data flow diagram.

Should we explode any of the processes shown in Fig. D.6 to an even lower level? Probably not. Why? How do you know when you have gone far enough? Consider process 1.1, *accept transaction*. Try to imagine breaking down that process without thinking about how you are going to accept transactions, or where (physically) those transactions are going to come from. When you reach the point where further subdivision forces you to think about how you are going to implement the process, you have gone far enough.

How detailed should a data flow diagram be? The example presented in this module is relatively trivial; what if the data flow diagram contained dozens of processes and data stores? Such a diagram would be very difficult to follow, thus defeating a primary purpose of the data flow diagram—communication. A number of studies suggest that human beings find it difficult to follow a data flow diagram containing more than *7±2* processes. These studies suggest a strategy. Start with a high level diagram. Explode it to a functional level. If exploding all the functions to their next level of detail would cause you to exceed the 9 process limit, don't explode the data flow diagram. Instead, take each function, one at a time, and develop a sub-diagram, showing only the explosion of the single process. Repeat this step for each process. The functional level data flow diagram can then be used to provide a logical overview of the entire system. If a user or a manager wants to know what happens within a given process, the appropriate sub-diagram can be shown.

CHECKING THE DATA FLOW DIAGRAM

Once you have completed a data flow diagram, how can you be sure that it is a reasonable model of the system? The key is to check it with the user, but the analyst can do several things first. For example, remember that the data flows define the minimum contents of the data stores. For any given data store, check the data inflows against the data outflows. The data store should contain all data elements flowing in, and all data elements flowing out. What if it doesn't? What if an element of data that did not flow into a data store flows out? Something is missing. Perhaps the analyst has missed an interface with another system. Perhaps a process is missing. What if a data element flows into a data store, but does not flow out? That data element might be redundant or unnecessary. Perhaps the function that should process the element of data has been overlooked.

Consider, for example, the data store named *inventory*. It provides such data elements as the stock on hand and the reorder level to process 1.2; where did those data elements come from? Obviously, there must be a source for this data, and the source is not shown on Fig. D.6. The explanation may be very simple: inventory master data already exists to support another inventory system. Adding this detail to the data flow diagram may do more harm than good, but the point still remains: there is an inconsistency on the data flow diagram, and that inconsistency must be explained.

The *exception report* provides another example of how checking the data flows can help to improve the system definition. It contains such fields as a part description, a reorder quantity, current price, primary supplier, and secondary supplier. Where do these fields come from? Follow the data flows. Clearly, there is only one possible source: the data store named *inventory*. These data elements should be added to *inventory* and to the data flow named *inventory data*.

Another good idea is to cross-check the different data stores looking for data redundancies; often, two or more data stores can be combined. Although a data flow diagram does not imply physical implementation, some people do tend to view data stores as files. Don't let your data flow diagram even suggest a bad physical implementation.

Be prepared to modify your data flow diagram; in fact, you may need three or four drafts before you even begin to draw the final version. Start with freehand drawings on a note pad, and discuss them with the user. Only when you are confident that the data flow diagram is accurate should you consider using a flowcharting template and ink. People don't like to throw away what they have created. If you draw a data flow diagram using a template and ink, and then discover that something is wrong, there is a danger that you will try to change the system to fit the model. Even experienced analysts do this; it's a natural human reaction. Recognize this temptation, and fight it, even if the error is not your fault.

USING THE DATA FLOW DIAGRAM

The data flow diagram serves a variety of purposes. First, it helps the analyst to organize the information about a system. The very act of creating a data flow diagram forces the analyst to summarize information, extract key details, and consider the relationships among those details. Missing elements that might be overlooked in a massive narrative are often highlighted in the graphic structure of the diagram. Additionally, the contents of the data flows and the data stores represent a base for developing a data dictionary (see Module E).

A data flow diagram is an excellent communication tool. The limited number of symbols and the lack of physical implementation details makes a data flow diagram accessable to most users. In the early planning stages, a rough sketch of a data flow can help to summarize the results of an interview or the contents of formal documentation. Later, a completed data flow diagram can be used to explain the analyst's understanding of a system. An excellent formal presentation technique is to begin with a data flow diagram that has all the symbols in place, and then write the appropriate labels on the sources, destinations, processes, data flows, and data stores as you follow the flow through the system.

The data flow diagram can also be used as a design aid. Using the timing requirements of the various processes as a guide, it is possible to draw a number of different automation boundaries on a diagram, and each automation boundary might suggest a different physical system. Consider Fig. D.6. Transactions occur continuously; thus process 1.1, *accept transaction*, must be on-line. Purchasing wants its report once a day, and thus process 2 would logically run in batch mode. The other processes, however, are not constrained by the problem description. For example, what if we were to accept transactions on-line and enqueue them, updating inventory, processing reorders, and generating the reorder report in a batch mode (Fig. D.7)? We would, of course, need a new data store for the transactions, but our data flow diagram would certainly support this option.

Change the automation boundaries. Enclose processes 1.1, 1.2, and 1.3 within a common boundary (Fig. D.8). Now our system would accept transactions, update inventory, process reorders, and output a reorder record on-line; process 2 would then prepare a reorder report in batch mode. Can you imagine any other automation boundaries? Would it not be possible to group processes 1.1 and 1.2, and then draw a separate boundary around 1.3 and 2? Imagine designing the on-line inventory update module to flag reorders. Later in batch mode, *process reorder* would search *inventory*

Fig. D.7: *Automation boundaries suggesting a batch inventory update.*

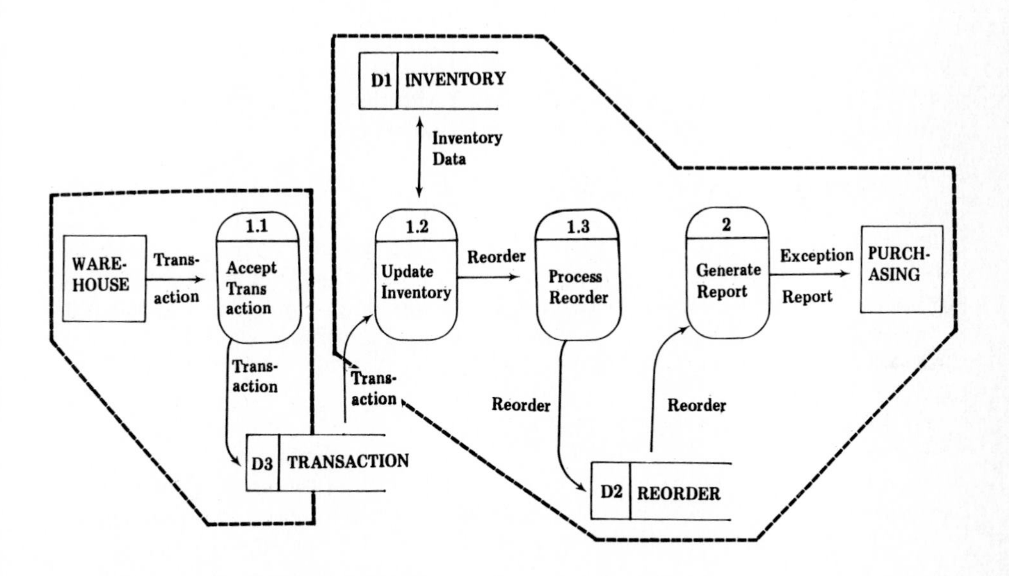

Fig. D.8: *Different automation boundaries suggesting an on-line inventory update.*

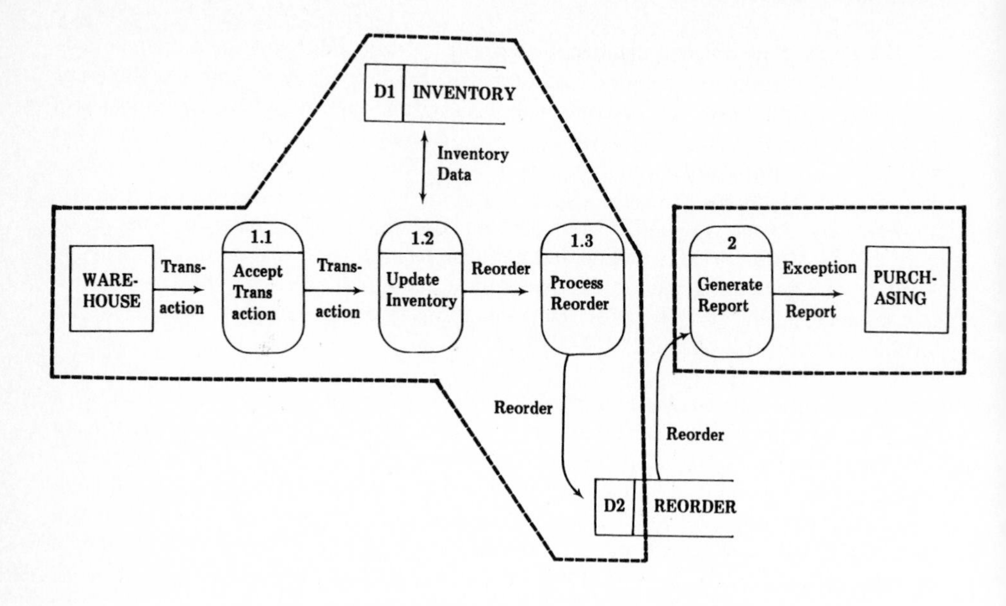

for flagged parts, and send them to *generate report*. Why not enclose all four processes within a single automation boundary and print reorder data in the purchasing department as the reorder condition occurs? Playing with the data flow diagram can allow the analyst to generate a number of reasonable alternative solutions to the problem.

Data flow diagrams can be used even in the detailed design stage. Assume, for example, that an on-line inventory update module will be developed as part of the system. What functions must be performed by this module? Although we are well beyond the logical design stage, we can still use the symbols and rules associated with data flow diagrams to develop a logical model of the program. It is relatively easy to move from a logical model of the data flow through a program to a high-level hierarchy chart of that program (see Module H).

Finally, the data flow diagram can be used to test the early physical design of a system. For example, imagine that a set of hierarchy charts and their associated input/process/output charts has been developed for each of the programs in a system, and that the necessary files have been specified. Each element of the system should have its logical equivalent on the data flow diagram. Do the necessary elements all exist? Are there any extra elements? Next, using the data flow diagram as a base, the analyst can actually trace the flow of data through the physical system. Is anything missing? Have any extra, unnecessary features been added between the analysis and design stages? The data flow diagram does not, of course, anticipate every detail of the physical system, but it is a very good guide to that system.

REFERENCES

1. Gane, Chris and Sarson (1979). *Structured Systems Analysis: Tools and Techniques.* Englewood Cliffs, New Jersey: Prentice-Hall, Inc.

2. Yourdon, Edward and Constantine (1979). *Structured Design.* Englewood Cliffs, New Jersey: Prentice-Hall, Inc.

Data Dictionaries

OUTLINE

WHAT IS A DATA DICTIONARY?

A data dictionary is a collection of data about data. The basic idea is to provide information on the definition, structure, and use of each data element an organization uses. A data element is a unit of data that cannot be decomposed. Fig. E.1 summarizes the information that might be stored in a data dictionary.

Fig. E. 1: *A list of information typically recorded for each data element on a data dictionary.*

General
- Name
- Aliases or synonyms
- Description

Format
- Data type
- Length
- Picture
- Units (lbs/in^2, etc.)

Usage Characteristics
- Range of values
- Frequency of use
- Input/output/local
- Conditional values

Control Information
- Source
- Date of origin
- Users
- Programs in which used
- Change authorizations
- Access authorizations

Group Information
- Parent structure
- Subsidiary structures
- Repetitive structures
- Physical location
  - Record
  - File
  - Data Base

Why use a data dictionary? Perhaps the most obvious reason is documentation; a collection of data about data would be a valuable reference in any organization. Documentation, however, is often promoted for its own sake. What makes a data dictionary so valuable?

Throughout an organization, different people or groups may define a given element of data quite differently. All too often, the analyst and the user will seem to agree on the objectives and scope of a system, only to discover later that they were not really talking about the same thing. Consider, for example, the number of students attending your school. To some, this data element implies a simple head count. To others, it represents the result of a computation: divide total student credit hours by the normal full-time student load to get full-time equivalent students. These numbers can be quite different, particularly at a school with many commuters. Two people can discuss something as "obvious" as the number of students attending a school and have very different ideas in mind. This problem is common. A data dictionary can help to improve analyst/user communication by establishing a set of consistent definitions.

The data dictionary can have more far-reaching implications as well. Consider, for example, a large program involving the efforts of several programmers. If all programmers are required to develop data descriptions from a common data dictionary, a number of potentially serious module interface problems can be avoided. At an even higher level, different systems must often be linked or interfaced. In general, a data dictionary can help to improve communications between broad segments of an organization, simply by providing a set of consistent definitions for the data.

As a new application is being developed, the systems analyst can check the required data elements against the organization's central data dictionary. Some data elements will already exist, and using their established names and formats can save the analyst a great deal of work—why reinvent the wheel? By highlighting already existing elements, a data dictionary can help the analyst to avoid data redundancy, a problem that occurs when a given element of data is physically stored in several different places, under several different formats, and with slightly different levels of control.

Perhaps the most concrete, short run advantage of a data dictionary is derived from the control information that is maintained for each data element (Fig. E.1). Normally, all programs using a given element are cross-referenced in the data dictionary. Thus it becomes very easy to assess the impact of a change in the data. Consider, for example, the proposed nine-digit zip code. With a data dictionary, a complete list of all programs that use the zip code can be quickly compiled, and these programs can all be scheduled for revision. Without a data dictionary, however, there is no way even to identify all programs that use the zip code. Instead, except for a few obvious programs, necessary changes are made haphazardly at best. Often, the fact that a program uses the newly changed data element is not discovered until that program fails.

Finally, a data dictionary is a valuable first step in developing a data base. In fact, many data base management systems include a data dictionary as a standard feature.

DATA DICTIONARY SOFTWARE

A number of data dictionary software packages are commercially available. Some are associated with a specific data base management system. Others are more general; many offer optional links to a variety of data base management systems. Some firms have even written their own customized data dictionary software.

What facilities might you expect to find in a typical data dictionary package? Since creating the data dictionary can be tedious, most contain data entry support. A few are designed to prepare at least part of the entry from programmer source code, a most valuable option when an established computer center decides to install a data dictionary. Another common approach is to display a dummy data dictionary entry on a CRT screen and allow the operation to enter missing elements; other systems use a conversational approach to help the analyst, programmer, or technical writer enter the data about the data.

Many data dictionary systems are designed to generate source code for application programs. The analyst normally checks each data element against the data dictionary. If the element already exists, its name and format will be on the data dictionary, and generating the source code to describe that data element is a relatively simple task. If the data element is new, it can be added to the data base as part of the crosschecking process; once again, generating the source code should be easy. Often, the output from the data dictionary system will be in the form of COBOL DATA DIVISION entries, PL/1 DECLARE statements, or similar data descriptions. The analyst might place this code on a source statement library so that the programmers can simply insert it into their programs.

Earlier, we discussed the possibility of using the control information recorded on a data dictionary to generate a list of all the programs that access a particular data element. Several other data usage reports and cross reference checks might be imagined as well. Many data dictionary software packages include a query feature that allows a user to request specific information about data use. This ability to have the computer investigate how data are used may well be the most valuable feature of a data dictionary.

A SIMULATED DATA DICTIONARY

Ideally, you should use a real data dictionary system in solving your class assignments. Frequently, however, students do not have access to such software. Thus a simulated data dictionary is suggested; we will use this simulation throughout the text.

Many advantages associated with a data dictionary are derived from the ability to process information about each data element separately. Simply listing all the data elements on a sheet of paper will not do, as it is difficult to deal with the individual elements when they are presented in list form. To maintain a semblance of direct access, information related to each data element will be recorded on a separate 3x5 filing card.

A minimum amount of information will be recorded for each data element (Fig. E.2). The data name will come first. It is a good idea to establish a set of conventions for assigning data names; we'll borrow from the COBOL language, although any other programming language would do as well. Often, a given element of data will be known by more than one name; thus any aliases or synonyms will be recorded. Next comes a brief description or definition of the data element, followed by its type and format. Finally, information related to the physical location of the data element will be recorded. Control information and usage characteristics (if available) will be noted on the back of the card. This control information is more important than its position would seem to indicate, but realistically, we will not be able to perform a large scale cross reference analysis on data recorded on 3x5 cards.

It is easy to overdo the data dictionary. Too many systems analysts, when faced with such a data collection task, tend to lose sight of the real objective. During the early feasibility study and analysis stages, not all the information demanded by a data dictionary will be available. If the analyst "hides behind" the requirement that he or she complete a data dictionary, nothing may really get done. Use the data dictionary as a structure for collecting data about the data. As key bits of information become available, add them to the developing data dictionary. Don't try to predefine everything; you won't be able to do it. In fact, many systems analysts use something like the simulated data dictionary as a note pad to record details as they are uncovered. When

Fig. E.2: *The minimum information to be collected for each data element on the simulated data dictionary.*

Name:

Aliases:

Description:

Format:

Location:

the time comes to add the information to the computerized data dictionary, these notes will contain most of the necessary information.

AN EXAMPLE

In Module D, a number of data flow diagram concepts were illustrated through a simple example. We might use the same example to illustrate several data dictionary concepts as well. Purchasing needs a daily inventory exception report listing, by part number, all items to be reordered to replenish stock. For each reorder item, purchasing needs: the part number, part description, reorder quantity, most current price, primary supplier, and secondary supplier. An item must be reordered when the stock on hand drops below a critical level. Inventory transactions (additions and deletions) are reported as they happen, through CRT terminals located in the warehouse. Purchasing requires its list once a day, at the start of business.

Let's concentrate on the data elements making up the inventory exception report. We cannot derive all necessary information from the problem definition, but we can at least lay out a skeleton data dictionary (Fig. E.3). The analyst might know a few details from personal experience. Discussions with the user might help to fill in many of the blanks. We might search the existing organizational data dictionary to see if a given element is used in another application; if it is, we can use the established information to describe the element. Missing information serves to alert the analyst that work remains to be done. The data dictionary eventually will be completed, as a direct result of the systems analysis and design process.

Fig. E.3: *A portion of the simulated data dictionary for the inventory example.*

Name: INVENTORY-EXCEPTION-REPORT
Aliases: REORDER-REPORT. PURCHASING-REPORT
Description: Group item representing daily list of parts to be reordered. Sent to purchasing. See back of card for structure.

Format: Group item.

Location: output to printer

Name: PART-NUMBER
Aliases:
Description: Key field that uniquely identifies a specific part in inventory.

Format: Alphanumeric; 8 characters.
PIC X(8).

Location: INVENTORY-EXCEPTION-REPORT
INVENTORY
REORDER

Name: REORDER-QUANTITY
Aliases:
Description: The numer of units of a given part that are to be reordered at a single time.

Format: numeric; 5 digits.
PIC 9(5)

Location: INVENTORY-EXCEPTION-REPORT
INVENTORY
REORDER

REFERENCES

1. Atre, S. (1980). *Data Base: Structured Techniques for Design, Performance, and Management.* New York: John Wiley and Sons.

2. Kroenke (1977). *Database Processing.* Chicago: SRA.

3. Lomax, J.D. (1977). *Data Dictionary Systems.* Rochelle Park, New Jersey: NCC Publications. Also distributed by Hayden Book Co.

System Flowcharts

OUTLINE

SYSTEM FLOWCHARTS

In the structured approach to systems analysis and design, we begin by constructing a logical model, often using data flow diagrams as a tool. During system design, this logical model must be converted to physical form. How do we describe the physical stystem? Remember that we are just beginning to move from the logical to the physical, and are not yet ready to begin specifying details. What we need is an overview.

A system flowchart is a traditional tool for describing a physical system. The basic idea is to provide a symbol to represent, at a black box level, each discrete component in the system—programs, files, forms, procedures, and so on. While a few of the symbols are the same, system flowcharts are quite different from program logic flowcharts (see Module J). A system flowchart is a high-level picture of a physical system.

Fig. F.1: *Basic flowcharting symbols.*

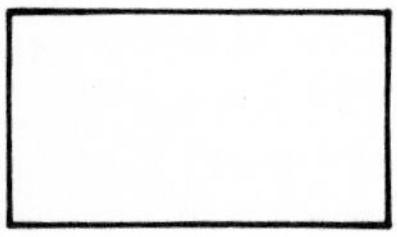

Process — **A process or component that changes the value or location of data. Examples include a program, a processor, and a clerical process.**

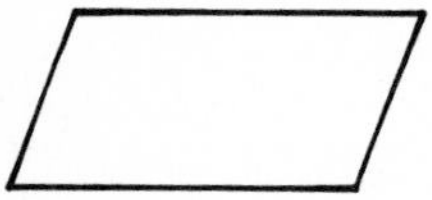

Input/output — **A generalized, device independent symbol for input, output, or both.**

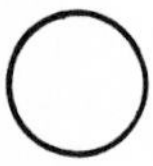

Connector — **Indicates an exit to or entry from another part of the flowchart, usually on the same page.**

Off-page connector — **Indicates an exit to or entry from another page of the chart.**

Flowline — **Used to link symbols. The flowlines define both sequence and direction of flow.**

FLOWCHARTING SYMBOLS

When a system flowchart is drawn, a separate symbol is used for each discrete component in the system. The basic symbols are shown in Fig. F.1. At first, the analyst may work almost exclusively with the basic symbols. Eventually, however, the general input and output operations must become specific files or data bases stored on specific devices, and the processes must become specific programs or manual procedures. The system flowcharting symbols (Fig. F.2) allow the analyst to represent the actual devices or processes that make up the system.

Fig. F.2: *System flowcharting symbols.*

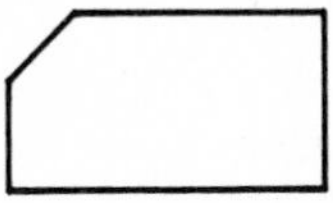

Punched card — **Input or output using punched cards. Also, a punched card file.**

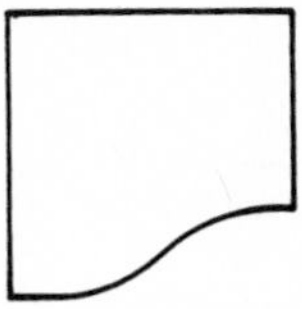

Document — **Normally used for printed output. Can also be used to designate data entry via printing terminal.**

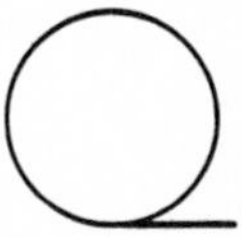

Magnetic tape — **Magnetic tape input or output, or a magnetic tape file.**

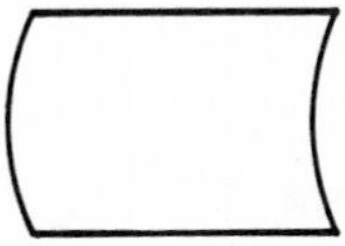

Online storage — **A generalized symbol for any type of online storage, including disk, magnetic drum, mass storage device, diskette, and so on.**

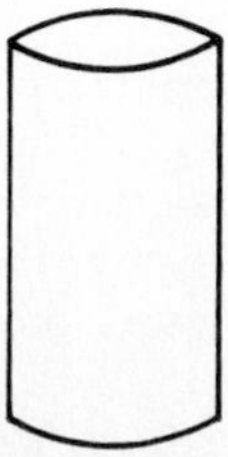

Magnetic disk — **Magnetic disk input or output, or a file or data base stored on magnetic disk. Note the a vertical symbol implies disk, while a horizontal symbol implies drum.**

Fig. F.2: *Continued.*

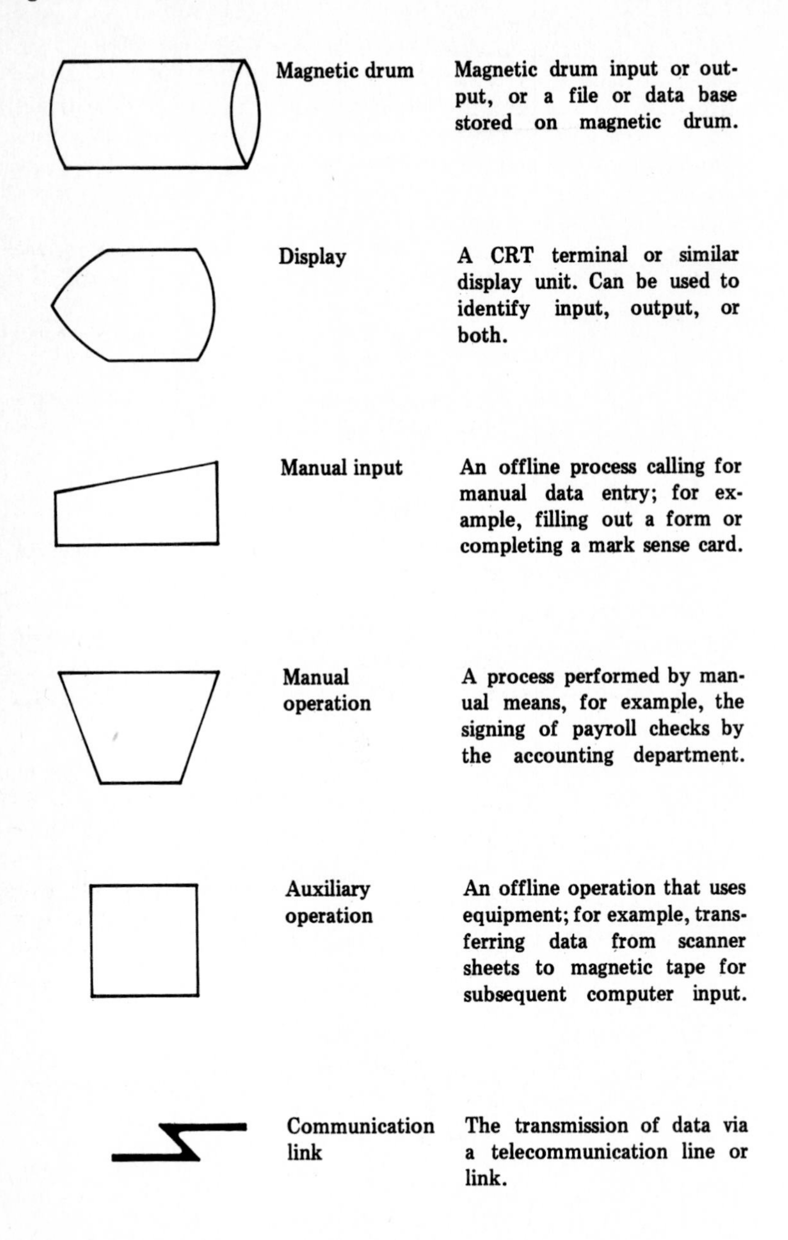

Magnetic drum
Magnetic drum input or output, or a file or data base stored on magnetic drum.
Display
A CRT terminal or similar display unit. Can be used to identify input, output, or both.
Manual input
An offline process calling for manual data entry; for example, filling out a form or completing a mark sense card.
Manual operation
A process performed by manual means, for example, the signing of payroll checks by the accounting department.
Auxiliary operation
An offline operation that uses equipment; for example, transferring data from scanner sheets to magnetic tape for subsequent computer input.
Communication link
The transmission of data via a telecommunication line or link.

AN EXAMPLE

Perhaps the best way to introduce system flowcharts is through a simple example. In Modules D and E, a logical model of a system to solve an inventory problem was developed. One alternative physical solution is illustrated in the system flowchart of Fig. F.3. Read the chart from top to bottom. Transactions enter the system through CRT terminals (the display symbol), and are processed by an inventory program.

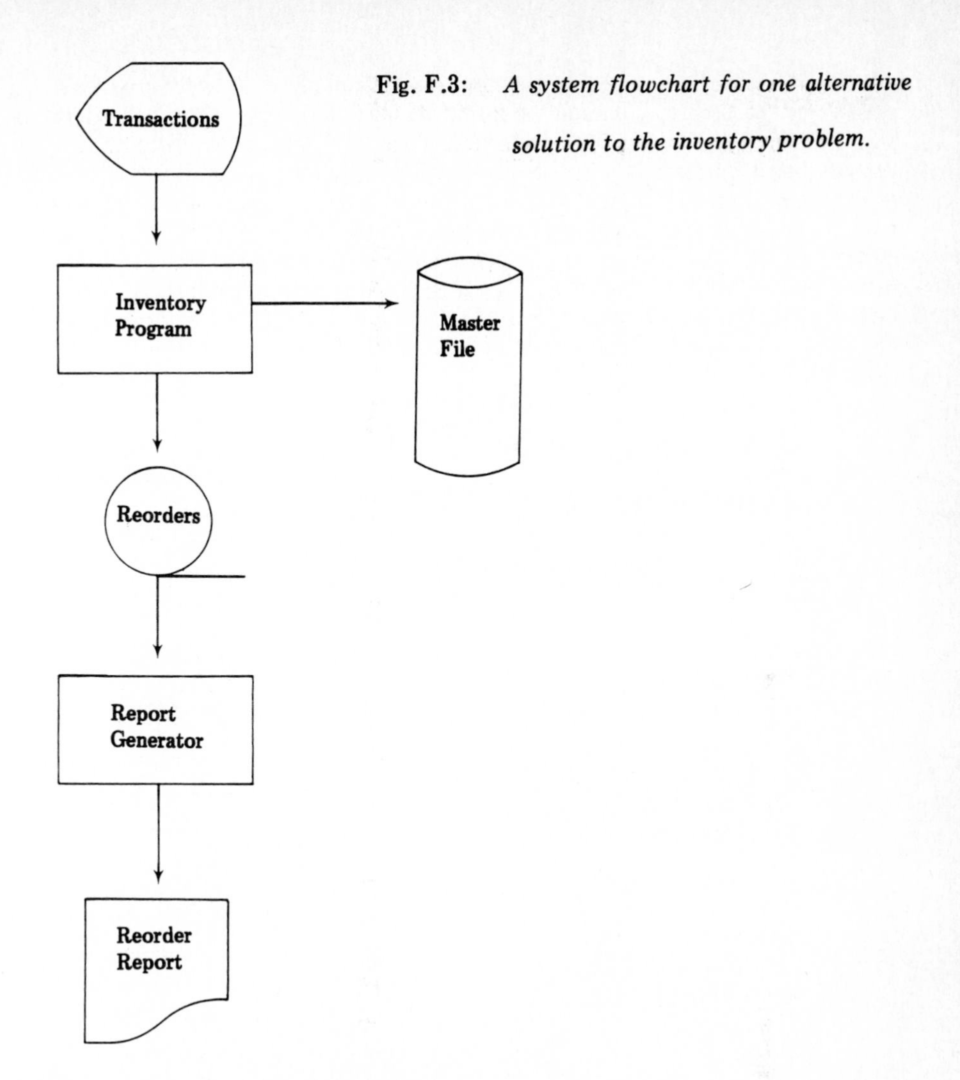

Fig. F.3: *A system flowchart for one alternative solution to the inventory problem.*

The inventory program updates the inventory master file (on disk), and writes necessary reorder information to magnetic tape. Eventually, this tape is read by a report generator program, and the reorder report is printed. Note how the system flowchart graphically illustrates the physical system. Note how each symbol defines, at a black box level, one of the discrete components that make up this system. Note also how the flowlines define the logical path through the system.

Generally, the path of flow through a system flowchart is from top to bottom. However, many analysts prefer a left to right flow; for example, we could start with the display symbol at the left of the page and the reorder report at the right, and align the two programs and the reorder tape horizontally across the page. It's basically a matter of personal choice.

We have labeled each of the symbols in Fig. F.3, thus providing documentation. Many analysts prefer to add more detailed notations to a system flow diagram; some even add a separate page of explanation.

The flowchart for a complex system can be quite large. The offpage connector symbol can be used to continue the flowchart on a subsequent page, but a multiple-page flowchart can be very difficult to read. When faced with a complex system, many analysts begin by drawing a high level flowchart outlining the key functions; on subsequent pages, each of these functions is exploded, one at a time, to the appropriate level of detail. In the inventory system, for example, a first flowchart might show only three symbols: an inventory process, a report generation process, and the magnetic tape file that links them (Fig. F.4). A second page would then concentrate on the inventory process, showing (Fig. F.5) the CRT terminal, the inventory program, the disk master file, and the tape. Page three would explode the report generation process (Fig. F.6).

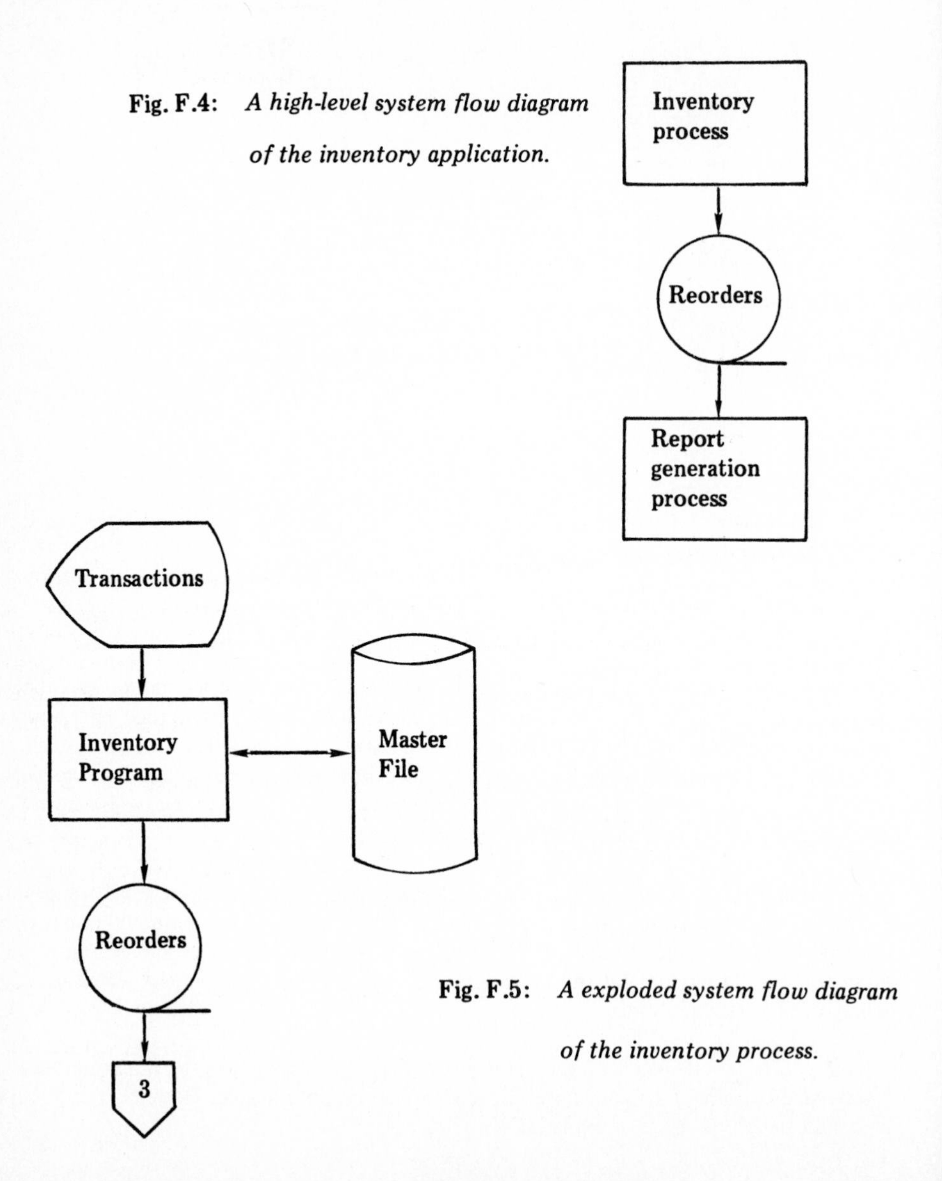

Fig. F.4: *A high-level system flow diagram of the inventory application.*

Fig. F.5: *A exploded system flow diagram of the inventory process.*

Fig. F.6: *An exploded system flow diagram of the report generation process.*

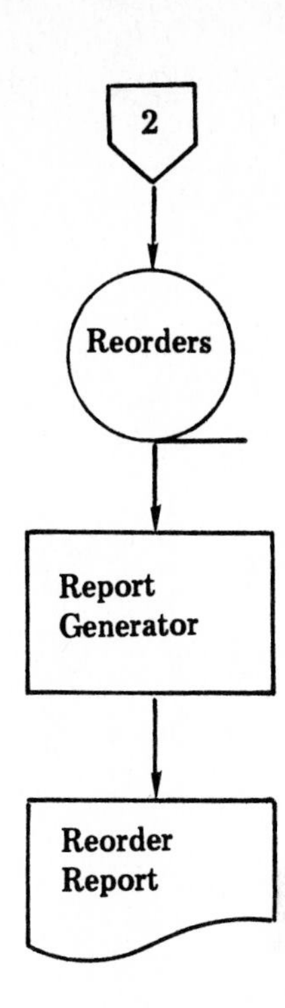

WHY USE SYSTEM FLOWCHARTS?

A data flow diagram presents a rather abstract picture of the system. In contrast, the system flowchart is more concrete. Specific programs or procedures replace the generalized processes, and specific files or data bases replace the data stores. Given the flowchart, it is possible to visualize how the system will be implemented. Such clear communication is particularly important at the end of the system design phase, when the user or management is asked to commit funds to implement the system.

The system flow diagram identifies each of the discrete components of the system. For planning purposes, these components can be attacked one at a time—the divide and conquer approach. Cost estimates are much more accurate when they are based on discrete, physical elements. A realistic implementation schedule can be devised, and parts of the work can be assigned to various groups in the organization. A system flow diagram is an excellent planning tool.

The further into the system life cycle we move, the more likely we are to encounter a group approach, with different people working independently on different aspects of the problem. A system flowchart represents a common reference point. By providing a visualization of the overall system, the system flowchart gives each independent group a sense of how their efforts fit into the bigger picture.

Fig. F.7: *A system flow diagram showing the hardware in a typical computer center.*

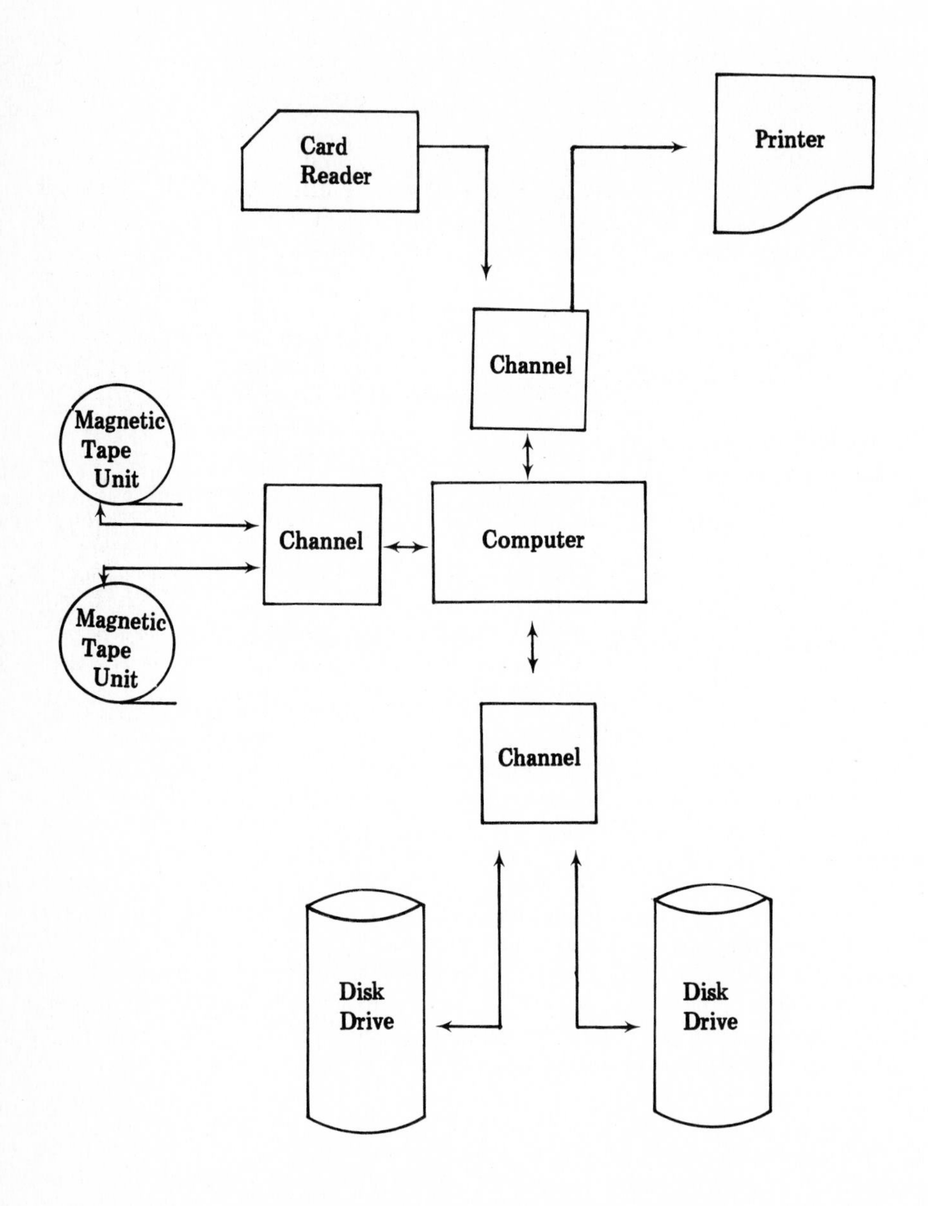

Some analysts use system flowcharts as a model to help develop job control language specifications. If IBM's job control language is used as an example, each program in the system calls for one EXEC statement, and each device or file that is linked to a program must be identified by a DD statement. A system flowchart clearly identifies each program and its associated input and output devices; thus each program symbol represents an EXEC statement and each file symbol implies a need for one DD statement. In effect, the system flowchart serves as a memory aid, helping the analyst to generate an accurate and complete set of job control statements.

OTHER USES

Often, the analyst must study the existing physical system. The objective is not to document the system fully, but to understand it. Too often, the documentation is unclear or incomplete. Drawing a flowchart is an excellent way to summarize a great deal of technical information and to highlight missing pieces.

A system flowchart can be used to map a hardware system (Fig. F.7); as an exercise, develop a flowchart showing the equipment in your school's computer center. Paperwork flows, procedural flows, and work flows can be described, too. Any time the analyst must summarize a number of physical facts concerning a system, a system flowchart might prove useful.

SYSTEM FLOWCHARTS AND DATABASES

The discrete, physical files tend to disappear on a database system. Rather than accessing a number of independent files to find information, a program accesses a central database; in effect, most online files are merged into that database. As a result, system flowcharts are somewhat less useful than they are when traditional files are used. Still, a system flowchart can provide a clear overview of a database application, showing key programs, data sources, and data destinations surrounding the database.

REFERENCES

1. Bohl, Marilyn (1978). *Tools for Structured Design.* Chicago: SRA.

2. Gore, Marvin and Stubbe (1979). *Elements of Systems Analysis for Business Data Processing, second edition.* Dubuque, Iowa: Wm. C. Brown Company Publishers.

3. Semprevivo, Philip C. (1982). *Systems Analysis: Definition, Process, and Design, second edition.* Chicago: SRA.

Module G

Cost/Benefit Analysis

OUTLINE

COST/BENEFIT ANALYSIS

There is a difference between spending and investing. We spend money to get what we need today. We invest money because we hope to get even more at some point in the future. Developing a system is an investment. Funds must be committed throughout the system life cycle. In return, certain future benefits are expected, often in the form of reduced operating costs, or new revenues. If the expected benefits are not greater than the cost, then that system is not worth doing.

Management has numerous alternatives for investing an organization's money. One option is to loan it to someone else through a bank or a money market. The advantage of this approach is a limited risk; for example, when you deposit funds in a bank, you are confident that, at the end of a specified period of time, you will get your money back with interest. Investing that same amount of money in the development of a new system carries a risk. The cost might be higher than anticipated, and the benefits less than expected. Is the system a good investment? Cost/benefit analysis anwers this question by giving management a reasonable picture of the costs, benefits, and risks associated with a given system, so they can compare this investment alternative with others.

COSTS AND BENEFITS

To compare costs and benefits, the analyst must first estimate them. Let's begin with the cost of developing the system. Each phase in the system life cycle has a cost. In estimating the total cost of a given step, the analyst should consider personnel, equipment, supplies, overhead, and such external factors as consulting fees; a list of specific cost elements that might be considered is shown in Fig. G.1.

Once a system is implemented, the organization must begin paying continuing operating costs (Fig. G.2). A cost is associated with using the hardware. Equipment must be maintained. Operators must be paid to run the system; clerks use the system; programmers maintain the code. Supplies are consumed. Overhead must be supported. What is the difference between a development cost and an operating cost? Development costs occur once. When the system is released, development ceases; at this point, operating costs begin, and continue over the entire life of the system. The development cost is a capital investment; operating costs are expenses.

When individuals invest their time or money, they expect something in return—a benefit. Note that the payoff need not be financial; for example, to many people, education is its own reward. In a business environment, however, the objective is usually to make a profit, and thus an ideal return on investment (or benefit) is an increase in profit. Consider the basic profit equation:

$$\text{PROFIT} = \text{REVENUE} - \text{COST}$$

Revenue is money flowing into the organization; for example, receipts from sales. Cost is money flowing from the organization; for example, salaries, material expenses, and rent. There are two ways to increase profit: increase revenue or decrease cost.

Fig. G.1: *A checklist of typical development cost elements.*

Personnel
- Analysts
  - Interviewing
  - Preparation of reports
  - Documentation
  - Contemplation
  - Preparation of procedures
  - System test
  - Inspections, walkthroughs
  - Training — operators
  - Training — clerical people
  - Consultations — users
  - Consultations — programmers
  - Supervision — pilot operation
  - Supervision — file conversion
  - Forms design
  - Formal presentations
- Programmers
  - Coding
  - Documentation
  - Debug
  - Inspections, walkthroughs
  - Customizing purchased pgms.
  - Consultations — analyst
  - Consultations — programmers
  - Formal presentations
- Operators
  - Conversion
  - Training
  - Programmer support
  - Consultations — analyst
- Clerical personnel
  - Conversion
  - Training
  - Consultations — analysts
- Management
  - Supervision
  - Consultation — analyst
- Other
  - Keypunching
  - Data entry
  - Art — forms design
  - Technical writer — documentation

Equipment
- Capital expenditures
  - New equipment
  - Packaged software
- Equipment installation
- Equipment test and debug
- Existing equipment use
  - Test and debug time
  - Disk space
  - Tapes
  - Other supplies
- File conversion
- System test

Materials and Supplies
- Publication of procedures
- Paper, forms, cards
- Preparation — new forms
- Copies

Overhead
- Management support
- Secretarial support
- Heat, light, space

External
- Consulting fees
- Special training

Fig. G.2: *A checklist of typical operating cost elements.*

Hardware costs
- Computer residency time
- Main memory space
- I/O operations
- Secondary storage space
- Maintenance

Personnel costs
- Operator support
- Clerical support
- Programmers — maintenance
- Direct management support

Materials
- Forms
- Paper, cards
- Tapes
- Disk packs
- Scrap
- Inventory carrying cost

Overhead

External costs
- Leases
- Rentals
- Subcontracting
- Auditing

Thus a cost/benefit analysis will generally focus on either cost reduction or revenue enhancement.

Revenue might be increased through a new or improved product that leads to more sales; thus benefits are computed by subtracting the operating costs associated with achieving new sales from this new revenue. When cost reduction is the justification for the system, benefits are derived from the difference between the operating costs of the old and the new systems. Generally, the cost of an existing system is well documented in various accounting reports. However, expected new revenue or the cost of a new, not yet implemented system must be estimated, and any estimate is subject to error. This potential for error is a risk. A well-done cost/benefit analysis will clearly identify the estimated costs, benefits, and risks.

PERFORMING A COST/BENEFIT ANALYSIS

The first step in performing a cost benefit analysis is to estimate the development cost, the operating cost, and the benefits associated with the proposed system. The benefits and operating costs occur over the entire life of the system; how long should this life be? On a research and development project, the finished system might not even begin returning benefits for several years, so an expected life of ten years or more is common. The longer the life of a system, however, the greater the risk of obsolescence, so estimates of benefits and costs for a time far in the future are risky. Our standard in this text will be a system life of five years.

For example, consider a proposal for modifying an inventory system to generate a daily list of parts to be reordered by purchasing. Assume that the user and the analyst, working together, have estimated that the proposed system could save $2500 per year by reducing the risk of inventory shortages. Since the benefits are realized every year, it is convenient to show them in the form of a table (Fig. G.3). If the system is to be implemented, the inventory program will have to be modified to flag items to be reordered, and a new program will be needed to generate the reorder report; the analyst has estimated the cost of developing this system at $5000. We'll use these numbers to illustrate a cost/benefit analysis.

Fig. G.3: *The cost and benefits of the inventory system.*

Development cost: $5000

Benefits:

Year	Amount
1	$2500
2	2500
3	2500
4	2500
5	2500

The Time Value of Money

How much would you be willing to invest today, to get $2500 in a year? If you think about it, the answer must be less than $2500. How much would you invest today to get $2500 in 5 years? Even less! $2500 today, $2500 a year from now, and $2500 in five years are not the same. Money has time value. Comparing present dollars to future dollars is a bit like comparing apples and oranges.

The time value of money is often expressed in the form of interest. A simple formula can be used to compute the future value of an investment assuming compound interest:

$$F = P\,(1+i)^{n}$$

where:

F is the future value of an investment,
P is the present value of the investment,
i is the interest rate per compounding period,
n is the number of compounding periods (usually years).

For example, if $5000 is invested in a certificate of deposit for 3 years at 12 percent interest, the value of that $5000 at the end of the 3 years would be:

$$F = 5000\,(1+0.12)^{3} = 5000\,(1.12)^{3} = 7024.64$$

Look again at Fig. G.3. We have an investment of $5000. Benefits are estimated at $2500 per year. The investment must be made "this year"; in other words, we have its present value. The benefits, however, occur in the future, and we cannot compare present values to future values without considering the time value of the money. Think, for example, of the $2500 we expect to receive at the end of the fifth year. How much would you be willing to invest right now, at current interest rates, to get $2500 at the end of 5 years? The answer is the present value of the benefit, which can be compared to the present value of the investment.

How do we compute the present value of the benefits? Start with the basic compound interest formula:

$$F = P\,(1+i)^{n}$$

Solve this equation for P, the present value:

$$P = \frac{F}{(1+i)^{n}}$$

Using this equation, we can compute the present value of a $2500 benefit 5 years from now (assuming 12 percent interest) as:

$$P = \frac{2500}{(1+0.12)^{5}} = 1418.57$$

In other words, if we invest $1418.57 today at 12 percent interest, we can expect to withdraw $2500 after 5 years. Converting a future value to its present value equivalent is known as discounting.

This computation can be repeated for each year's expected benefit; the answers are summarized in Fig. G.4. We now have the present values of both the investment and the benefits, and can begin comparing them.

An interest rate of 12 percent was used above, purely for illustration. In practice, the interest rate used to discount future values should reflect a realistic, low-risk investment option. If a local bank is paying 14 percent on guaranteed certificates of deposit, 14 percent might be an excellent choice. Many business organizations use the widely publicized prime rate; a large company can always invest its money at or near this rate with little risk. Developing a new system is risky. If the proposed system does not promise a higher return, then the low-risk, "sure thing" is a better investment.

Fig. G.4: *The present values and accumulated present values of the annual benefits.*

Year	Future value	$(1+i)^n$	Present value	Cumulative Present value
1	2500	1.12	2234.14	2234.14
2	2500	1.25	1992.98	4225.12
3	2500	1.40	1779.45	6004.57
4	2500	1.57	1588.80	7593.37
5	2500	1.76	1418.57	9011.94

The Payback Period

One common measure of the relative value of a project is its payback period: How long does it take for the accumulated benefits to equal the initial investment? Obviously, the shorter this payback period, the sooner we begin realizing a profit; thus, the more desirable the investment becomes.

Look, for example, at Fig. G.4; the last column on the right shows the cumulative present values of the benefits. The investment is $5000. At the end of the first year, the accumulated present value of the benefits (only one year) is $2234.14, not nearly enough to offset the initial investment. At the end of the second year, the accumulated present value of the benefits amounts to $4225.12; still not enough to cover the investment. By the end of the third year, however, the accumulated benefits come to $6004.57; thus payback occurs somewhere in the third year. After two full years, we had $4225.12 in benefits. We need $5000 to reach payback; the difference is $774.88. The present value of the third year's benefits is $1779.45; thus we need 44 percent of that third year's benefits. The payback period is 2.44 years.

The payback period is a very conservative economic measure; it should not be used alone. It is, however, quite valuable when used in combination with other measures, particularly when the risk of technological obsolescence is significant.

The Net Present Value

Another useful measure is net present value: the difference between the present value of the benefits and the present value of the investment. The inventory application (Fig. G.4), for example, yields a cumulative benefit of $9011.94 on an investment of $5000; that's a net present value of $4011.94. Earlier, future benefits were discounted back to their present values using a simple formula; in our example, we assumed a 12 percent interest rate. The net present value can be viewed as the amount of benefit over what could have been earned on an imaginary, risk-free investment. If the net present value is zero, the project will return the same amount as that risk-free investment, and thus is probably not worth doing. If the net present value is negative, the project is definitely not worth doing, as the imaginary risk-free investment would earn more.

In an effort to compare alternative investment opportunities, the net present value is sometimes expressed as a percentage of the investment, for example:

$$\frac{4013.94}{5000.00} = 0.80 \text{ or } 80\%$$

Given the 5 year life of the project, this number might be restated as 16 percent per year. Be careful in using such statistics, however; that 16 percent figure cannot be compared with the annual percentage rates offered by a bank or paid on a loan. The number does provide a relative measure of a project's return on investment, and can prove useful in comparing alternatives as long as the "rate of return" is computed in the same way, but it is not an absolute measure of the project's true return on investment.

The Internal Rate of Return

It is possible to compute an internal rate of return that can be compared with the prime rate and other common financial market statistics. To do this, we treat the interest rate as an unknown. We have the present value of the investment, and a series of estimated future benefits. Imagine that we were to place the initial investment in a savings account, and withdraw the appropriate estimated future value at the end of each year; for example, in the inventory problem, we would deposit $5000 now, and withdraw $2500 each year for 5 years. At the end of the time period, there would be nothing left in the account. What interest rate would we have to earn to be able to make this pattern of withdrawals and have nothing left after 5 years?

The key to solving the problem is the fact that, at the end of the specified time period, nothing is left. Thus, as long as the point in time is consistent, the investment and the benefits must be equal. Choosing the present as our common time period, we can write the following equation:

$$P = F_1\left[\frac{1}{(1+i)^1}\right] + F_2\left[\frac{1}{(1+i)^2}\right] + F_3\left[\frac{1}{(1+i)^3}\right] + \ldots + F_n\left[\frac{1}{(1+i)^n}\right]$$

Rearranging the terms, we get:

$$0 = -P + F_1\left[\frac{1}{(1+i)^1}\right] + F_2\left[\frac{1}{(1+i)^2}\right] + F_3\left[\frac{1}{(1+i)^3}\right] + \ldots + F_n\left[\frac{1}{(1+i)^n}\right]$$

which is a polynomial. Many scientific subroutines are available to solve polynomials, and many programmable calculators offer a polynomial routine. Figure G.5 shows a simple BASIC program that estimates, to the nearest full percentage point, the internal rate of return for a project with a 5 year life; you may want to generalize this program. It should be noted that other techniques might be used to generate more accurate results, or to iterate to a solution in fewer steps; this program is presented because of its simplicity.

The internal rate of return is analogous to the annual percentage rate, the number that banks and financial institutions use when advertising an investment opportunity or a loan; thus it is possible to use this number to compare or to rank alternatives both inside and outside the organization. For this reason, it is the preferred measure in any cost/benefit analysis. The internal rate of return for the inventory problem (as estimated by the program of Fig. G.5) lies between 41 and 42 percent (Fig. G.6).

RISK

Consider three different investment alternatives: a bank deposit, a new inventory system, or the fourth horse in the third race. The bank is virtually a sure thing. The inventory system may seem very straightforward, but estimates of cost and benefit can be wrong, and if they are, the rate of return will be wrong; thus an element of risk is involved. The horse race is a pure gamble, no matter how good your inside information may be. It is possible that the computed rate of return for all three alternatives could be the same, but the degree of risk is quite different. An investor expects to be compensated for risk; the greater the risk, the higher the rate of return that will be required.

The analyst must communicate risk to management. One common technique is to provide an optimistic estimate, a pessimistic estimate, and a most likely estimate for each questionable cost or benefit. Often, these three estimates are weighted: 20 percent, 20 percent, and 60 percent (for the most likely), and the result of this weighting process is used in all cost/benefit computations. Another option is to perform a complete cost/benefit analysis using only the worst case numbers (maximum cost and minimum benefit); the actual outcome is not likely to be any worse. Given a reasonable assessment of the risk, management can make an intelligent decision.

ESTIMATING COSTS

In discussing cost/benefit analysis, we have essentially treated the cost and benefit estimates as given. We presented two checklists of typical cost factors (Figs. G.1 an G.2), but said little about how the analyst might use this information to estimate costs. An in-depth treatment of cost estimating is well beyond the scope of this book, but we can present a number of general guidelines.

It is difficult to estimate the cost of a system. It is much easier to estimate the costs of each of the system components, and then sum these results; the total cost of the system is, after all, nothing more than the sum of the costs of its parts. Thus, the first step in developing a cost estimate is to break the problem into as many pieces as possible. During the feasibility study, this subdivision might be limited to the steps in

Fig. G.5: *A BASIC program to estimate the internal rate of return.*

```
10                                   REM  * * * * * * * * * * * * * * * * *
20                                   REM  * Program to compute the internal   *
30                                   REM  * rate of return for a project with a *
40                                   REM  * five year expected life.            *
50                                   REM  *    By:  W.S. Davis                  *
60                                   REM  *         1/15/83                     *
70                                   REM  * Variables:                          *
80                                   REM  *    F  =  future value, benefits     *
90                                   REM  *    P  =  present value, investment *
100                                  REM  *    X  =  net present value          *
110                                  REM  *    N  =  year (maximum of 5)        *
120                                  REM  * * * * * * * * * * * * * * * * *
130  DIM F(5)
140  PRINT "Enter initial investment amount";
150  INPUT P
160  FOR N = 1 TO 5
170    PRINT "Enter benefit amount for year  ";N;
180    INPUT F(N)
190  NEXT N
200                                  REM  * * * * * * * * * * * * * * * * *
210                                  REM  * Once all the data have been      *
220                                  REM  * entered, compute net present value.*
230                                  REM  * If it is negative, the investment is *
240                                  REM  * larger than the sum of the benefits, *
250                                  REM  * and there is no return on invest-  *
260                                  REM  * ment.                               *
270                                  REM  * * * * * * * * * * * * * * * * *
280  LET X = F(1)+F(2)+F(3)+F(4)+F(5)−P ,
290  IF X > = 0 THEN 420
300  PRINT  "The amount of the investment is greater than or equal to"
310  PRINT  "the sum of the benefits. There is no return."
320  STOP
330                                  REM  * * * * * * * * * * * * * * * * *
340                                  REM  * If the net present value is positive, *
350                                  REM  * the value of the polynomial is    *
360                                  REM  * computed with a 1% rate. The      *
370                                  REM  * value is then re-computed for     *
380                                  REM  * increments of 1% until the net    *
390                                  REM  * present value becomes negative or *
400                                  REM  * zero.                              *
410                                  REM  * * * * * * * * * * * * * * * * *
420  FOR I = .01 TO 1 STEP .01
430    LET X = (−P)
440    FOR N = 1 TO 5
450       LET X = X + F(N) / (1+I) ↑ N
460    NEXT N
470    IF X < = 0 THEN 580
480  NEXT I
490  PRINT "The rate of return exceeds 100%."
500  STOP
510                                  REM  * * * * * * * * * * * * * * * * *
520                                  REM  * The internal rate of return is less *
530                                  REM  * than or equal to the rate that    *
540                                  REM  * caused the net present value to   *
550                                  REM  * become negative or zero. Thus,    *
560                                  REM  * both (I−0.01) and I are printed.  *
570                                  REM  * * * * * * * * * * * * * * * * *
580  PRINT "The internal rate of return lies between:"
590  PRINT I−.01; "or"; (I−.01)*100; "percent; and"
600  PRINT I; "or"; I*100; "percent"
610  END
```

Fig. G.6: *The BASIC program of Fig. G.5 can be used to estimate the internal rate of return for the inventory example.*

```
Enter initial investment amount? 5000
Enter benefit amount for year  1 ? 2500
Enter benefit amount for year  2 ? 2500
Enter benefit amount for year  3 ? 2500
Enter benefit amount for year  4 ? 2500
Enter benefit amount for year  5 ? 2500
The internal rate of return lies between:
   .4099999 or 40.99999 percent; and
   .4199999 or 41.99999 percent
```

the system life cycle, yielding an estimate that is perhaps accurate to ±50 percent. Later, the personnel, equipment, and material costs associated with each step might be estimated separately. During detailed design those personnel costs might be subdivided into programmers, operators, clerks, and others. Eventually, the analyst may be able to break the work of the programmers down to individual modules on a hierarchy chart, and generate a much more accurate cost estimate; the smaller the module, the easier it is to estimate the cost.

On most systems, the bulk of the cost is concentrated in a few of the system components. By evaluating these few key elements carefully, it is possible to develop a very accurate cost estimate with a minimum of effort. If the handful of factors that account for 80 or even 90 percent of a system's cost can be identified and accurately estimated, ballpark figures or even guesses may be perfectly adequate for the other factors.

Historical data are an important source of information. Try to find the actual expenses for a past project of similar scope; they can provide a very good set of guidelines or targets. Of course, the actual cost of the existing system (as recorded by accounting) is the standard against which the new system is measured, and can prove invaluable in estimating the cost of a new system. Consider also the inspection process as described in Module A. As part of that process, rework time is estimated; later the actual rework time is recorded. If properly maintained, such data can be extremely useful.

Standards provide another source of cost information. Many firms add a fixed percentage to the sum of all other costs to cover overhead. Maintenance might be estimated as a fixed percentage of the implementation cost. We might use a number like 10 percent of programmer cost to estimate computer charges during the program test and debug period.

Personnel costs are a function of time; thus if we can estimate the time required to do a job, we can often estimate its cost. Let's assume that, based on historical data, our programmers average 10 lines of tested, documented, debugged code per day.

Given this standard, a good way to estimate the cost of a program is to start with the hierarchy chart. Each block represents one module, and each module should be limited to roughly one page of code—about 50 lines. That's an average of five programmer days per module; a moderate program containing 50 modules would be expected to take roughly 250 days. If programmers cost $100 per day, that's $25,000. Rather than using an average of five days per module, the analyst could, of course, evaluate each module independently, and thus develop more accurate estimates; the sum of these estimates would then represent the total time needed to write the program. Most organizations have standard personnel costs, and many have established coding standards based on measures such as lines of code per day.

When you prepare a cost estimate, remember to break the system into pieces first. Estimate the cost of each piece separately; to get the system cost, sum the components that represent the bulk of the system cost. Take advantage of historical cost figures and standards whenever possible. Cost estimating is not easy, but with a little work you can learn to do a good job.

REFERENCES

1. Teichroew, D. (1964). *An Introduction to Management Science.* New York: John Wiley and Sons, Inc.

2. Numerous books are available on such topics as engineering economic analysis, business economic analysis, management science, and investment analysis. Most contain at least a chapter or two on cost/benefit analysis.

HIPO with Structured English

WITH: *Laurena Burk*

OUTLINE

THE HIPO TECHNIQUE

Once we have completed the analysis stage of the system life cycle, we should have a logical module of the system documented by a data flow diagram, a data dictionary, and a description of key algorithms. During system design, a system flowchart may be created; it typically identifies, at a black box level, one or more programs. In this module, we will develop a set of specifications for those programs using the *HIPO (Hierarchy plus Input/Process/Output)* technique.

A completed HIPO package has two parts. A *hierarchy chart* is used to represent the top down structure of the program. For each module depicted on the hierarchy chart, an *IPO (Input/Process/Output) chart* is used to describe the inputs to, the outputs from, and the process performed by the module; the data dictionary is the source of the inputs and outputs, and the algorithm descriptions define the processes.

The HIPO documentation serves a number of purposes. By using it, designers can evaluate and refine a design, and correct flaws prior to implementation. Given the graphic nature of HIPO, users and managers can easily follow a program's structure. Finally, programmers can use the hierarchy and IPO charts as they write, maintain, or modify the program.

AN EXAMPLE

In Module F, we developed a system flowchart for an inventory application. The flowchart identified two programs; we'll use the HIPO technique to prepare specifications for the inventory update program. What basic functions are necessary to update the inventory master file? We might begin by listing the following major steps:

1. Get a transaction.
2. Get the master record.
3. Process the transaction.
4. Rewrite the master record.
5. Write a reorder record (if necessary).

We can summarize these steps in a high-level hierarchy chart (Fig. H.1). At the top is the main control module, *Update Inventory*, which controls the order in which the lower level modules are invoked. *Get Transaction* is the first subordinate module to be called, followed by *Get Inventory*, *Process Transaction*, *Rewrite Inventory*, and *Write Reorder*. As each subordinate module completes its task, control is passed back to the main control module, which invokes the next subordinate module. The arrangement of modules graphically represents the program hierarchy.

Note the names chosen for the modules. Each consists of a strong verb followed by a clear subject. The name tells you what the module does, and thus adds to the documentation.

Fig. H.1: *The first-level hierarchy of the on-line inventory update program.*

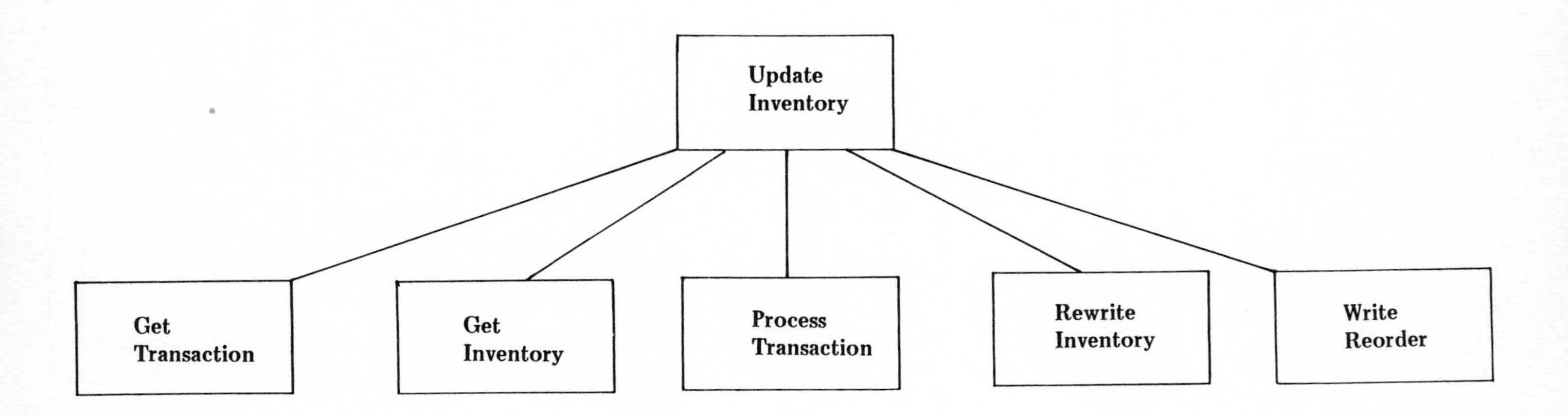

A separate IPO chart supports each module on the hierarchy chart (see Figs. H.2 through H.6). For example, consider the IPO chart for Get Inventory (Fig. H.3). The top lines on the chart identify the system, the module, and the author. Next, we show how this module is related to other modules: *Get Inventory* is called by *Update Inventory*, and calls no lower level modules. The part number and an error flag flow into *Get Inventory* from *Update Inventory*; the inventory record and the error flag flow back to *Update Inventory*. The logic performed by *Get Inventory* is clearly identified in the process block. Finally, there is space near the bottom of the chart for a list of local data elements and notes.

Many sources use a more elaborate IPO chart than the one described in this module. Often, large arrows are used to illustrate data flows graphically. Our intent is to concentrate on the essence of the HIPO technique, however, and we can do this with a minimum of detail. Another advantage to our simplified IPO diagrams is ease of maintenance, a topic we will return to later.

Fig. H.2: *An IPO chart for the Get Transaction module.*

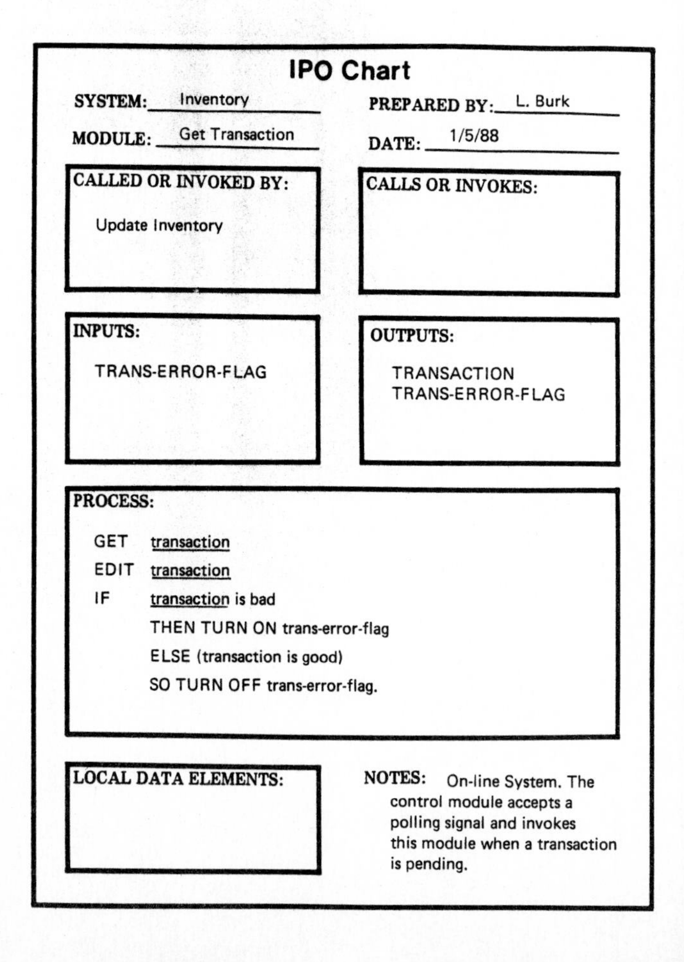
IPO Chart

SYSTEM: Inventory PREPARED BY: L. Burk

MODULE: Get Transaction DATE: 1/5/88

CALLED OR INVOKED BY:

Update Inventory

CALLS OR INVOKES:

INPUTS:

TRANS-ERROR-FLAG

OUTPUTS:

TRANSACTION
TRANS-ERROR-FLAG

PROCESS:

GET transaction
EDIT transaction
IF transaction is bad
THEN TURN ON trans-error-flag
ELSE (transaction is good)
SO TURN OFF trans-error-flag.

LOCAL DATA ELEMENTS:

NOTES: On-line System. The control module accepts a polling signal and invokes this module when a transaction is pending.

Fig. H.3: *The Get Master IPO chart.*

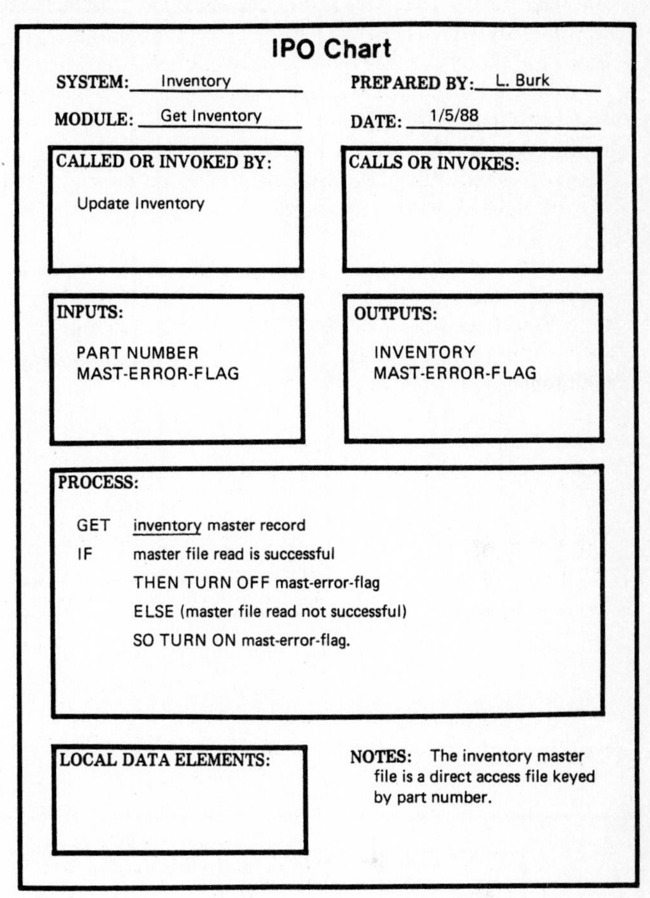

IPO Chart

SYSTEM: Inventory

PREPARED BY: L. Burk

MODULE: Get Inventory

DATE: 1/5/88

CALLED OR INVOKED BY:
Update Inventory

CALLS OR INVOKES:

INPUTS:
PART NUMBER
MAST-ERROR-FLAG

OUTPUTS:
INVENTORY
MAST-ERROR-FLAG

PROCESS:

GET inventory master record
IF master file read is successful
THEN TURN OFF mast-error-flag
ELSE (master file read not successful)
SO TURN ON mast-error-flag.

LOCAL DATA ELEMENTS:

NOTES: The inventory master file is a direct access file keyed by part number.

Functional Decomposition

A hierarchy chart is developed from the top, down. High level modules are general, performing control functions, while detailed computations are normally performed at the lower levels. Should we break any of the modules shown in Fig. H.1 down to a lower level? Ideally, each module should perform a single, complete logical function. Consider, for example, *Process Transaction*. What functions make up this module? Basically, we might consider four transaction types:

1. Increase the stock on hand as items are received at the warehouse.

2. Decrease the stock on hand as items are shipped from the warehouse.

3. Add a new record to the inventory file.

4. Delete an old record from the inventory file.

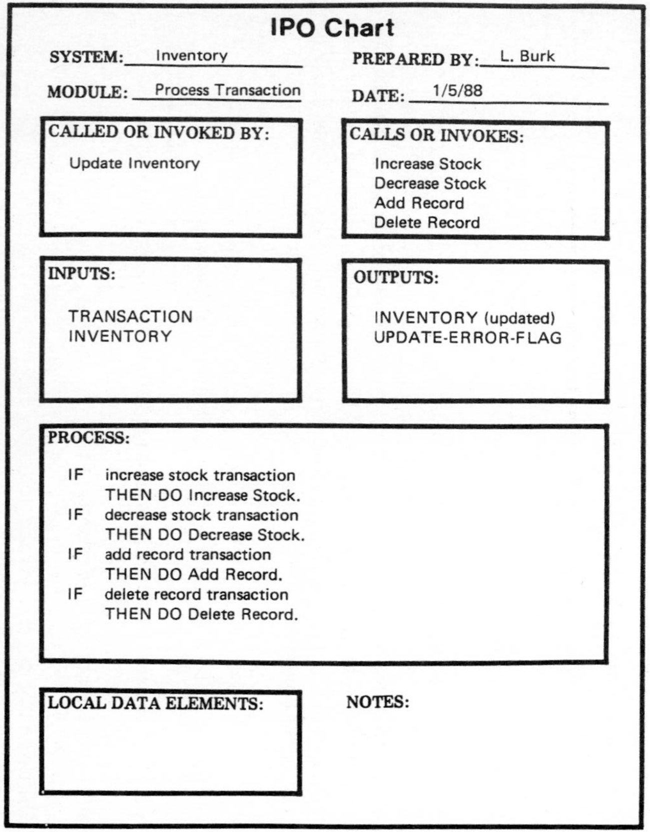

IPO Chart

SYSTEM: Inventory PREPARED BY: L. Burk

MODULE: Process Transaction DATE: 1/5/88

CALLED OR INVOKED BY:

Update Inventory

CALLS OR INVOKES:

Increase Stock
Decrease Stock
Add Record
Delete Record

INPUTS:

TRANSACTION
INVENTORY

OUTPUTS:

INVENTORY (updated)
UPDATE-ERROR-FLAG

PROCESS:

```
IF    increase stock transaction
      THEN DO Increase Stock.
IF    decrease stock transaction
      THEN DO Decrease Stock.
IF    add record transaction
      THEN DO Add Record.
IF    delete record transaction
      THEN DO Delete Record.
```

LOCAL DATA ELEMENTS:

NOTES:

Fig. H.4: *The Process Transaction IPO chart.*

Fig. H.5: *The Rewrite Inventory IPO chart.*

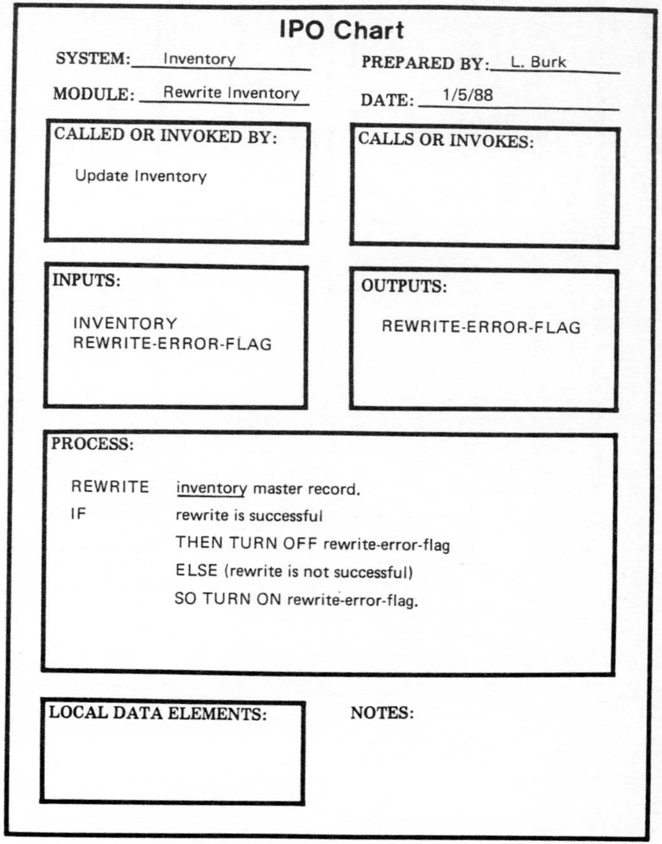

IPO Chart

SYSTEM: Inventory PREPARED BY: L. Burk

MODULE: Rewrite Inventory DATE: 1/5/88

CALLED OR INVOKED BY:

Update Inventory

CALLS OR INVOKES:

INPUTS:

INVENTORY
REWRITE-ERROR-FLAG

OUTPUTS:

REWRITE-ERROR-FLAG

PROCESS:

```
REWRITE   inventory master record.
IF        rewrite is successful
          THEN TURN OFF rewrite-error-flag
          ELSE (rewrite is not successful)
          SO TURN ON rewrite-error-flag.
```

LOCAL DATA ELEMENTS:

NOTES:

Fig. H.6: *The Write Reorder IPO chart.*

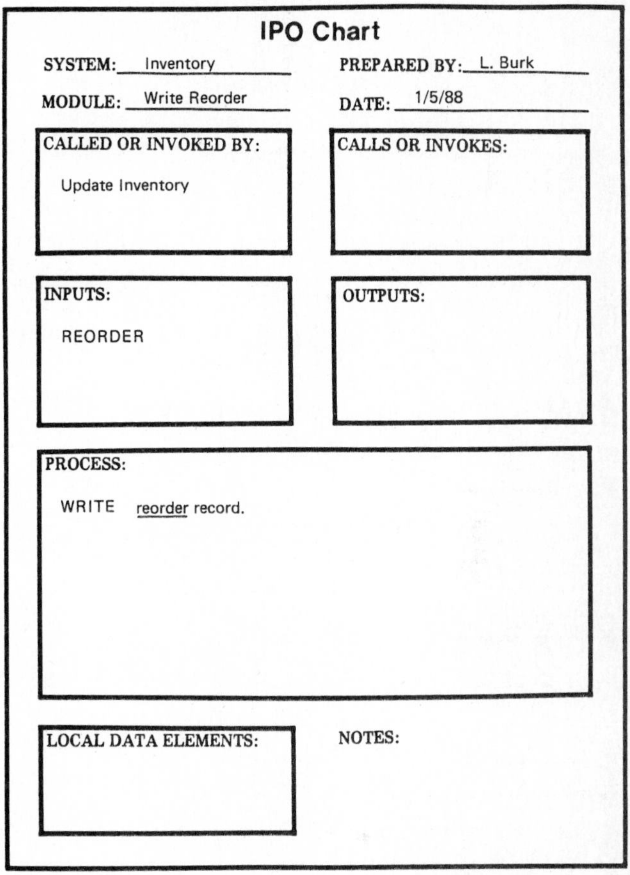

IPO Chart

SYSTEM: Inventory PREPARED BY: L. Burk

MODULE: Write Reorder DATE: 1/5/88

CALLED OR INVOKED BY:	CALLS OR INVOKES:
Update Inventory	

INPUTS:	OUTPUTS:
REORDER	

PROCESS:

WRITE reorder record.

LOCAL DATA ELEMENTS:

NOTES:

Processing transactions to update the inventory master file requires four different algorithms, one for each transaction type. A given transaction will call for only one of these algorithms. Each transaction type, each algorithm, represents an independent function; thus the *Process Transaction* module should be subdivided into four lower level modules (Fig. H.7). Breaking a module into its subtasks is called *functional decomposition*. A module that performs a single, complete logical function is said to be *cohesive*.

Should any of these modules be further decomposed? Consider *Increase Stock*; its IPO chart is shown in Fig. H.8. The process block identifies only a single function—calculate stock on hand. Since the module performs a single, complete logical function, it is already cohesive, and further decomposition is unnecessary.

Fig. H.7: *A functional decomposition of the Update Inventory program.*

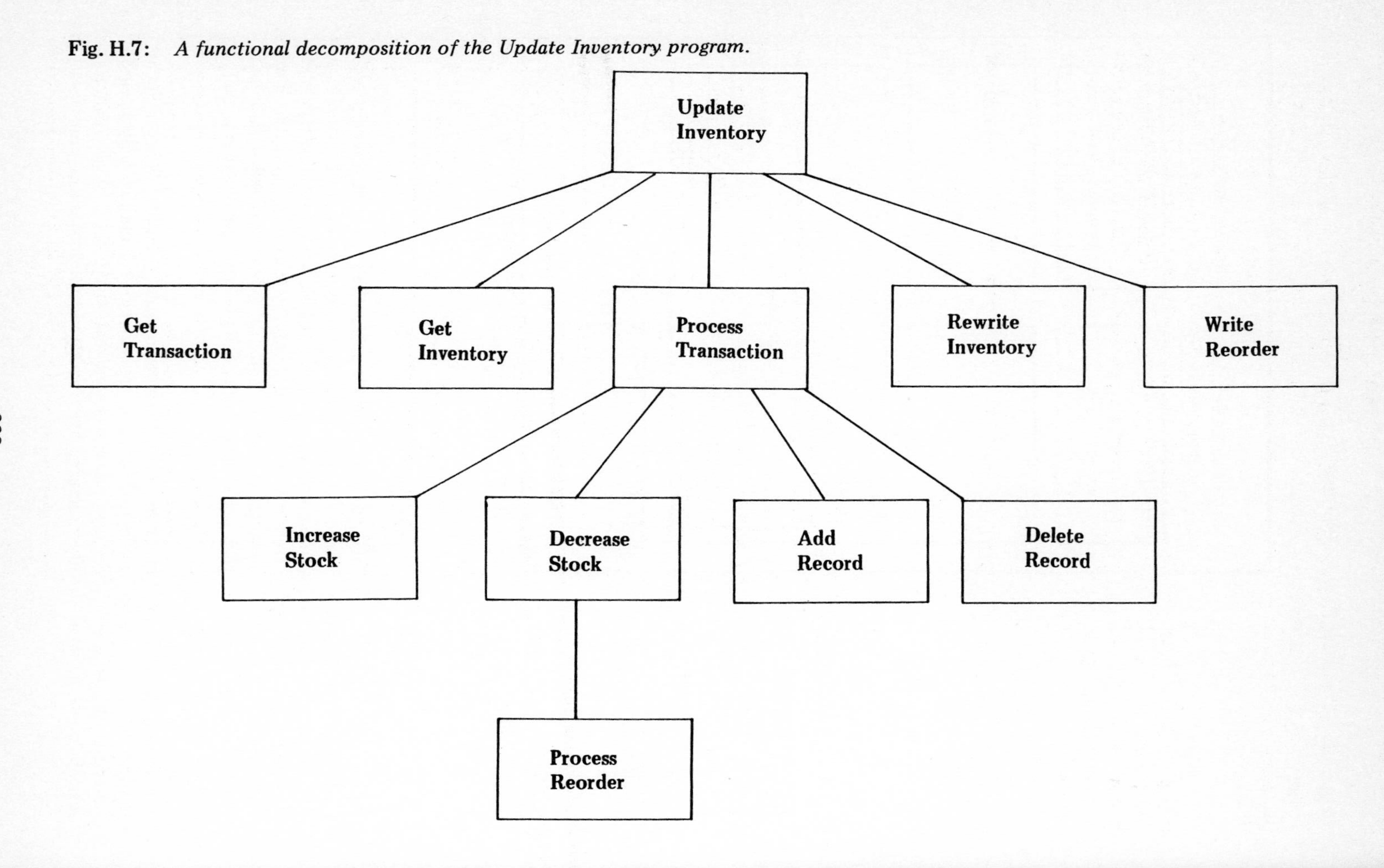

Next, consider *Decrease Stock* (Fig. H.9). Two functions can be identified; we always calculate stock-on-hand, but sometimes find it necessary to process a reorder as well. It might be reasonable to define *Process Reorder* as a subtask of *Decrease Stock* (Fig. H.7).

Is cohesion enough? Not really. It might be argued that *Process Transaction* is a complete logical function, and thus should not be decomposed. Without decomposition, however, *Process Transaction* would be so large that it would be difficult to follow. As a rule, no module should exceed a single page of code; as a planning guideline, decompose a module when its process cannot be completely defined on a single IPO chart. A well designed program is composed of a hierarchy of small, cohesive modules.

In the interest of brevity, we will not decompose the other modules of Fig. H.7.

Fig. H.8: *The Increase Stock IPO chart.*

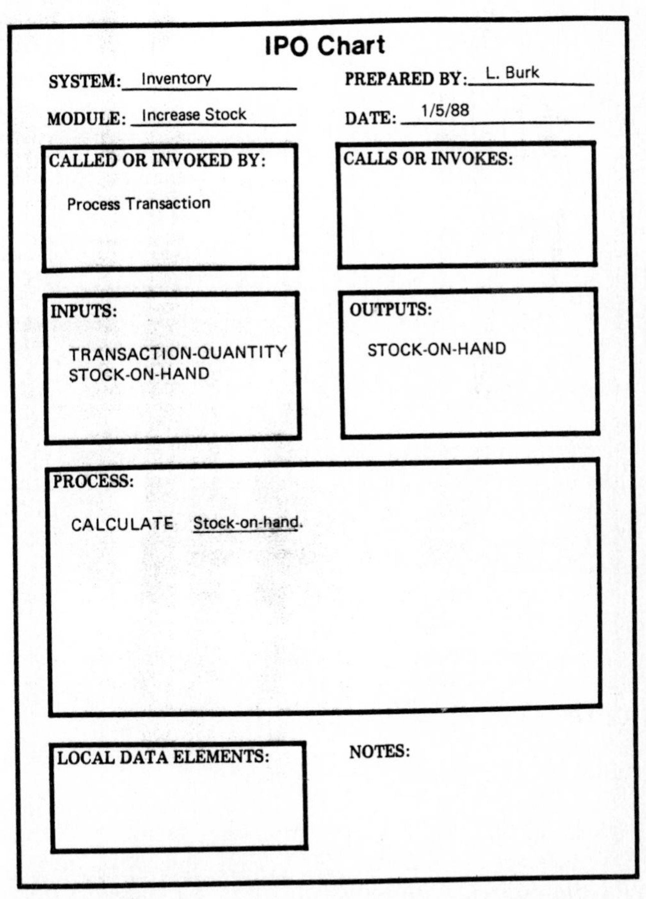

IPO Chart

SYSTEM: Inventory — PREPARED BY: L. Burk

MODULE: Increase Stock — DATE: 1/5/88

CALLED OR INVOKED BY:
Process Transaction

CALLS OR INVOKES:

INPUTS:
TRANSACTION-QUANTITY
STOCK-ON-HAND

OUTPUTS:
STOCK-ON-HAND

PROCESS:
CALCULATE Stock-on-hand.

LOCAL DATA ELEMENTS:

NOTES:

Fig. H.9: *The Decrease Stock IPO chart.*

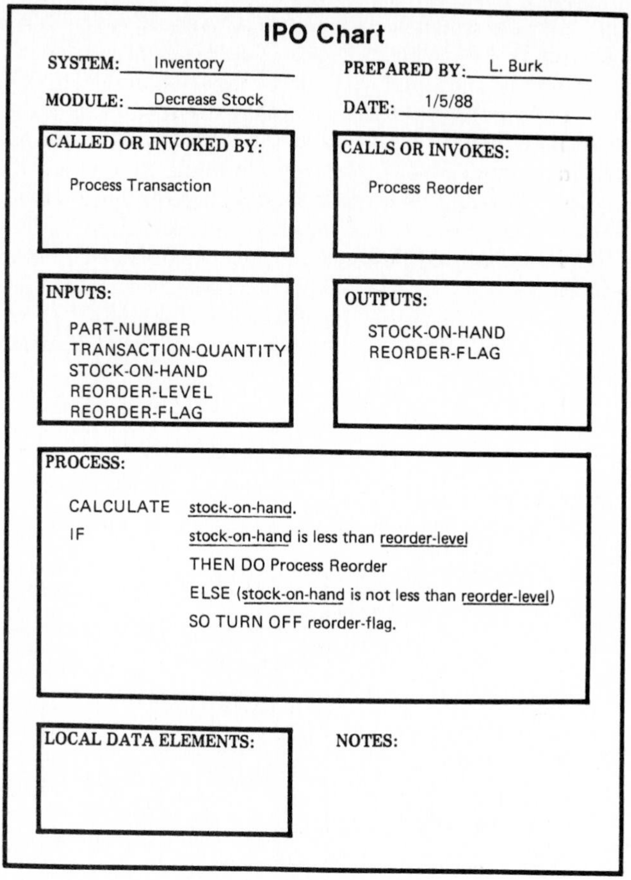

IPO Chart

SYSTEM: Inventory PREPARED BY: L. Burk

MODULE: Decrease Stock DATE: 1/5/88

CALLED OR INVOKED BY:
Process Transaction

CALLS OR INVOKES:
Process Reorder

INPUTS:
PART-NUMBER
TRANSACTION-QUANTITY
STOCK-ON-HAND
REORDER-LEVEL
REORDER-FLAG

OUTPUTS:
STOCK-ON-HAND
REORDER-FLAG

PROCESS:

CALCULATE stock-on-hand.
IF stock-on-hand is less than reorder-level
THEN DO Process Reorder
ELSE (stock-on-hand is not less than reorder-level)
SO TURN OFF reorder-flag.

LOCAL DATA ELEMENTS:

NOTES:

Checking the Flow of Control

Once we have created the functionally decomposed hierarchy chart, we must review the flow of control through the program. Control normally flows from the top, down, following the lines that link the modules. The scope of control of a given module extends only over those modules lower in the hierarchy chart; a low-level module never tells its parent what to do. In Fig. H.7, for example, *Process Transaction* should control only itself and its subordinates. *Update Inventory* may tell *Rewrite Inventory* what to do, but *Process Transaction* may not. The analyst should carefully check the processes on each IPO chart to make sure that the rules concerning the flow of control have not been violated. Since the modules are ultimately linked by data flows, the next step is to check for data independence.

Data Coupling and Data Independence

We know that data must pass between modules; thus relationships among the modules are essential. What data should be allowed to flow into a module? Only those elements that are essential to the module's function. What data should flow out from a module? Once again, only essential data should move. Why? One reason is ease of maintenance. If a data element is altered in form, then every module that accesses that data element is a candidate for maintenance. Why maintain a module just because a nonessential data element changes? Why waste time? Another problem, the ripple effect, occurs when a bug in one module propagates bugs in other modules. Modules are linked by data flows. An unnecessary data flow represents an unnecessary risk. Finally, consider program simplicity. What is easier to understand, a subroutine with a parameter list of three elements, or one with six? Unnecessary data flows mean unnecessary complexity.

Data coupling is a measure of the links between modules. The more loosely modules are coupled, the more independent they are. We achieve module independence and reduce data coupling by minimizing the number of connections, or the number of data flows, linking a module to the rest of the program. The IPO chart lists the data links in the input and output blocks; thus we must study the IPO charts to be sure that coupling is loose, by verifying that all input to a module is essential to its function, and that no superfluous data are passed to or returned by a module.

The Structure Chart

Following data from IPO to IPO is difficult. An excellent way to keep track of the data flows is to develop a *structure chart*. Start with a single IPO chart. Locate the module associated with this IPO on the hierarchy chart, and copy each of the input and output data element names alongside the line linking this module to its parent. Use small arrows to indicate the direction of flow: into the module or back to the higher level. Repeat this process for each IPO chart, and you have a structure chart (Fig. H.10). (Note that two new IPO charts, Figs. H.11 and H.12, have been added to the six presented earlier in the chapter.)

Given a structure chart, we can easily evaluate data coupling. Remember the key rule: only those data elements that are required by a module should be passed to that module. Note that two structures flow into *Process Transaction*. (Remember that a structure contains a number of data elements). Are all data elements contained in both the inventory and transaction structures needed by *Process Transaction*, or by a module controlled by *Process Transaction*? Yes. Consider, for example, *Add Record* (Fig. H.12). The only way to add a complete record to the inventory file is to have a complete record; the only way *Add Record* can get a complete record is if *Process Transaction* passes it a complete record.

In contrast, note *Increase Stock* (Fig. H.8). Why are only selected data elements sent to this module? Does *Increase Stock* need every element in both the transaction and the inventory records? No. Since *Increase Stock* does not need such data elements as the part number, part description, and reorder quantity, there is no need to pass these elements to *Increase Stock*. Generally, a program should be designed to keep entire data structures in management modules, and to pass to the subordinate modules only those elements essential to their function. Structures are helpful in grouping related fields, but bundling data just to shorten a parameter list leads to confusion, and not simplification.

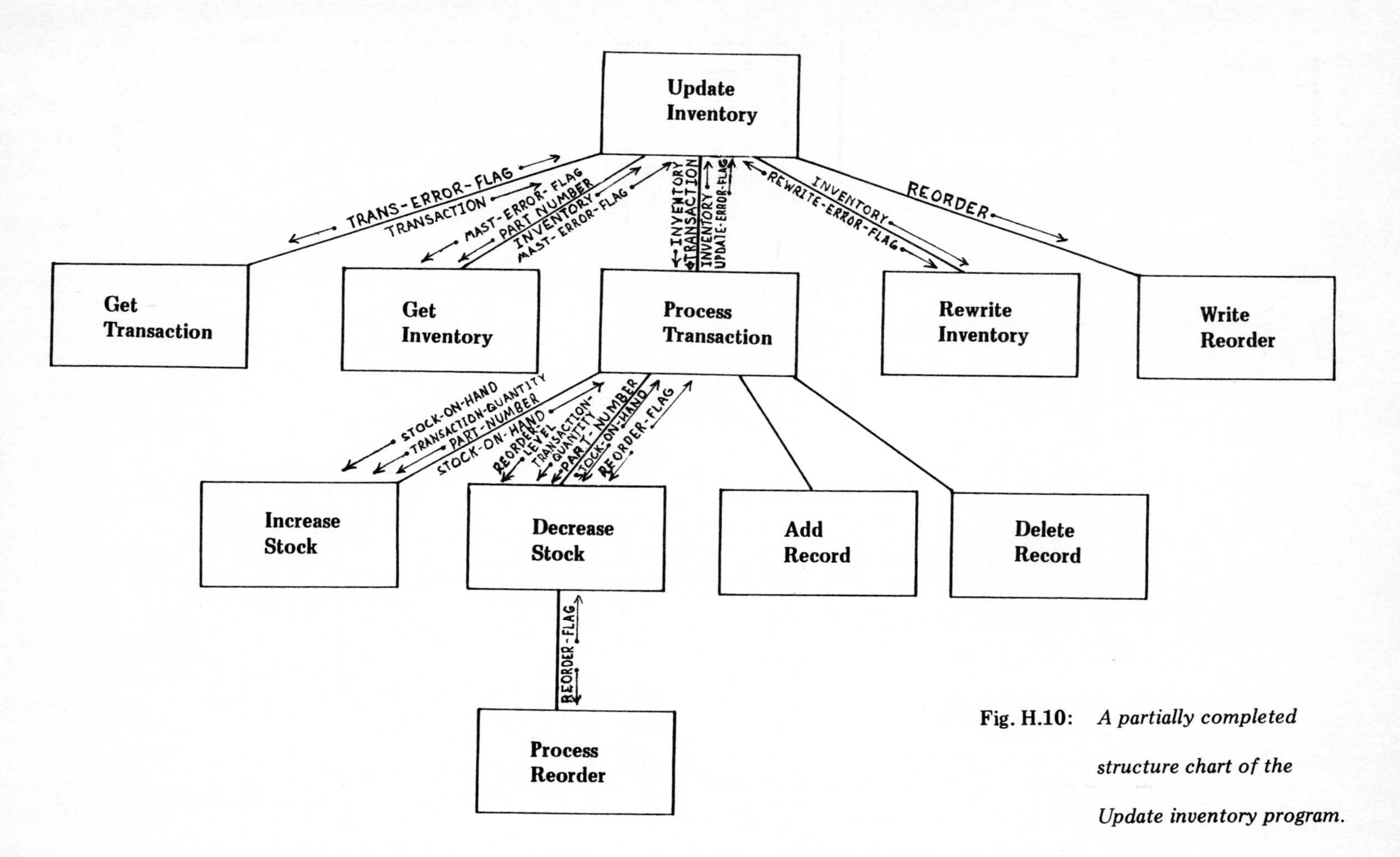

Fig. H.10: *A partially completed structure chart of the Update inventory program.*

For documentation purposes, adding data flows to a hierarchy chart is probably counterproductive: a structure chart is hard to maintain and difficult to read. A structure chart should be used to verify a design. Take the time to review each module in the HIPO, and make sure that it is independent and loosely coupled.

STRUCTURED ENGLISH

In this module, we have used *structured English* to define the process steps on each IPO chart. Structured English is a very limited, highly restricted subset of the English language. In some ways, it resembles a programming language; thus programmers tend to find it easy to understand. The base for structured English is, of course, English, so users find it easy to follow, too.

Fig. H.11: *The Process Reorder IPO chart.*

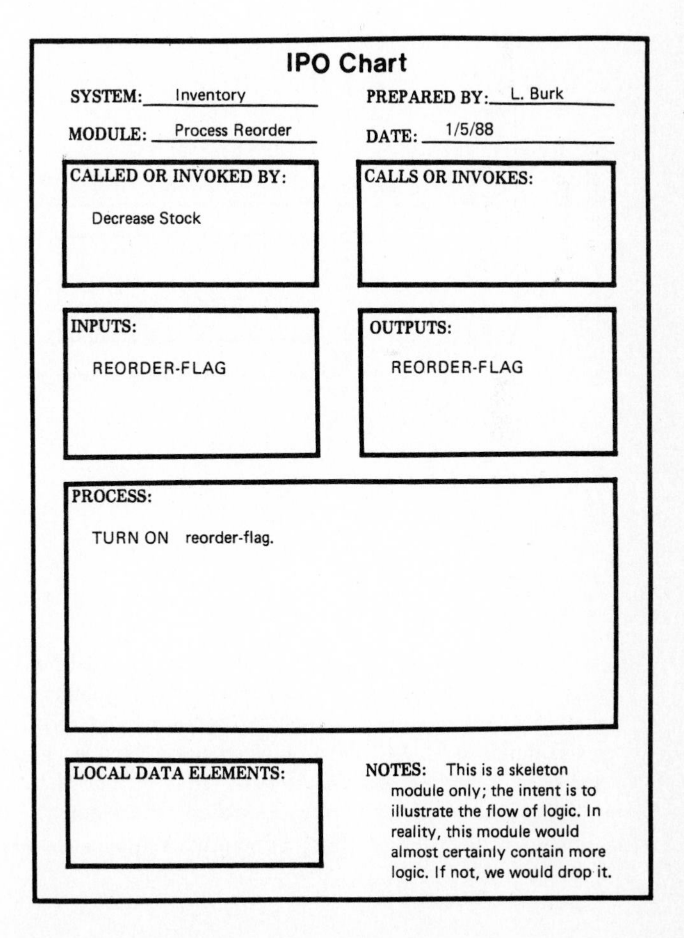

IPO Chart

SYSTEM: Inventory
PREPARED BY: L. Burk
MODULE: Process Reorder
DATE: 1/5/88

CALLED OR INVOKED BY:
Decrease Stock

CALLS OR INVOKES:

INPUTS:
REORDER-FLAG

OUTPUTS:
REORDER-FLAG

PROCESS:
TURN ON reorder-flag.

LOCAL DATA ELEMENTS:

NOTES: This is a skeleton module only; the intent is to illustrate the flow of logic. In reality, this module would almost certainly contain more logic. If not, we would drop it.

Fig. H.12: *The Add Record IPO chart.*

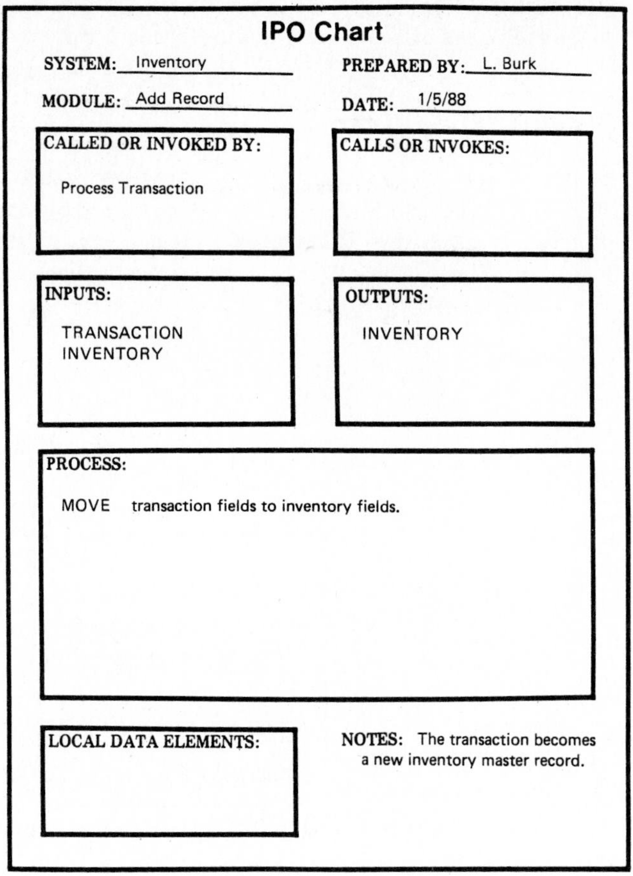

IPO Chart

SYSTEM: Inventory PREPARED BY: L. Burk

MODULE: Add Record DATE: 1/5/88

CALLED OR INVOKED BY:	CALLS OR INVOKES:
Process Transaction	

INPUTS:	OUTPUTS:
TRANSACTION INVENTORY	INVENTORY

PROCESS:

MOVE transaction fields to inventory fields.

LOCAL DATA ELEMENTS:

NOTES: The transaction becomes a new inventory master record.

There are several variations of structured English, and none even approaches the status of a standard; we can, however, offer a number of guidelines. Three basic instruction types are used: sequence, decision, and repetition. Sequence instructions are the simpliest; CALCULATE new stock-on-hand is a good one. Sequence statements begin with commands such as MOVE, GET, WRITE, READ, or COMPUTE, followed by the name or names of the associated data elements. A good structured English statement reads like a short imperative sentence; several examples are shown in Fig. H.13.

Fig. H.13: *Some Typical Structured English imperative statements.*

COMPUTE gross-pay.

ADD 1 to counter.

MULTIPLY hours-worked by pay-rate to get gross-pay.

GET master-record.

MOVE field-a to output-record.

WRITE new master-record.

TURN ON reorder-flag.

It is often convenient to group several structured English statements into a block, assign a name to the block, and treat it as a single sequence statement. For example, all the instructions required to compute gross pay might be grouped in a block under the name *compute gross pay*. Once this has been done, we can write:

DO *compute gross pay*.

and reference the entire block. Note that the gross pay logic might well contain decisions and even repetitive code; a block can contain any combination of code.

Decision logic follows the IF-THEN-ELSE structure. For example:

IF stock-on-hand is less than reorder-point

THEN turn on reorder-flag

ELSE (stock-on-hand not less than reorder-point)

SO turn off reorder-flag

The key word IF is followed by a condition. If the condition is true, the block following THEN is executed. ELSE identifies the negative of the condition; SO preceeds the block to be executed if the initial condition is false. In general write:

```
IF condition

    THEN block-1

    ELSE (not condition)

    SO block-2.
```

Indenting makes the IF-THEN-ELSE logic easier to read.

Earlier, we mentioned that a block of structured English code can contain any combination of sequence, decision, and repetition logic. Thus, in the general structure of the decision logic described above, block-1 or block-2 (or both) could contain another decision, for example:

```
IF condition-1

    THEN IF condition-2

             THEN block-a

             ELSE (not condition-2)

             SO block-b

    ELSE (not condition-1)

    SO block-c.
```

This is a nested decision. Note that any or all of the three logic blocks could contain yet another decision.

Repetitive logic defines a block of structured English that is executed repeatedly until a terminal condition is reached. For example, such instructions as:

```
REPEAT UNTIL condition-1

        block-1

FOR EACH TRANSACTION

        block-a
```

imply both repetitive logic and the condition used to terminate that logic.

By convention, only key words such as IF, THEN, SO, REPEAT, UNTIL, DO, and so on are capitalized; data names and the general English needed to complete a sentence or a phrase should be lower case. Many sources recommend that a data name defined in a data dictionary be underlined. Indentation should always be used to show the relationship between the parts of a block.

Structured English is excellent for describing an algorithm, particularly when user communication is necessary. If the main concern is communication with the programmers, however, pseudocode may be a better choice (see Module I). Both tools are less effective at describing the logic of a high level control structure or an algorithm in which numerous decisions must be made. In these cases, a logic flowchart (Module J), a decision table, or a decision tree (Module O) may be a better choice. No single tool is always best. Choose the technique that best fits the application.

HIPO AND LONG-TERM DOCUMENTATION

Ideally, the hierarchy chart and the IPO charts should form the core of a long-term documentation package. As the program changes (and it will), the documentation should change, too. Unfortunately, this is not always the case. In this module, we have considered only the essence of the HIPO technique; most sources recommend a more elaborate procedure for developing the IPO charts, with large arrows graphically depicting the data flows. Unless an organization has professional documentators or a dependable, on-line, computer-aided documentation system, it is unlikely that such charts will be maintained for very long. A distinct advantage to the "essential elements only" approach of this module is the fact that an IPO chart can be imbedded into the source code as a series of comments. Source code comments are perhaps the easiest form of documentation to maintain.

REFERENCES

1. Gane, Chris and Sarson (1979). *Structured Systems Analysis: Tools and Techniques.* Englewood Cliffs, New Jersey: Prentice-Hall, Inc.

2. IBM Corporation (1974). *HIPO—A Design Aid and Documentation Technique.* White Plains, New York: IBM Corporation. Publication Number GC20-1851.

3. Katzan, Harry Jr. (1976). *Systems Design and Documentation: An Introduction to the HIPO Method.* New York: Van Nostrand Reinhold.

4. Peters, Lawrence J. (1981). *Software Design: Methods and Techniques.* New York: Yourdon Press.

5. Yourdon, Edward and Constantine (1979). *Structured Design.* Englewood Cliffs, New Jersey: Prentice-Hall, Inc.

Module I

Pseudocode

OUTLINE

WHAT IS PSEUDOCODE?

In Module H, structured English was used to describe the logic of a process on an IPO chart. Pseudocode is an alternative to structured English. *Pseudo-* means similar to; thus pseudocode is similar to real code. In fact, the structure of pseudocode is often based on a real programming language such as COBOL, FORTRAN, or Pascal. When structured English is used, such details as opening and closing files, initializing counters, and setting flags are often ignored or implied; with pseudocode, they are explicitely coded. The analyst using pseudocode would not, however, be concerned with language-dependent details such as the distinction between subscripts and indexes in COBOL, or the difference between real and integer numbers in FORTRAN. The idea is to describe the executable code in a form that a programmer can easily translate.

There is no standard pseudocode; many different versions exist. Most, however, incorporate the three structured programming conventions: sequence, decision, and repetition. We will use a hybrid pseudocode in this text, borrowing from COBOL, FORTRAN, and Pascal.

SEQUENCE

Perhaps the easiest way to define sequential logic is by coding a FORTRAN-like expression, for example:

```
COUNT = 0
```

or:

```
STOCK = STOCK + QUANTITY
```

Such details as the sequence of operations and the rules for using parentheses are also borrowed from FORTRAN; data names should be taken from the data dictionary and/or the list of inputs and outputs on the IPO diagram. If you feel more comfortable with another language, the conventions of that language would serve just as well.

Input and output instructions are explicitely defined in pseudocode. We will use:

```
READ data FROM source
```

and

```
WRITE data TO destination
```

where "data" is a list of variables, a data structure, or a record, and "source" and "destination" refer to a file or a data base.

Blocks of Logic

It is possible to write a number of pseudocode instructions and treat them as a single block; for example:

```
COUNT = 0

ACCUMULATOR = 0
```

might be assigned the block name INITIALIZE. Once this has been done, we can refer to the entire block as a single sequence instruction:

```
PERFORM INITIALIZE
```

A block can contain any set of sequence, decision, and/or repetition logic. The general form of a PERFORM is:

```
PERFORM block
```

To distinguish between formal subroutines and internal procedures, some analysts use:

```
PERFORM block USING list
```

for a subroutine; "list" designates a list of variables passed between the calling routine and the subroutine.

To simplify our discussion of decision and repetition logic, we will refer to blocks of instructions rather than to the individual commands. Remember that a pseudocode block consists of one or more pseudocode instructions.

DECISION

The general format of a decision pseudocode block is:

```
IF condition
   THEN
           PERFORM block-1
   ELSE
           PERFORM block-2
ENDIF
```

For example:

```
IF HOURS > 40
   THEN
            PERFORM OVERTIME
   ELSE
            PERFORM REGULAR
ENDIF
```

A standard IF-THEN-ELSE structure is used, with the THEN block executed if the condition is true, and the ELSE block executed if the condition is false. Note the ENDIF. A feature of most pseudocodes is that each block of logic is clearly delimited. A decision block always begins with IF and ends with ENDIF; there is no ambiguity. Note also the use of indentation; it makes the block easy to read.

A pseudocode block can contain any combination of code, including a decision block. Thus it is possible to nest decision logic; for example:

```
IF condition-1
   THEN
            IF condition-2
               THEN
                        PERFORM block-a
               ELSE
                        PERFORM block-b
            ENDIF
   ELSE
            PERFORM block-c
ENDIF
```

Note how indentation highlights the relationship among these instructions. Note also how IF and ENDIF clearly delimit both decision blocks.

REPETITION

In structured English, we did not distinguish between the various forms of repetitive logic; in pseudocode, we do. The basic idea of repetitive code is that a block is executed again and again until a terminal condition occurs. DO WHILE logic tests for this terminal condition at the top of the loop; for example:

```
WHILE condition DO
   PERFORM block
ENDWHILE
```

Note the use of indentation and the way WHILE and ENDWHILE delimit the block. As an alternative, the REPEAT until structure tests for the terminal condition at the bottom of the loop:

```
REPEAT
    PERFORM block
UNTIL condition
```

Some analysts use a pseudocode structure much like a FORTRAN DO loop to define a count-controlled loop:

```
DO index = initial TO limit
    PERFORM block
ENDDO
```

Again, note the indentation and the ENDDO.

THE CASE STRUCTURE

A common programming problem involves selecting from among several alternative paths. For example, a program might be designed to process four different types of inventory transactions; imagine a control module that accepts a transaction and, based on a code, passes control to one of four lower-level routines. Although nested decision statements could be used to define this logic, IF-THEN-ELSE blocks tend to be difficult to follow when nesting goes beyond three or four levels. The case structure is a better option; its general form is:

```
SELECT variable
    CASE (value-1) block-1
    CASE (value-2) block-2
    .
    .
    .
    DEFAULT CASE block-n
ENDSELECT
```

The decision made by a case structure depends on the value of the variable specified by the word SELECT. If it contains "value-1", then block-1 is executed; if it contains "value-2", then block-2 is executed, and so on. The DEFAULT CASE is coded in the event that the variable contains none of the listed values. A case structure is delimited by ENDSELECT. Once again, indentation makes the structure easy to read. For example, the control module for the inventory program described above might be coded as:

```
SELECT TRANSACTION TYPE
    CASE (increase) PERFORM INCREASE STOCK
    CASE (decrease) PERFORM DECREASE STOCK
    CASE (add)      PERFORM ADD RECORD
    CASE (delete)   PERFORM DELETE RECORD
    DEFAULT CASE    PERFORM TRANSACTION ERROR
ENDSELECT
```

A NOTE OF CAUTION

Pseudocode is very much like real code; thus programmers find it easy to use and to understand. This fact does, occasionally, lead to a problem, however. Many systems analysts were once programmers. These former programmers feel comfortable with pseudocode; in fact, they may feel too comfortable with pseudocode. Rather than using this tool as an aid for planning or designing a program, they write the code. As a result, the program is written twice: once in pseudocode and once in real code. Aside from being an obvious waste of time, double coding misses the point of program design; if the analyst worries about coding details, crucial design considerations may well be overlooked. Also, programmers tend to resent such overspecification—they want to be told what to code, and not how to code it.

Even if the pseudocode is well done, the programmers sometimes resent it. Specifying algorithms in what is essentially a high-level programming language does limit the programmer's flexibility. Often, the programmer will fail to distinguish between the analyst's coding technique and the analyst's design; the result may be criticism (even rejection) of a perfectly good design based on inappropriate criteria. If your programmers dislike pseudocode, consider using structured English or some other technique instead.

Pseudocode is an excellent tool for defining computational algorithms. It is not a good tool for describing control structures, particularly when several nested decisions are involved. In this text, we will use structured English or pseudocode primarily for the low-level, computational modules of a hierarchy chart.

REFERENCES

1. Gane, Chris and Sarson (1979). *Structured Systems Analysis: Tools and Techniques.* Englewood Cliffs, New Jersey: Prentice-Hall, Inc.

2. Gillett, Will D. and Pollack (1982). *An Introduction to Engineered Software.* New York: Holt, Rinehart and Winston.

3. Peters, Lawrence J. (1981). *Software Design: Methods and Techniques.* New York: Yourdon Press.

Module J

Program Logic Flowcharts

OUTLINE

FLOWCHARTING

A flowchart is a graphical representation of program logic. Four standard symbols are used (Fig. J.1); they are linked by flowlines that show the sequence and direction of flow. By convention, logic flows from the top down, and from left to right; arrowheads are added to the flowlines to indicate deviations from this standard pattern. Since arrowheads make a flowchart easier to read we will always use them, even when the conventions are followed.

Fig. J.1: *Basic flowcharting symbols.*

Symbol	Meaning	Explanation
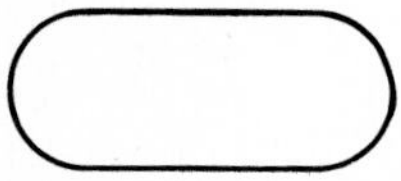 (rounded oval)	Terminal point	Marks the beginning or end of a program or program segment.
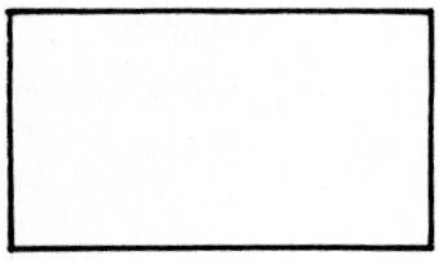 (rectangle)	Process	Indicates any arithmetic or data copy operation.
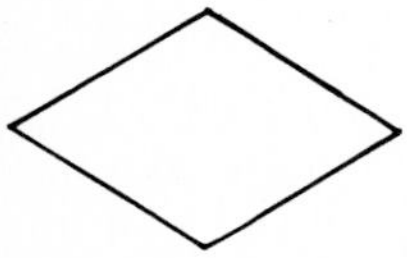 (diamond)	Decision	Indicates a yes/no decision to be made by the program.
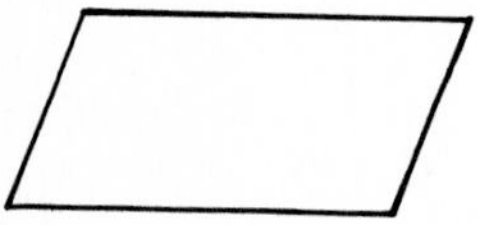 (parallelogram)	Input/output	Indicates any input or output operation.

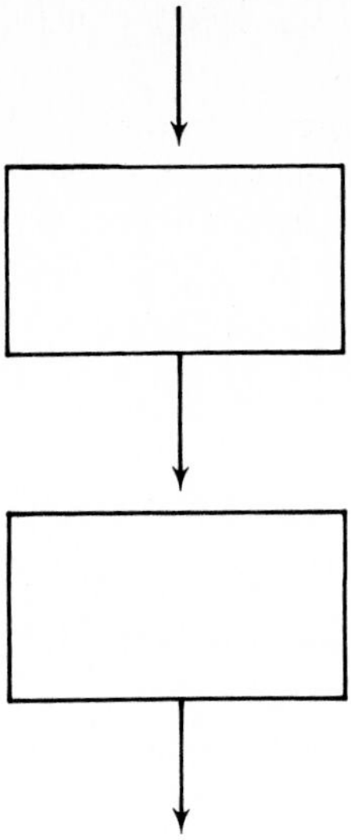

Fig. J.2: *Sequence logic.*

Program logic can be expressed as combinations of three basic patterns: sequence, decision, and repetition. The sequence pattern (Fig. J.2) implies that the logic is executed in simple sequence, one block after another. A sequence block on a flowchart is identified by a rectangle, and can represent one or more actual instructions.

A decision block implies IF-THEN-ELSE logic (Fig. J.3). A condition (the diamond symbol) is tested. If the condition is true, the logic associated with the THEN branch is executed, and the ELSE block is skipped. If the condition is false the ELSE logic is executed and the THEN logic is skipped.

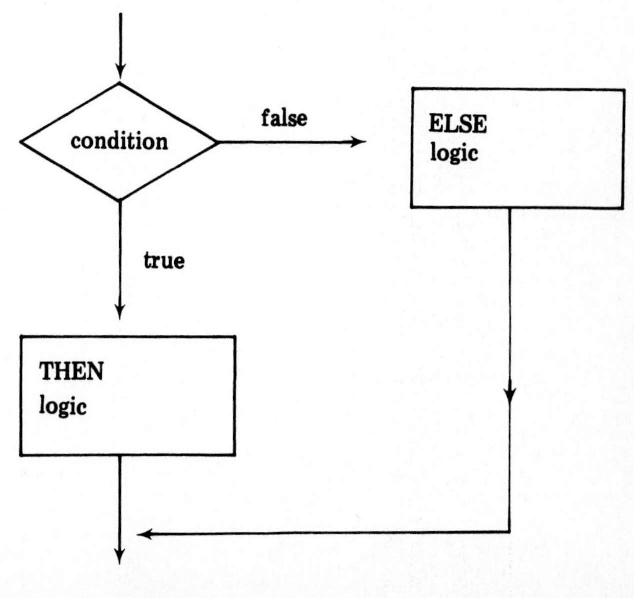

Fig. J.3: *Decision or IF-THEN-ELSE logic.*

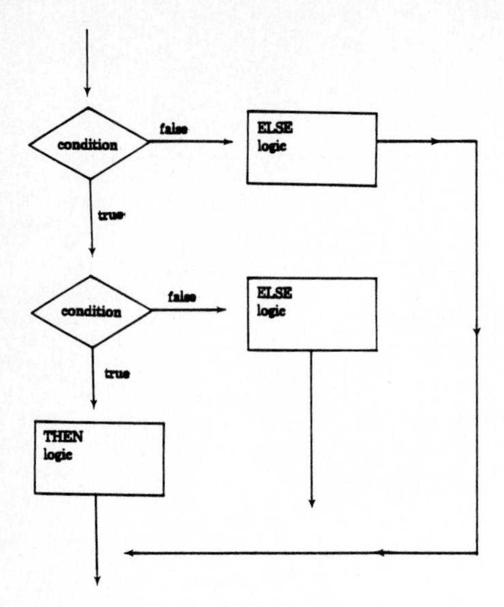

Fig. J.4: *Nested decision logic.*

Each decision block contains two sequence blocks: one for the THEN logic and one for the ELSE logic. A sequence block can contain one or more actual instructions. These instructions need not be limited to simple sequential logic; for example, on a high-level flowchart, a sequence block might be used to represent a subroutine containing sequence, decision, and repetitive logic. Taking advantage of this fact, we can expand the THEN branch of a decision block to contain another decision block (Fig. J.4): a nested IF. (Note that dashed lines have been added to Fig. J.4 to identify the position of the original THEN block.) Decision logic can be nested on the ELSE branch too (Fig. J.5), and nesting can occur on both paths.

Fig. J.5: *Another example of nested decision logic.*

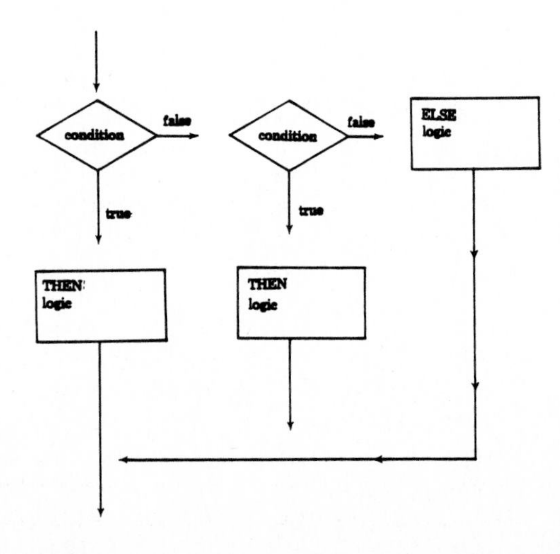

Fig. J.6: *DO WHILE logic.*

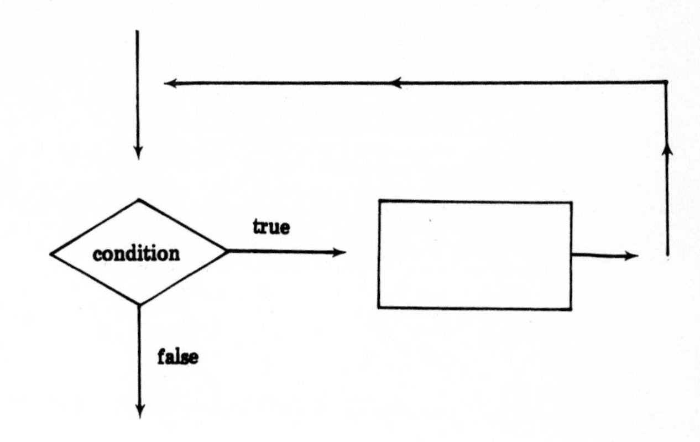

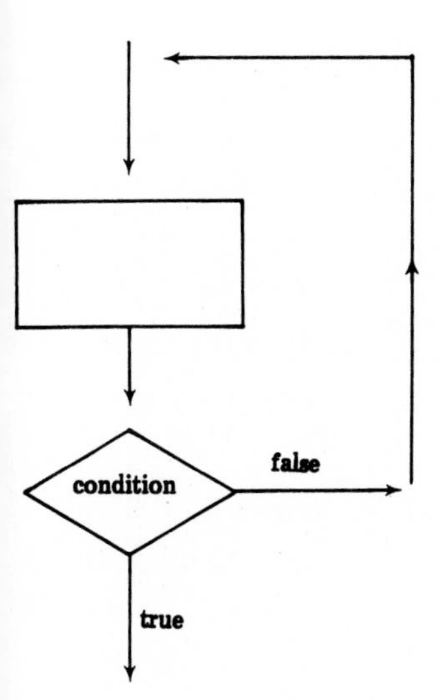

Fig. J.7: *DO UNTIL logic.*

Two basic patterns exist for showing repetitive logic. In the DO WHILE pattern (Fig. J.6), a test is performed at the top of a loop. If the condition tested is true, then the logic of the loop is executed, and control is returned to the top for another test; if the condition is false, the logic of the loop is skipped, and control is transferred to the block following the DO WHILE. The DO UNTIL pattern (Fig. J.7) is different in that the test is performed at the bottom of the loop; otherwise the rules are similar.

A program is ultimately composed of combinations of these three basic structures. Program logic flows from block to block. Each block should have a single entry point (at the top) and a single exit; in other words, once the logical flow enters a block, it stays there until the function is completed. Simple sequence, decision, and repetition blocks can be combined to form larger blocks; for example, each module on a hierarchy chart can be viewed as a block of logic that performs a single, complete function. Even at this level, however, a block should have a single entry point and a single exit.

Fig. J.8: *A hierarchy chart of the inventory program.*

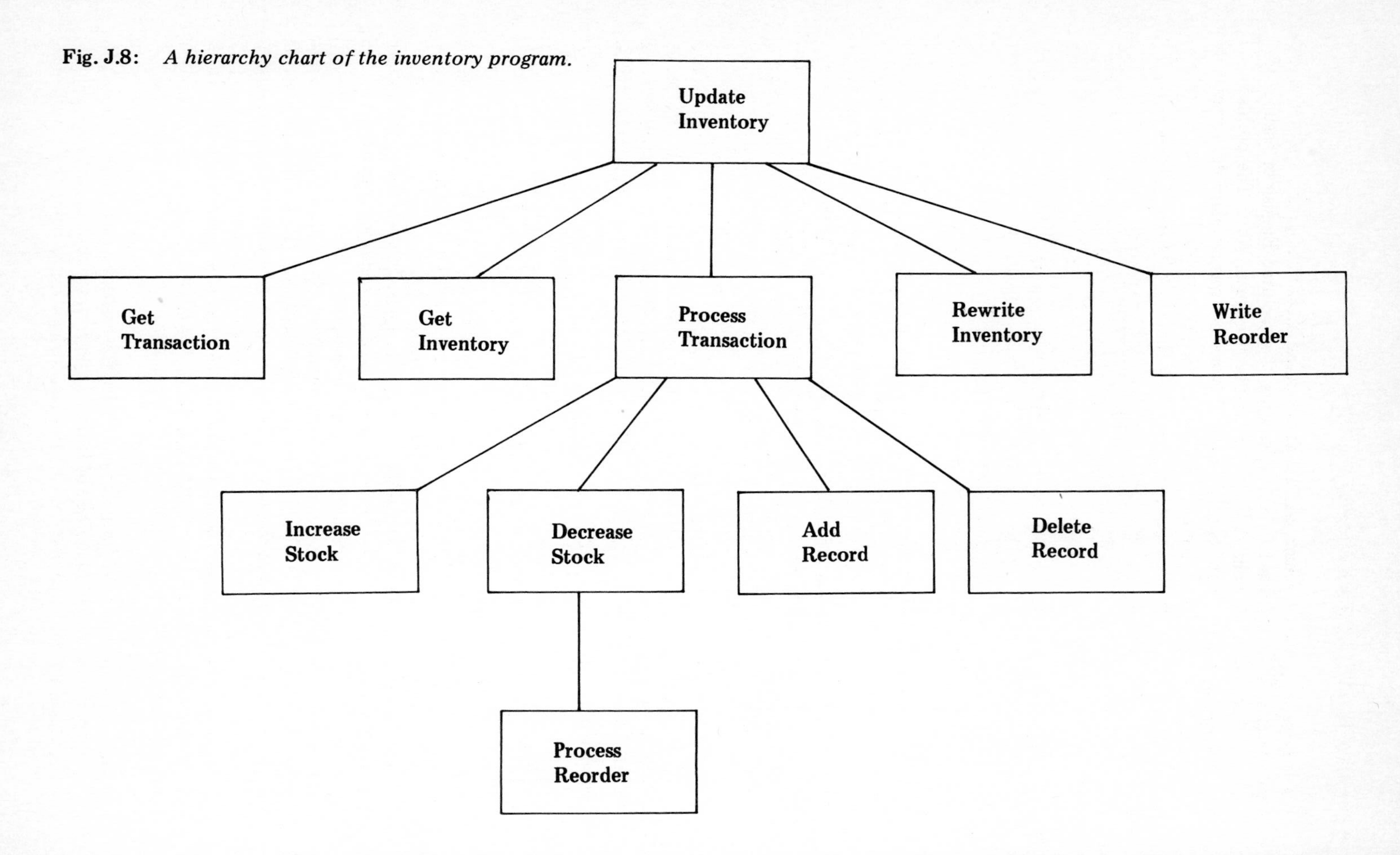

FLOWCHARTING: USES AND LIMITATIONS

A properly prepared flowchart can illustrate logical flow at a glance; in fact, many texts use simple flowcharts to clarify elements of structured English or pseudocode. In spite of this fact, flowcharting has fallen into disrepute. Why? The problem is a long history of misuse. For years, a detailed flowchart was standard documentation for a program, and programmers were required to prepare them. All too often, the flowchart was drawn after the program was written, and did little more than echo the code. A single flowchart tracing the complete logic of a program at a detailed level is too complex to follow, and impossible to maintain; such flowcharts are a waste of time and effort.

In this text, we will not attempt to develop a flowchart of the logic of a complete program. Instead, a hierarchy chart will be used to provide an overview of the program. Each block on the hierarchy chart represents one logical function, and an IPO chart is used to describe the details. For certain logical structures, a flowchart may well be the simplest, clearest way to define a process at the level of the IPO chart. Flowcharts are particularly good for decision-based algorithms, where the number of alternative paths does not exceed two or three. Flowcharts are a waste of time on algebraic algorithms, where no decisions are involved; algebraic expressions, structured English, or pseudo code are much better. (In other words, if you must write the algebraic expression anyway, why draw a box around it?) Complex case structures are better defined with a decision tree or a decision table. We will limit the use of flowcharts to relatively simple decision or control structures, usually at a high level in the hierarchy chart.

Incidently, most flowcharting standards mention two additional symbols: the connector and the off-page connector. We won't use them. If a flowchart is sufficiently complex to require connector symbols, it almost certainly describes too much logic for a single IPO chart. If the problem involves a series of arithmetic steps, then algebra, structured English, or pseudo code are better choices.

SOME EXAMPLES

In Module H, a hierarchy chart for an inventory program was developed; it is reproduced as Fig. J.8. The main control module, Update Inventory, determines the order in which the second level modules are executed; Fig. J.9 shows a flowchart of the process. Note that the first four modules are always executed in simple sequence, but that the Write Reorder module is conditional; a flowchart allows the reader to determine at a glance the logic of such control structures.

The Process Transactions module (Fig. J.10) represents a somewhat more complex control structure; had more than four decisions been involved, we would not have used a flowchart. As we move down to Decrease Stock, we see that the flowchart (Fig. J.11) clearly shows an arithmetic operation followed by a decision. We would probably not use a flowchart for Increase Stock, however, since only the arithmetic operation is required.

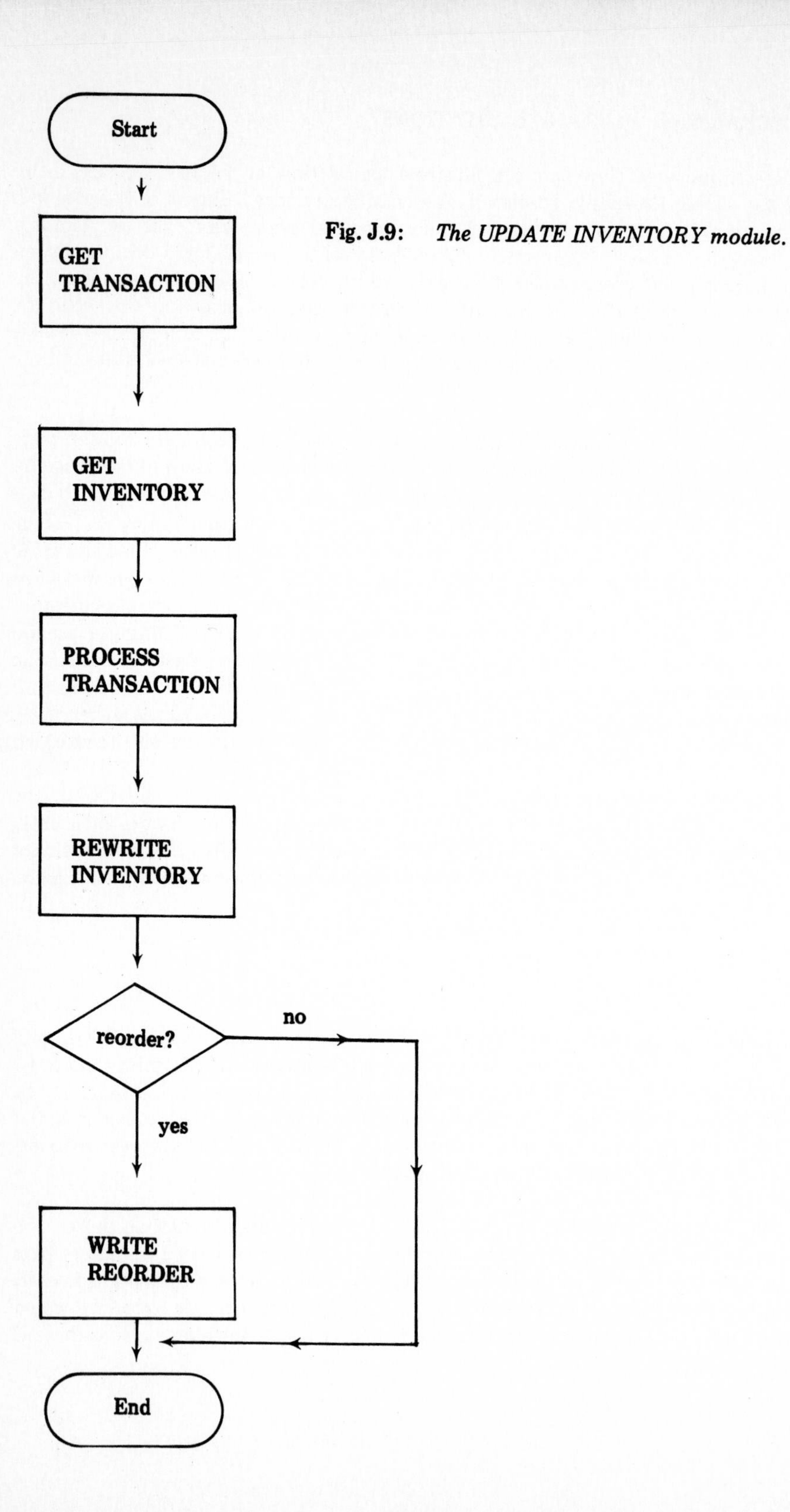

Fig. J.9: *The UPDATE INVENTORY module.*

Fig. J.10: *The PROCESS TRANSACTIONS module.*

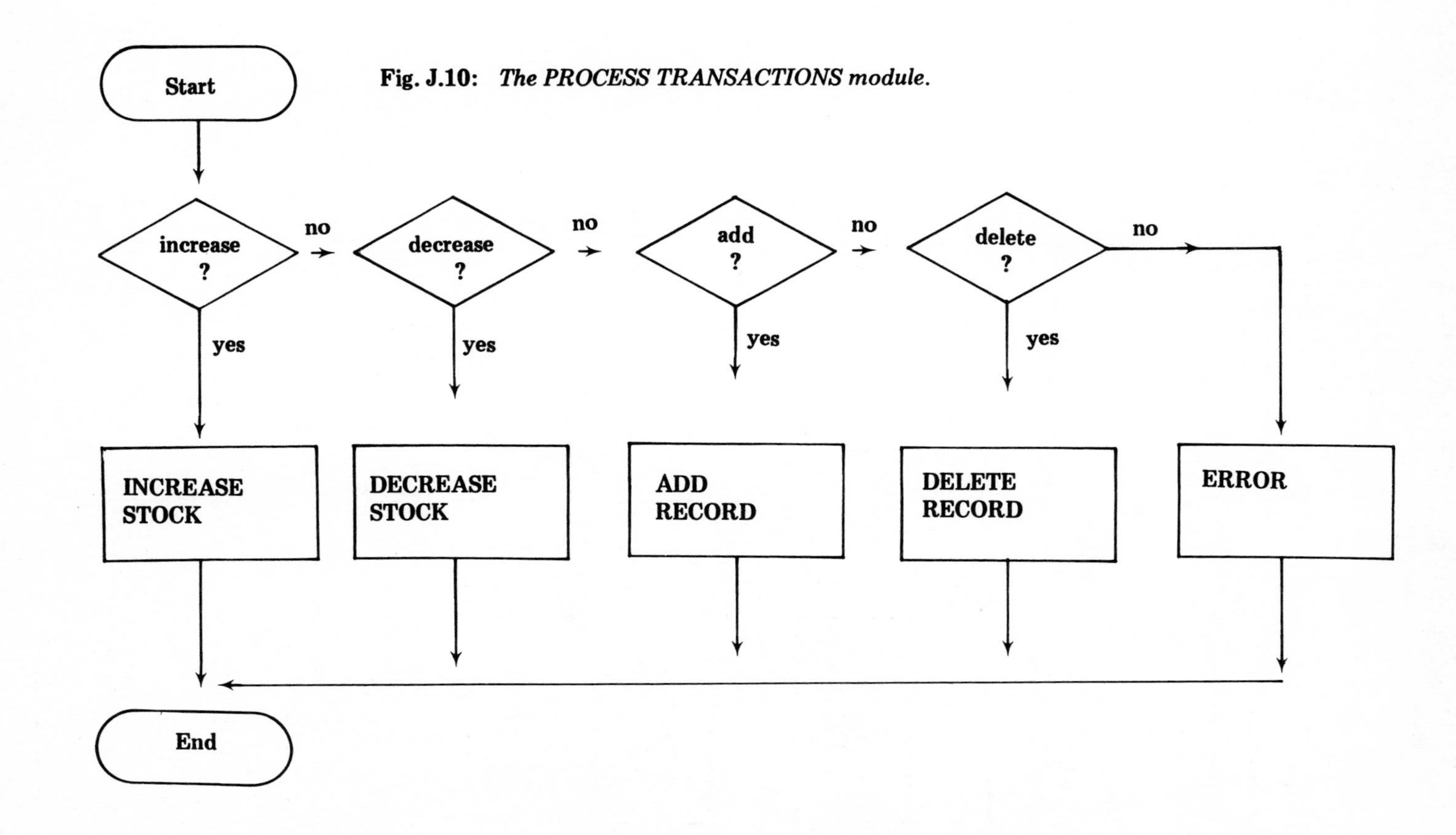

Fig. J.11: *The DECREASE STOCK module.*

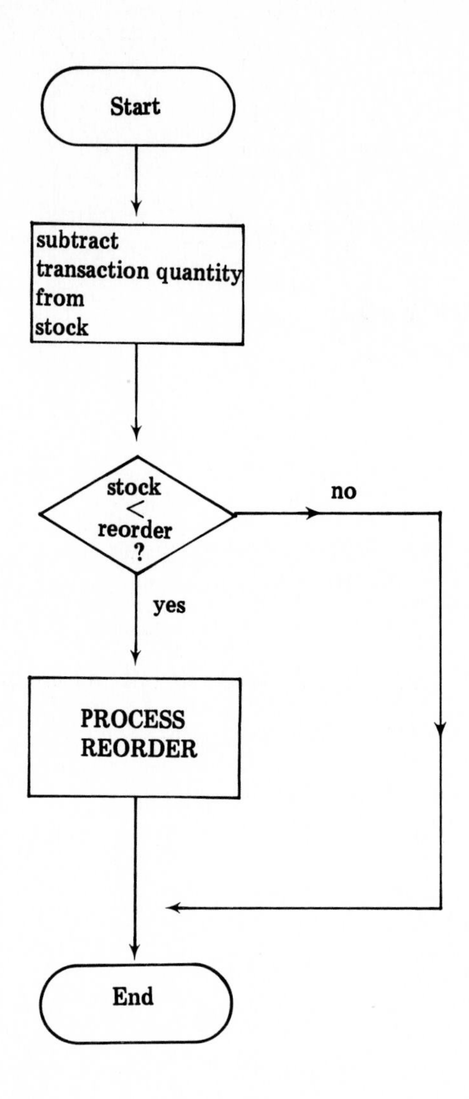

REFERENCES

1. Bohl, Marilyn (1978). *Tools for Structured Design.* Chicago: Science Research Associates, Inc.

Warnier/Orr Diagrams

WITH: *David C. Haddad*

OUTLINE

WARNIER-ORR DIAGRAMS

As an alternative to HIPO, many analysts prefer the Warnier-Orr design methodology. Developed in the early 1970s by Jean-Dominique Warnier and extended to system design by Ken Orr, this technique has enjoyed wide acceptance in France and, recently, in the United States. This introduction is merely an overview of the technique; additional details can be found in the references listed at the end of the module.

At first glance, a Warnier-Orr diagram seems like a hierarchy chart turned on its side. For example, Fig. K.1 shows a portion of a hierarchy chart for a simple program, while Fig. K.2 shows the equivalent Warnier-Orr diagram. Read the diagram from left to right. The program is divided into five high level modules: *Get Transaction*, *Get Master Record*, *Process Transaction*, *Write New Master*, and *Produce Report*. One of these modules, Get Transaction, is itself broken into a number of lower-level modules. The symbol { that groups these low-level modules with their parent is called a brace, though some would call it a bracket or a parenthesis. Note how the structure closely resembles the hierarchy chart.

A hierarchy chart describes only the program structure; it says nothing about the logical flow. A Warnier-Orr diagram does. Consider the sequence in which modules are invoked. Read the Warnier-Orr diagram (Fig. K.2) from top to bottom; *Get Transaction* is executed first, *Get Master Record* second, and so on. How often is each module executed? Note the numbers under the module names in Fig. K.2; they indicate the number of times the module is repeated on each program cycle. For example, each time the program is repeated, *Get Transaction* is executed once, and each time *Get Transaction* is invoked, *Read Transaction* is executed once.

A Warnier-Orr diagram can be used to describe a data structure, a set of detailed program logic, or a complete program structure; it is a flexible tool. Before we illustrate how Warnier-Orr diagrams can be used to help design a system or a program, we will begin with the basics, and discuss data structures and program logic; then we'll consider an example.

Defining Data Structures

A key principle in the Warnier-Orr methodology is that the design and structure of a well-written program are tied to the structure of the data; thus the designer typically begins with the data structures. For example, assume that an application calls for processing library circulation data. Each of a large number (N) of potential borrowers might have several books checked out. For each book, we must keep the call number, title, and author. Figure K.3 shows a Warnier-Orr diagram describing the hierarchical structure of these data. The numbers under the data element names show that data are maintained for N borrowers; that each borrower has K (a variable number of) books; and that for each book, one call number, one title, and one author are kept. A similar structure can be developed for most data.

Defining Program Logic

At a detailed level, a program consists of logic instructions. Any programming task can be defined by combinations of three elementary constructs: sequence, decision, and

Fig. K.1: *A typical hierarchy chart.*

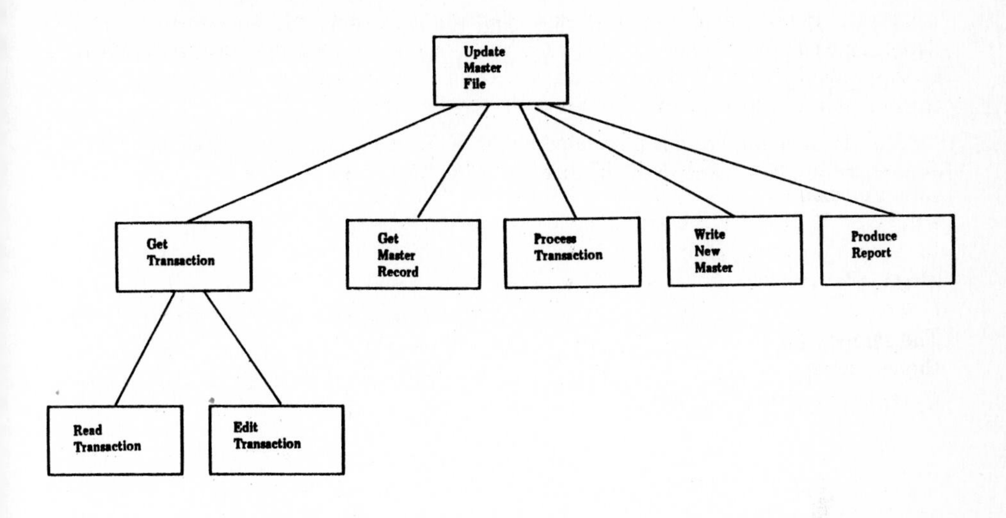

Fig. K.2: *An equivalent Warnier-Orr Diagram.*

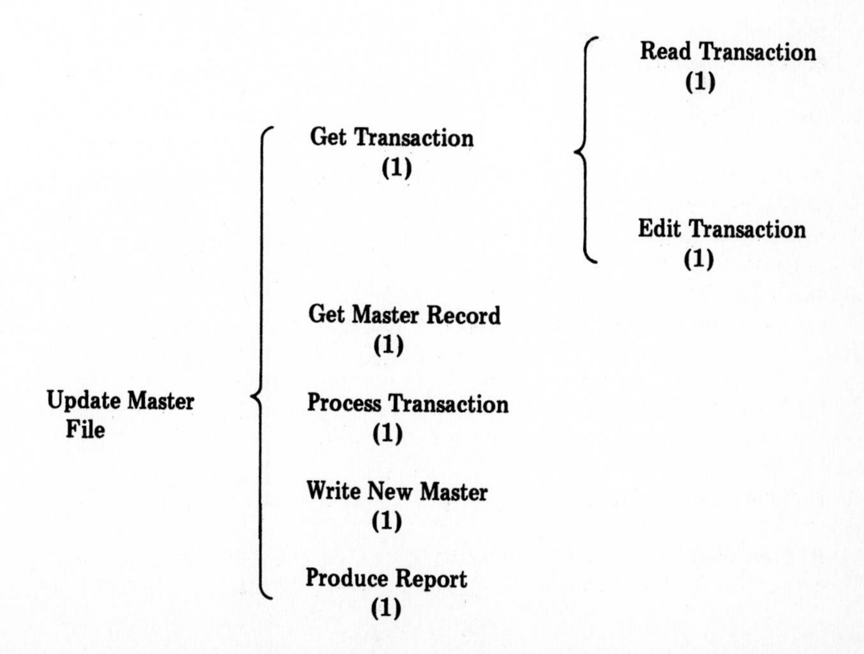

repetition. Let's consider how Warnier-Orr diagrams can be used to show each of these basic program logic blocks.

Figure K.4 shows both a flowchart and the equivalent Warnier-Orr diagram for a sequence operation. The (1) under READ A on the Warnier-Orr diagram indicates that this instruction is executed once.

The decision structure is illustrated in Fig. K.5; once again, a flowchart is included for comparison. Near the top of the diagram is the line:

$$A>0 \left\{ B=\sqrt{A} \right.$$

Under this line are the numbers (0,1), which tell us that the expression $B=\sqrt{A}$ is executed either 0 or 1 times, depending on whether the condition, $A>0$, is true or false. Near the bottom is another line:

$$\overline{A>0} \left\{ \text{WRITE ERROR} \right.$$

Fig. K.3: *The hierarchical structure of library circulation data.*

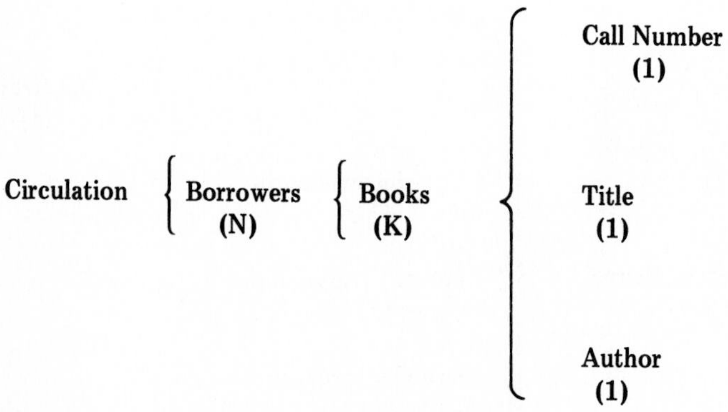

$$\text{Circulation} \left\{ \begin{matrix}\text{Borrowers}\\ \text{(N)}\end{matrix} \left\{ \begin{matrix}\text{Books}\\ \text{(K)}\end{matrix} \left\{ \begin{matrix}\text{Call Number}\\ \text{(1)}\\ \\ \text{Title}\\ \text{(1)}\\ \\ \text{Author}\\ \text{(1)}\end{matrix} \right. \right. \right.$$

Fig. K.4: *Sequence.*

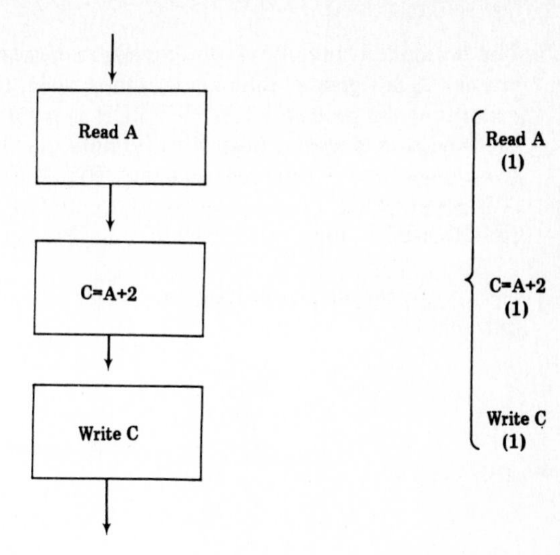

Fig. K.5: *Decision.*

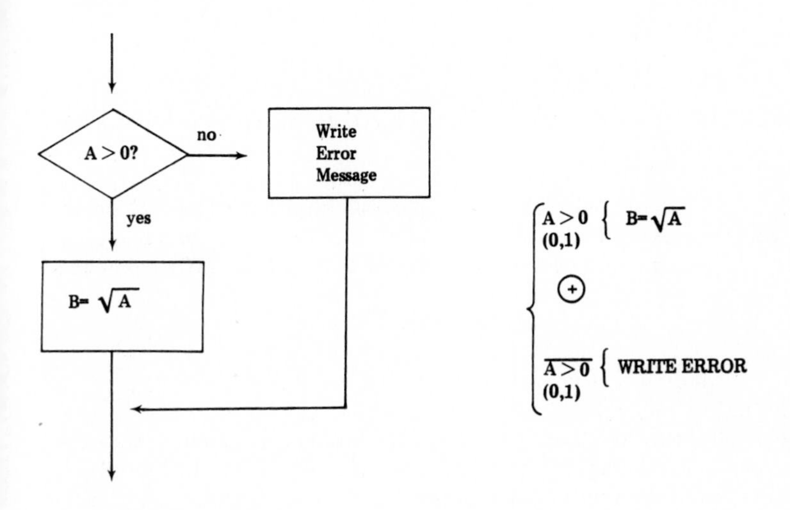

Fig. K.6: *Repetition.*

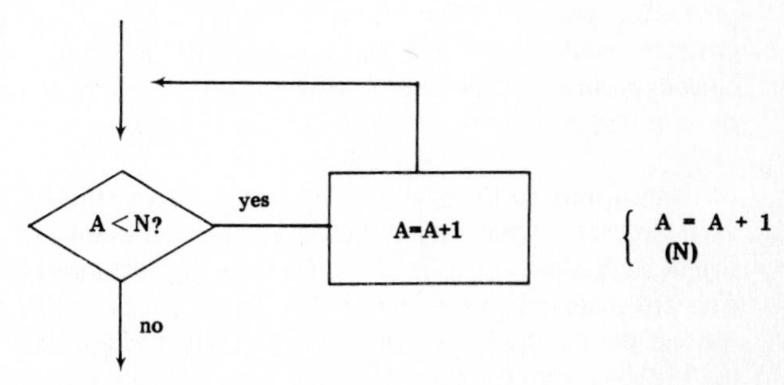

The horizontal line above the condition negates the condition; in other words, $\overline{A>0}$ means "A not greater than zero." Once again, the numbers (0,1) are found below the condition; the process WRITE ERROR is invoked 0 or 1 times depending, on whether the condition is true or false. Finally, note the "plus sign in a circle" (⊕) between the two lines of the Warnier-Orr diagram (Fig. K.5). This is an *exclusive or* symbol. Either A is greater than 0, or it isn't; the two conditions are mutually exclusive. This is how IF-THEN-ELSE logic is represented on a Warnier-Orr diagram.

It is possible to indicate that multiple expressions follow a given condition, for example:

$$A>0 \left\{ \begin{array}{l} B=\sqrt{A} \\ \quad (1) \\ \\ \\ B=B+1 \\ \quad (1) \end{array} \right.$$

Note that a group of statements is enclosed by the brace or bracket that follows the condition.

Figure K.6 illustrates the repetition (DO WHILE) structure. The (N) below A=A+1 tells us that it is executed a variable number of times; if the program logic called for repeating a set of logic 10 times, we would have written (10) below the expression.

AN EXAMPLE

Now that certain basic ideas have been introduced, let's use an example to illustrate how the Warnier-Orr methodology might support the design process. In Module H, the HIPO documentation for an inventory program was developed; we'll use this same example again. Let's assume that the physical implementation chosen for the system calls for two programs, an on-line inventory update program and, an exception report generation program. Five major functions can be identified for the inventory update program, while the report generation program simply reads, formats, and writes the reorder records; these functions are summarized in the system-level Warnier-Orr diagram of Fig. K.7.

According to the Warnier-Orr methodology, the key to designing a program is the data structure. What major data flows are associated with the functions described in Fig. K.7? Basically, transaction and inventory data are read and processed, and reorder data are generated; these major data flows can be added to the diagram (Fig. K.8). If we can define the structure of each of these major data flows, we should be able to use the structures to design the programs.

Fig. K.7: *The high-level structure of the inventory system.*

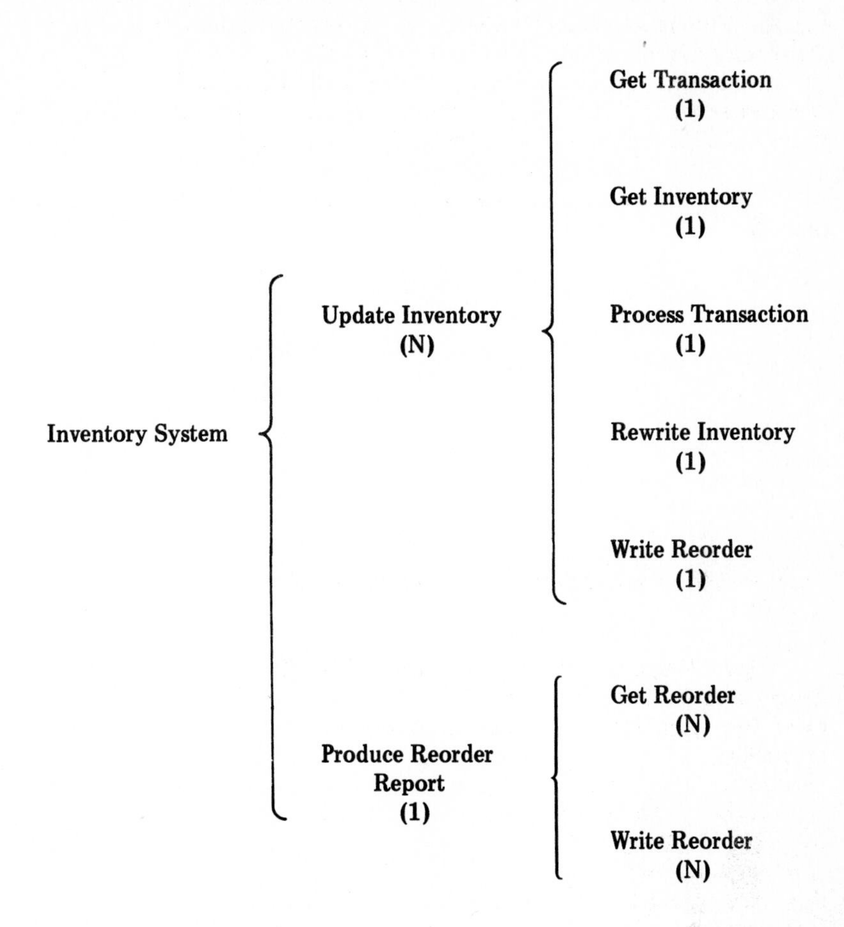

Defining the Data Structure

It is best to start with the output data. Why? Because the objective of the program (or the system) is to produce that output. Any data elements in the output must enter the system through one of the inputs. Thus, by beginning with the output, we define the minimum data the system must contain. The ultimate output from the system is an inventory exception report. What data elements does this report contain? (Note: check the data dictionary.) As a minimum, we can identify the part number, part description, reorder quantity, price, primary supplier, and secondary supplier; the structure of this report is summarized in Fig. K.9.

What is the source of the exception report? The *Produce Reorder Report* program. Where does this program get its data? Figure K.8 shows the input to this program is *Reorder*, so *Reorder* must contain all the data elements found in *Exception Report*. Note that there is one exception report (or one reorder file) containing an unknown

Fig. K.8: *The high-level structure of the inventory system with major inputs, outputs, and data flows.*

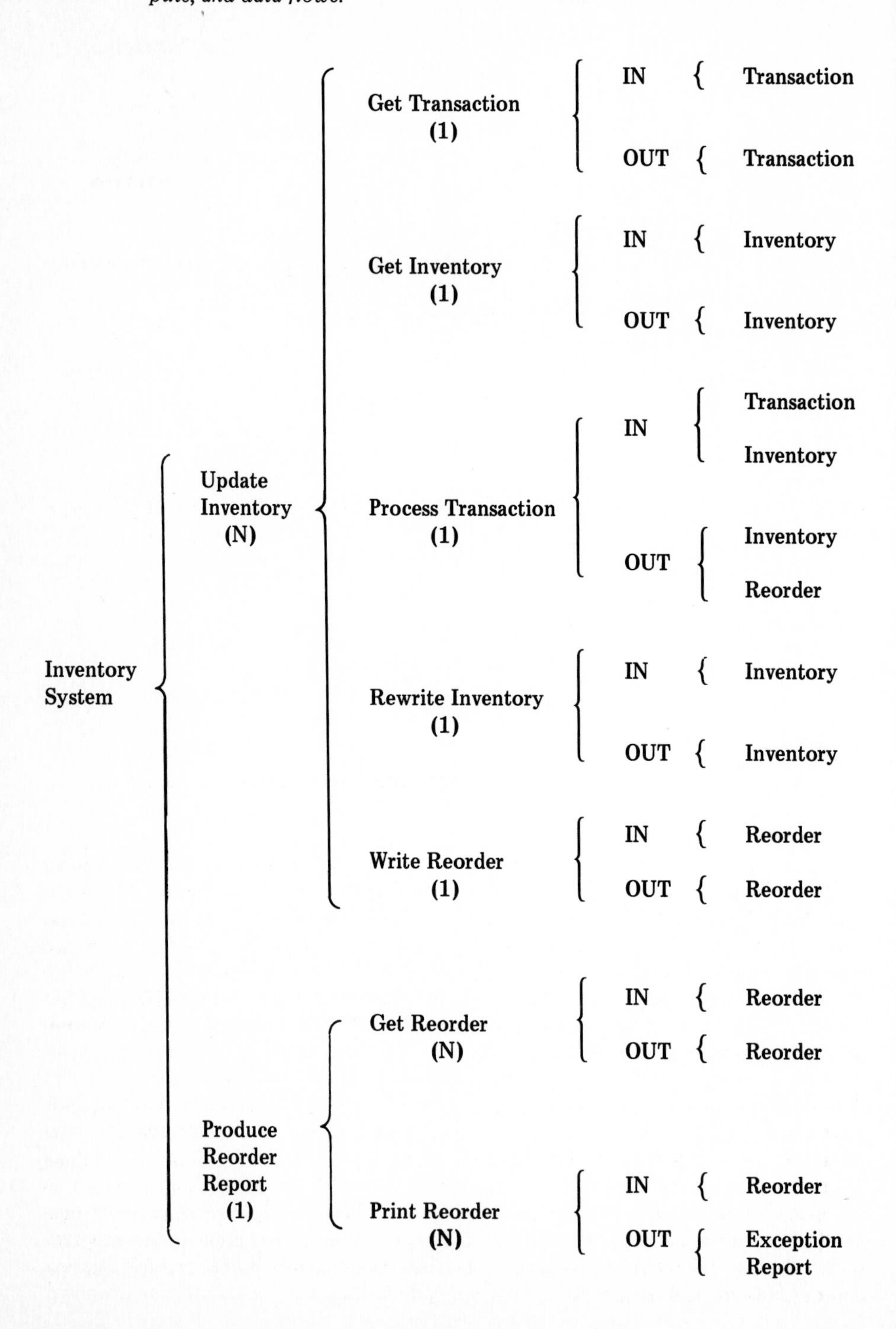

Fig. K.9: *The structure of the Exception Report. Note that the Reorder Data flow has the same structure.*

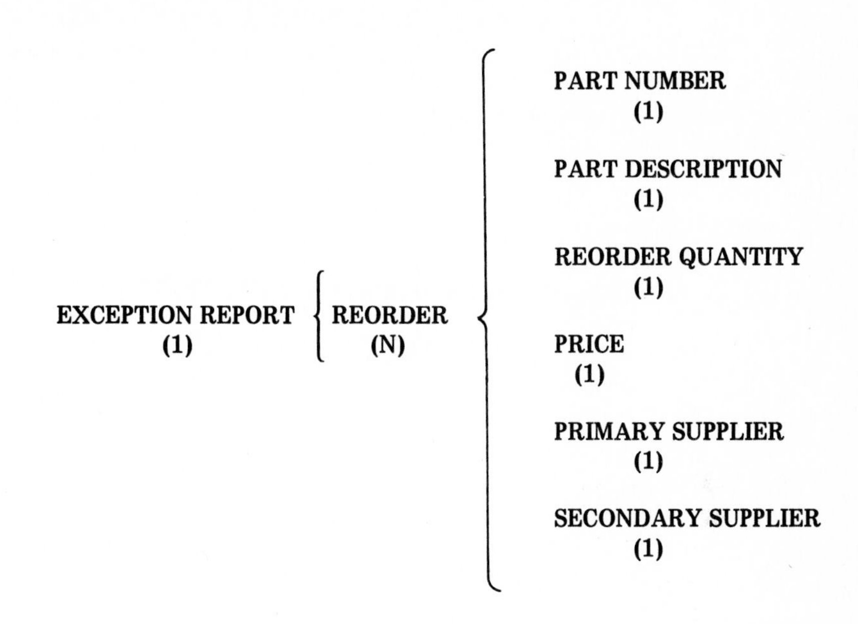

number of reorder records, and that each reorder record contains one part number, one part description, and so on.

What is the source of the reorder flow? Keep reading up the rightmost column of the inventory system diagram until you find a module that produces *Reorder* as its output. We can ignore *Write Reorder*, as its only input is *Reorder*. Look at *Process Transaction*, however. The inputs are *Transaction* and *Inventory*; clearly these two data flows must supply the data elements of *Reorder*. The structure of an inventory record is shown in Fig. K.10; note that the first six data elements match Reorder. Additionally, the system needs the stock on hand and the reorder point, and *Inventory* seems the only reasonable place to put them.

The structure of a transaction is a bit more complex. The system must process four different types of transactions: an increase to stock, a decrease from stock, an addition of a new part, or a deletion of an existing part. The data elements contained in a transaction vary with the transaction type (Fig. K.11). Note that there is one transaction file containing Q transactions. A transaction can be any one of four types; the (0,1) below each transaction type indicates that the substructure can occur either 0 or 1 times, and the exclusive or symbols imply that for any given transaction, one and only one of these structures can exist.

Fig. K.10: *The structure of an INVENTORY record.*

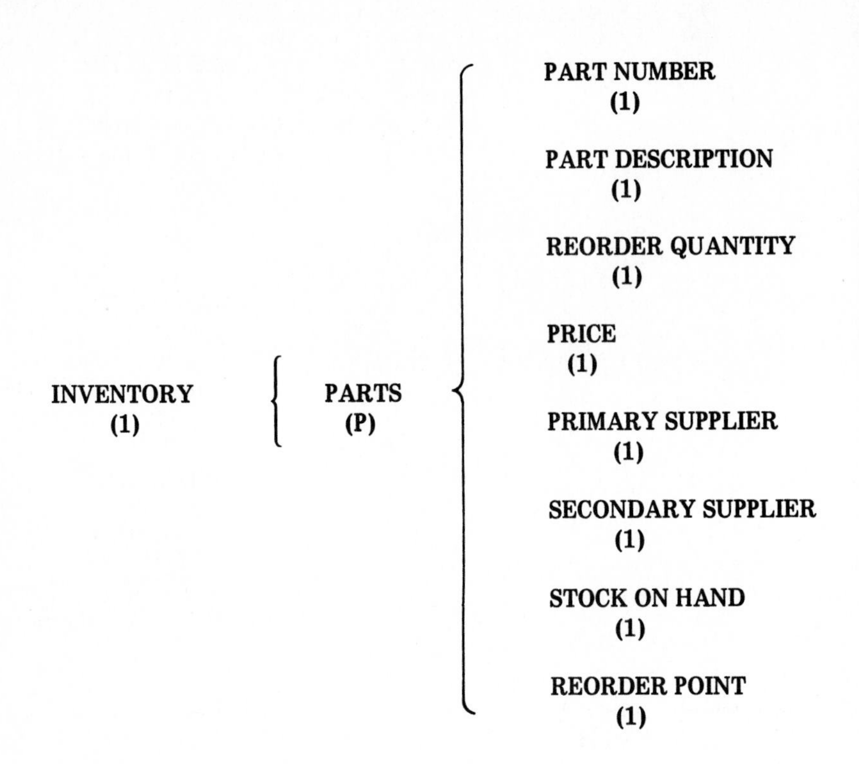

The Report Generation Program

Now that the data structures have been defined, we can begin to design the programs. We'll start with the report generation program. Ideally, the design of a program should be based on the structure of its input data. The only input to the report generation program is *Reorder*, which has the same structure as the output exception report (Fig. K.9). There is one reorder file containing N reorders; that suggests a repetitive data structure. Move over to the decomposition of *Reorder*, and note that all the data elements occur only once. The data structure suggests a very simple program that repeats until there are no more reorders; we are looking at a read, format, and write loop.

The program itself is described in Fig. K.12. It is divided into three primary functions—begin, process, and end. The processing loop repeats N times, once for each reorder. Read the Warnier-Orr diagram from left to right and from top to bottom; you should have little trouble relating the structure to a simple report generation program. As an exercise, write the pseudocode or structured English for this program.

Fig. K.11: *The Transaction structure.*

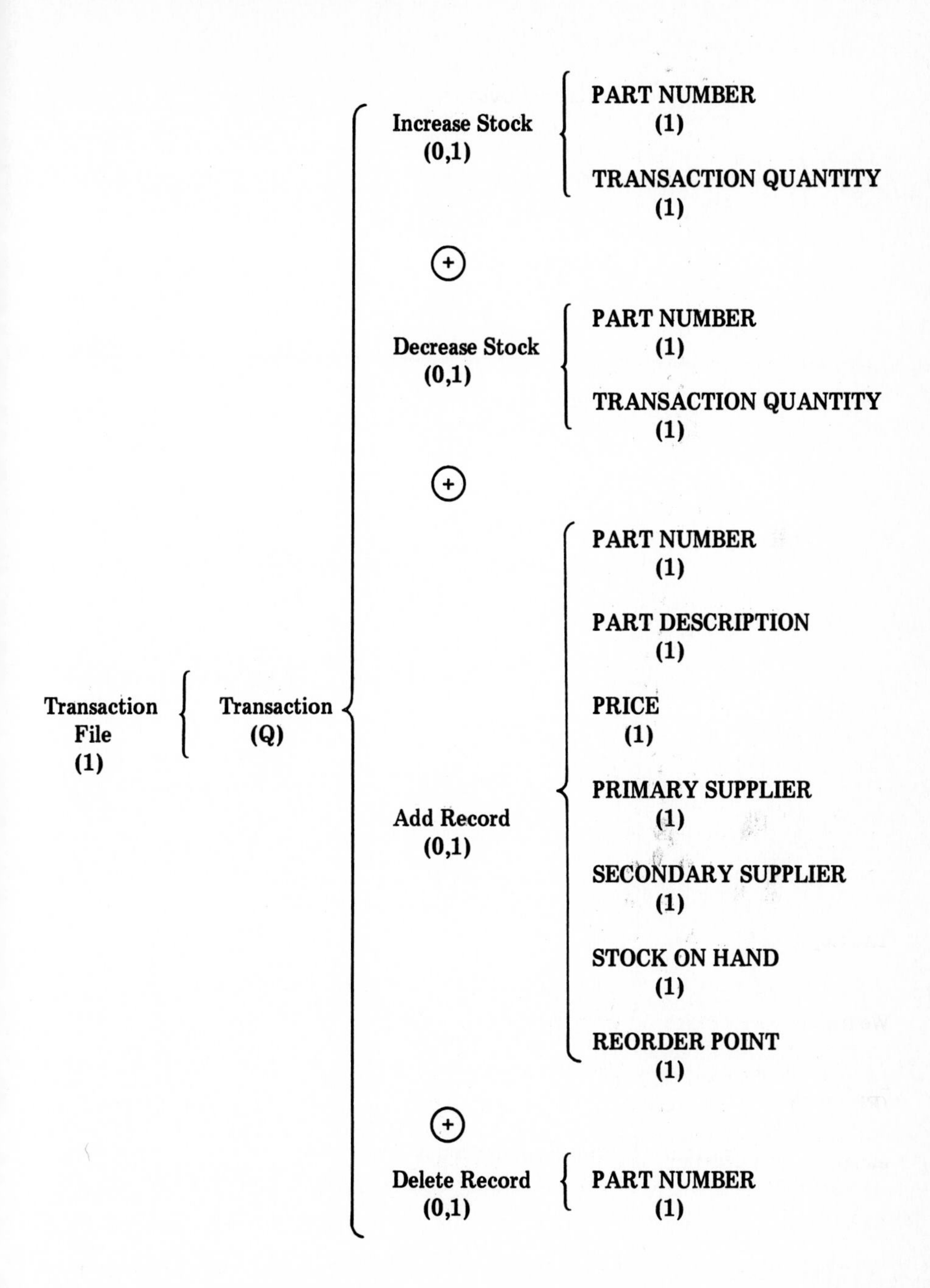

Fig. K.12: *The Produce Reorder Report program.*

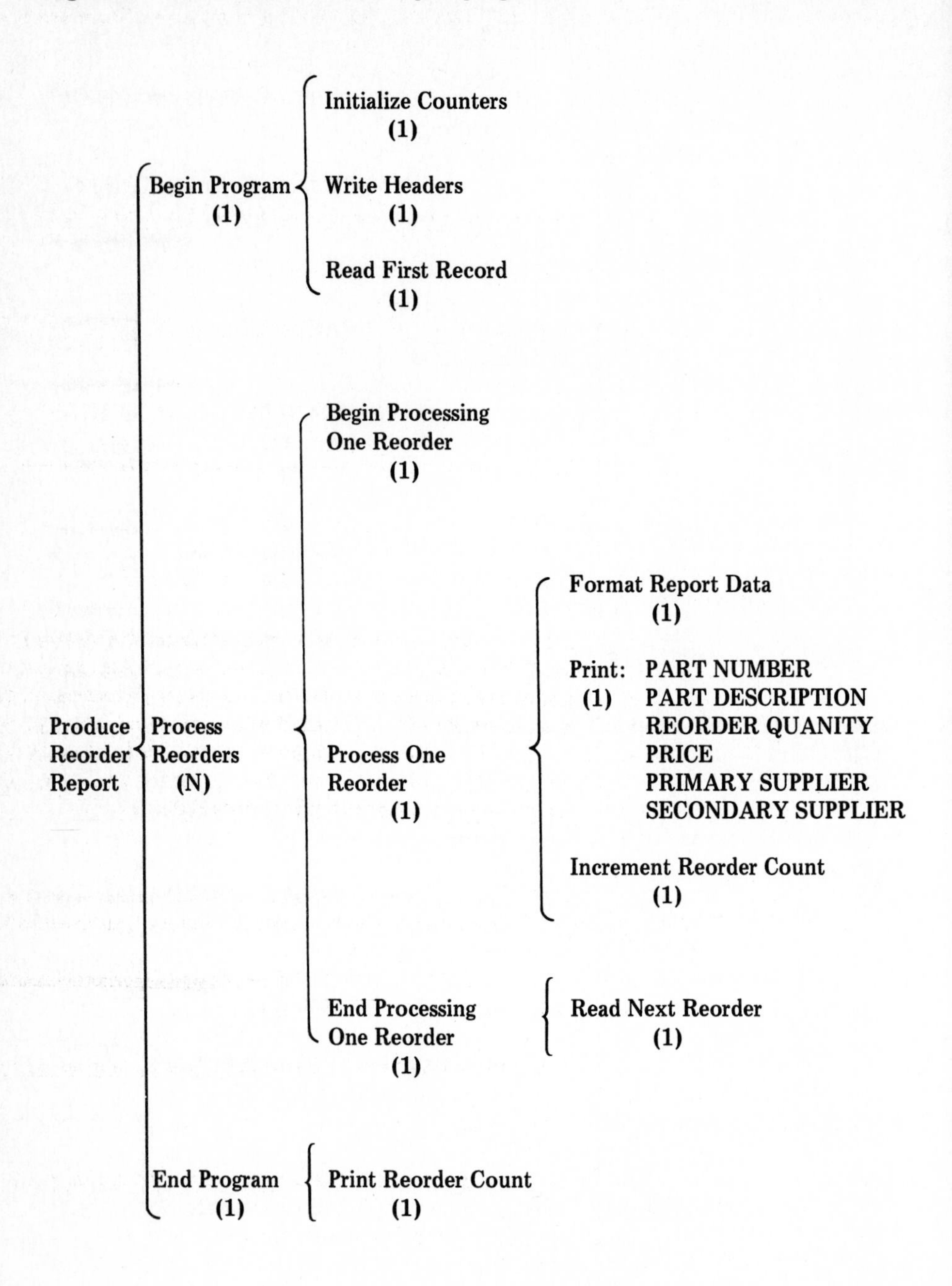

Incidently, if there is a difference in the structures of the input and the output data, which structure should be used as a model for the program? Normally, the input data structure is used. Why? Outputs are frequently subject to change. If the output data structure is used as a model for the program, a minor change in the data could require a major program modification. Generally, the input data are more stable.

The Inventory Update Program

Next we turn to the inventory update program. This on-line program will read a transaction, update the inventory master record associated with that transaction, and, where appropriate, identify and write reorder records. The primary input is a transaction (Fig. K.11). Look at the structure of this data flow. At the highest level, the number of transactions is unknown, suggesting that the main control module will contain repetitive logic and loop until there are no more transactions. The next level involves a decision, with a link to one of four lower-level modules, depending on the transaction type. (Note: a typical on-line program resides in main memory. A polling signal alerts the program when a transaction is ready to be processed; you might view the logic as a continuous loop, with the program sitting in core and waiting for input.)

Are there any other inputs to the inventory update program? Yes, *Inventory*. Look at the structure of an inventory record (Fig. K.10). Except for the lack of a decision block, it resembles the *Transaction* structure; there is no apparent incompatibility in the two data flows. The output is *Reorder*, and its structure also resembles *Transaction*. Thus we can safely use the structure of *Transaction* as a model in designing the program (Fig. K.13).

Let's follow the Warnier-Orr diagram as a single transaction is processed. (Remember to read the diagram from top to bottom and from left to right.) We'll assume that any necessary initialization functions have been done; thus we begin near the middle of the second column from the left (Fig. K.13) with *Transaction*. What steps are involved in processing a transaction? Move one column to the right. We begin processing by getting the transaction. Next, assuming that the transaction does not add a new record to the inventory file (note the decision logic), we read an inventory record. We then process the transaction; let's assume it's a decrease in stock. The decision logic (moving to the right) determines the transaction type; a decrease stock transaction means that we subtract the transaction quantity from the stock on hand and then check to see if the stock has fallen below the reorder point. Based on this test, the reorder flag is set to either 0 or 1. Once the transaction has been processed, the next major step (note that we are back to column three) is to rewrite the inventory record. Finally, if the reorder flag is set to 1, we write a reorder record. This ends the processing of one transaction.

Obviously, many details have been ignored in developing the sample Warnier-Orr diagram; for example, we did not include a transaction type to modify any of the fields (other than STOCK ON HAND) of an existing master record. This was, after all, an introductory-level example, and we did not want to bury the concept in unnecessary details. How much detail should be included? It is a good idea to remember the purpose of a Warnier-Orr diagram: to design and develop the structure of the system or the program. The intent is not to include every programming detail. We want to map the system, not implement it. Given enough practice, you will develop your own style

Fig. K.13: *The UPDATE INVENTORY program.*

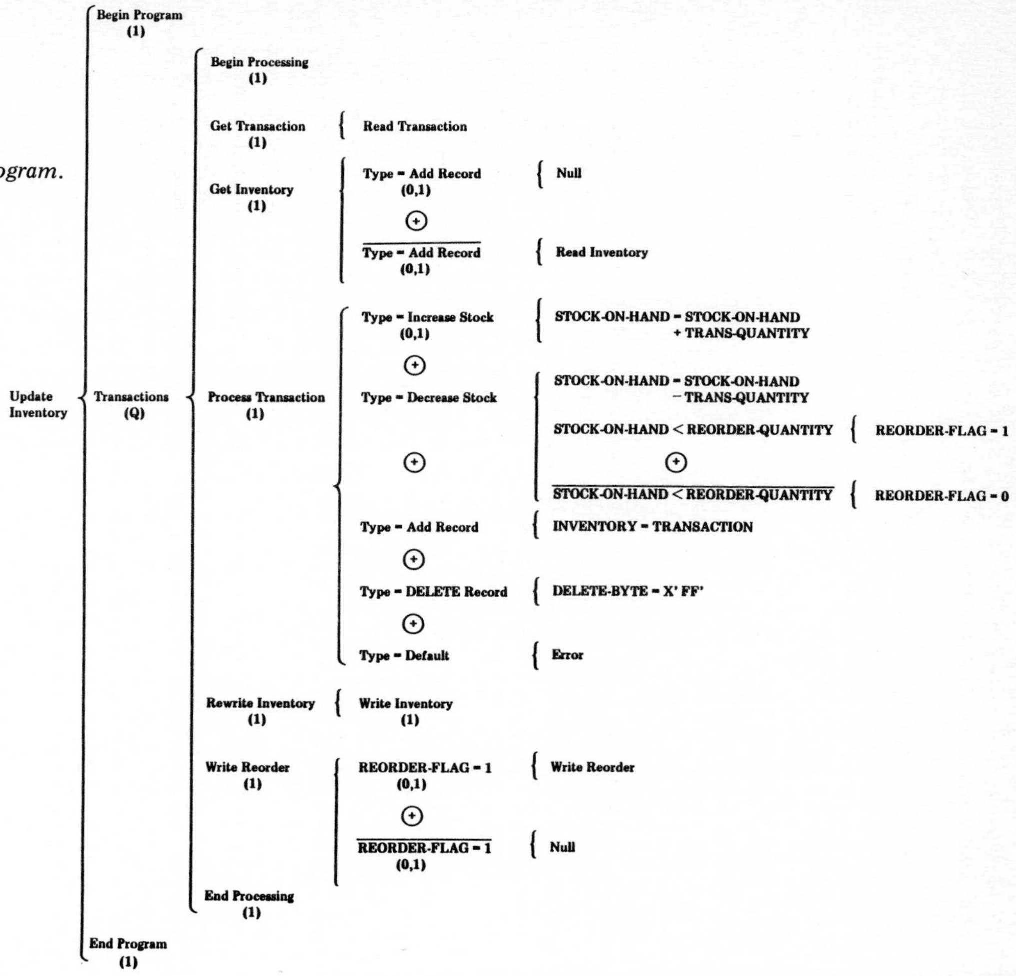

- **Update Inventory**
  - **Begin Program** (1)
  - **Transactions** (Q)
    - **Begin Processing** (1)
    - **Get Transaction** (1)
      - Read Transaction
    - **Get Inventory** (1)
      - Type = Add Record (0,1)
        - Null
      - ⊕
      - $\overline{\text{Type = Add Record}}$ (0,1)
        - Read Inventory
    - **Process Transaction** (1)
      - Type = Increase Stock (0,1)
        - STOCK-ON-HAND = STOCK-ON-HAND + TRANS-QUANTITY
      - ⊕
      - Type = Decrease Stock
        - STOCK-ON-HAND = STOCK-ON-HAND − TRANS-QUANTITY
        - STOCK-ON-HAND < REORDER-QUANTITY
          - REORDER-FLAG = 1
        - ⊕
        - $\overline{\text{STOCK-ON-HAND < REORDER-QUANTITY}}$
          - REORDER-FLAG = 0
      - ⊕
      - Type = Add Record
        - INVENTORY = TRANSACTION
      - ⊕
      - Type = DELETE Record
        - DELETE-BYTE = X' FF'
      - ⊕
      - Type = Default
        - Error
    - **Rewrite Inventory** (1)
      - Write Inventory (1)
    - **Write Reorder** (1)
      - REORDER-FLAG = 1 (0,1)
        - Write Reorder
      - ⊕
      - $\overline{\text{REORDER-FLAG = 1}}$ (0,1)
        - Null
    - **End Processing** (1)
  - **End Program** (1)

and learn to include enough detail to support writing the pseudocode or structured English, but not so much as to obscure the design.

Expect to complete several drafts of your Warnier-Orr diagram before an acceptable design emerges; no one, not even the most experienced systems analyst, can design a system in one shot. Start with rough sketches on a chalkboard or with paper and pencil; you will be more likely to throw away your false starts and begin again if your initial diagrams are rough. Only when the system begins to make sense should you attempt to develop neat Warnier-Orr diagrams.

STRUCTURE CLASH

In our sample program, the structures of the input and output data were similar, and thus the design of the program was straightforward. Often, with more complex programs, the structures of the input and output data differ radically; this is called a *structure clash*. When the structures clash, it is difficult to design a program. A common solution is to create an intermediate data store or data file with a structure that does not clash with either the input or the output. The program can then be split in two, with the input processed to intermediate form, and the intermediate data subsequently processed into the desired output.

SCIENTIFIC PROGRAMMING

A common problem in scientific programming is that neither the input nor the output data have a structure; the input is often an array of numbers, and the output a single value or a set of statistics. The key to such programs is usually a mathematical model or algorithm. The design of such programs should be based on the structure of the algorithm rather than the structure of the data; Warnier-Orr diagrams can still prove useful, however.

REFERENCES

1. Higgins, David A. (1979). *Program Design and Construction.* Englewood Cliffs, New Jersey: Prentice-Hall, Inc.

2. Orr, Kenneth T. (1981). *Structured Requirements Definition.* Topeka, Kansas: Ken Orr and Associates, Inc.

3. Orr, Kenneth T. (1977). *Structured Systems Development.* New York: Yourdon, Inc.

4. Warnier, Jean-Dominique (1976). *The Logical Construction of Programs.* New York: Van Nostrand Reinhold.

5. Warnier, Jean-Dominique (1978). *Program Modification.* London: Martinus Nijhoff.

Module L

PERT and CPM

WITH: *Teruo Fujii*

OUTLINE

PROJECT MANAGEMENT

Managing a complex project can be very difficult. Often, the best approach is to break it into a series of small tasks that are more easily controlled or managed. The danger with such subdivision is that it becomes too easy to focus on the individual tasks or activities, thus losing sight of the whole project. A tool is needed to support the subdivision into smaller tasks or activities, while preserving an overall view. PERT and CPM are two of the better known techniques.

PERT (Program Evaluation and Review Technique) gained prominence during the late 1950s when it proved invaluable in scheduling and controlling the highly complex Polaris missile project. It is particularly useful in research and development projects where the times required to complete the various activities are uncertain. CPM (Critical Path Method), on the other hand, is most useful in situations where the duration of the project can be controlled by increasing or decreasing resources; for example, by pouring more money into a car, it might be possible to finish a 500-mile race in a shorter time. Basically, PERT should be used when the activity times are highly uncertain, and CPM when prior history or experience makes the activity times relatively well-defined.

The key to both PERT and CPM is the project network. The basic idea is to break the project into logical steps or activities, and then to order these activities into a sequence. The project network is a graphic tool for describing this sequence.

AN EXAMPLE

Perhaps the easiest way to gain an understanding of how to develop and use a project network is through an example. Imagine that you have been assigned the task of painting an old-fashioned, rectangular, wooden army barracks. You'll have fifteen soldiers under your command. The job involves three distinct steps. First, the old, loose paint must be scraped from the walls. Next, the new paint is applied. Finally, razor blades are used to remove excess paint from the window panes. The problem is compounded by limited materials; you'll have only five scrapers, five paint brushes, and five razor blades.

One way to accomplish the task might be to scrape all the walls first, then paint all the walls, and finally remove the excess paint from all the windows. This would be very inefficient, however. We have fifteen soldiers, and only five of any one tool; thus, if we do one task at a time, ten soldiers will always be idle. Another, less obvious problem arises from the fact that the best time to remove paint from a window pane is shortly after it has been applied; after the paint has hardened, removing it becomes much more difficult. There must be a better way.

Why not work on one side at a time? The first logical activity is scraping the loose paint; thus we'll have five soldiers pick up the scrapers while the other ten relax. As soon as the first side has been scraped, the first five soldiers can move on to side two, while a second contingent picks up the paint brushes and begins to apply the paint. Now, only five soldiers are resting. Later, as the scrapers move to side three and the painters to side two, the last five soldiers can begin to remove the excess paint with the razor blades. With everyone working, the job should be finished in less time.

The barracks is rectangular, with the long sides twice as long as the short. Also, the jobs are not equivalent; painting takes longer than scraping, which takes longer than razoring. Let's assume that we have estimated the activity times shown in Fig. L.1, and develop a simple *Gantt chart* or *bar chart* (Fig. L.2) to illustrate a plan for completing the work. Note that scraping begins at time zero, and ends twelve hours later. Painting begins at the end of the second hour, after the first side has been scraped, and ends eighteen hours later at the end of hour twenty. Razoring is dependent upon painting; it makes little sense to remove paint from windows until after the paint has been applied. Note the intermittent activity of the razor wielders as indicated on the Gantt chart. The relationship between painting and razoring is not explicit, but implied.

Read the Gantt chart. A horizontal line following an activity shows when a given group of soldiers is active; the absence of a line denotes a rest period. The total elapsed time for the project is twenty-two hours. How long would it take without planning? Scraping would take twelve hours, painting eighteen, and cleaning six more, for a total time of thirty-six hours. Planning and coordination yield definite benefits.

Fig. L.1: *Time estimates (in hours) for painting a barracks.*

	Scraping	Painting	Razoring
Sides 1 and 3	2	3	1
Sides 2 and 4	4	6	2

Fig. L.2: *A Gantt chart or bar chart for the barracks painting project.*

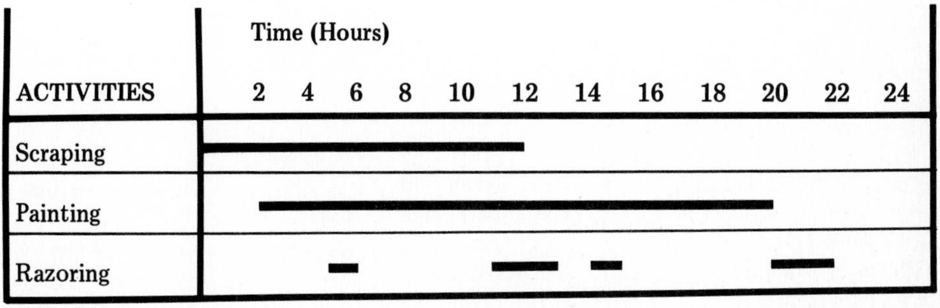

ACTIVITIES	Time (Hours) 2	4	6	8	10	12	14	16	18	20	22	24
Scraping												
Painting												
Razoring												

The Project Network

The *project network* (Fig. L.3) extends the capabilities of the bar chart. Like the bar chart, it depicts activities, and their starting and completion times. However, the project network explicitly illustrates how activities depend on each other, whereas the bar chart simply implies this activity dependence; this is what makes the project network so much more powerful as a tool for systems analysis and design.

A project network is quite easy to follow. The activities (painting, scraping, and razoring) are represented by lines or arrows, while *events* (the beginning or end of an activity) are represented by circles. For example, consider scraping the first side (Fig. L.3). It begins with event 1, and ends with event 2. The activity itself is identified by the numbers of its beginning and ending events; thus scraping the first side is activity 1-2. An event has no time duration and requires no resources; it is merely a definable instant in time. Activities, on the other hand, consume both time and resources.

Fig. L.3: *A project network for the barracks painting project.*

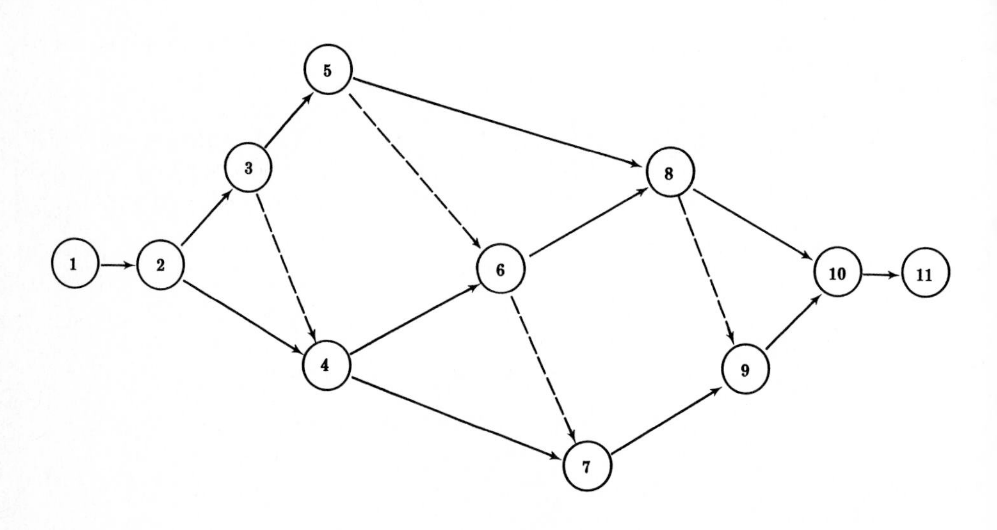

1-2 scrape side 1
2-3 scrape side 2
2-4 paint side 1
3-5 scrape side 3
4-6 paint side 2
4-7 razor side 1

5-8 scrape side 4
6-8 paint side 3
7-9 razor side 2
8-10 paint side 4
9-10 razor side 3
10-11 razor side 4

Dummy activities
3-4
5-6
6-7
8-9

Follow the project network of Fig. L.3. Begin at the left. Activity 1-2 represents scraping the first side. When this activity is completed, two other activities can begin: 2-3 represents scraping the second side, and 2-4 represents painting the first side. Note that event 2 marks the end of activity 1-2 *and* the beginning of activities 2-3 and 2-4.

Focus on event 4, and note the dashed line linking events 3 and 4. Event 3 marks the end of scraping side two. Event 4 marks the end of painting the first side. Event 4 also marks the beginning of activity 4-6, paint side two. Before painting can begin on side two, both scraping side two (event 3) and painting side one (event 4) must be completed; there is a dependency relationship between events 3 and 4. Activity 3-4 shows this relationship, but consumes no time and no resources. It is a *dummy activity*.

Read through the rest of Fig. L.3, using the activity descriptions printed below the graphics as a key. Try to explain the reason for each of the dummy activities.

Dummies can also be used to link parallel activities in a network. Consider, for example, Fig. L.4, which shows a simple project network for starting an automobile. Activity 5-6 represents fastening seat belts. When must the belts be fastened? On most cars, at any time. Near the top of Fig. L.4 is a project network that shows event 5 "hanging out there," with no apparent relationship to the best of the network. Below it is another version with a dummy activity, 2-5, added. It makes the network a bit easier to read.

Fig. L.4: *Dummy activities can be used to link parallel activities to the project network.*

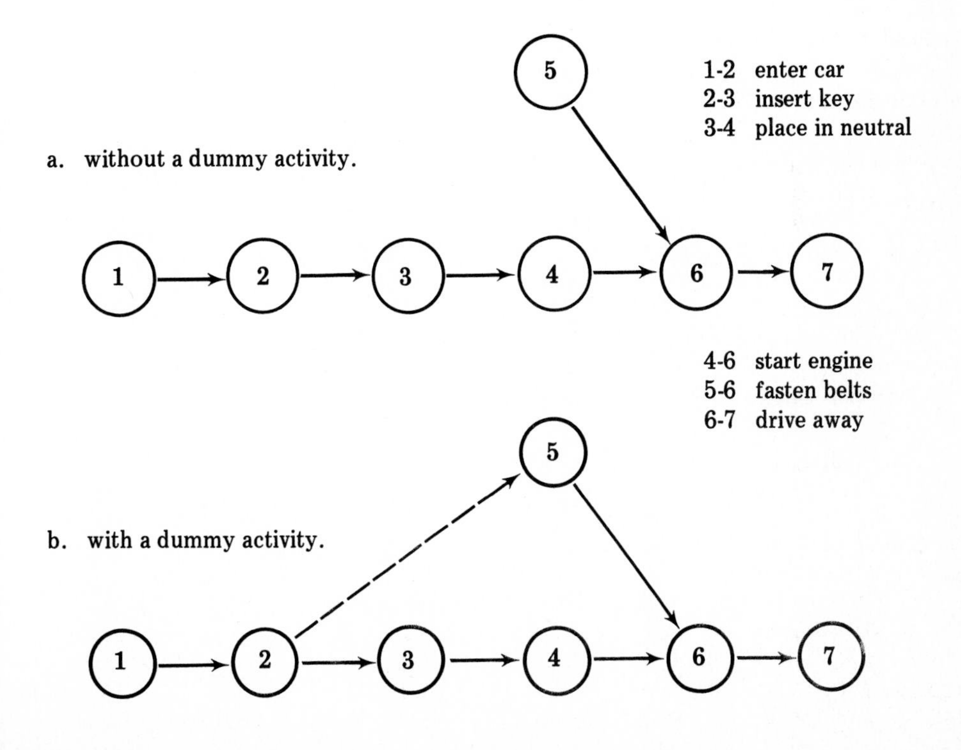

Estimating Activity Times

The project network shows the relationships between the various activities and events. At a glance, it is possible to see how a given event depends upon the completion of one or more activities. The next step is to estimate the activity times.

In most practical situations, an experienced systems analyst can make a reasonably accurate estimate of the probable duration of each activity. By laying out the project network, the analyst is breaking the project into a series of relatively brief, discrete steps, and estimates made on such small elements are often quite accurate. It is possible, however, that an analyst may feel uncertain about an activity's likely duration. When this happens, the PERT technique suggests using three different estimates: optimistic, most likely, and pessimistic. (Perhaps several different experts might be asked to supply a best guess, and the extremes used along with the average.) Once the three estimates have been generated, they can be plugged into the following formula:

$$\text{Event time} = \left[\frac{\textit{Optimistic} + (4 * \textit{most likely}) + \textit{pessimistic}}{6}\right]$$

to calculate the event times to be used in the project network.

Let's add time estimates (Fig. L.1) to the project network for the barracks painting problem; the new project network is shown in Fig. L.5. Note that there is no relationship between the length of an arrow representing an activity and the time estimate associated with it. The arrows show dependency relationships, while the number above the arrow represents activity time.

Scheduling the Project

Once the project network is completed, the analyst can begin to use it as an aid to scheduling. The first step is to compute two statistics for each event: the *earliest event time (EET)* and the *latest event time (LET)*.

The earliest event time represents the earliest time an event can possibly begin; conventionally, it is zero for the first event. For all others, the EET is computed by using three simple rules:

1. Select all activities entering the event.

2. For each entering activity, sum the activity's duration and the EET of its initial event.

3. Select the highest EET obtained.

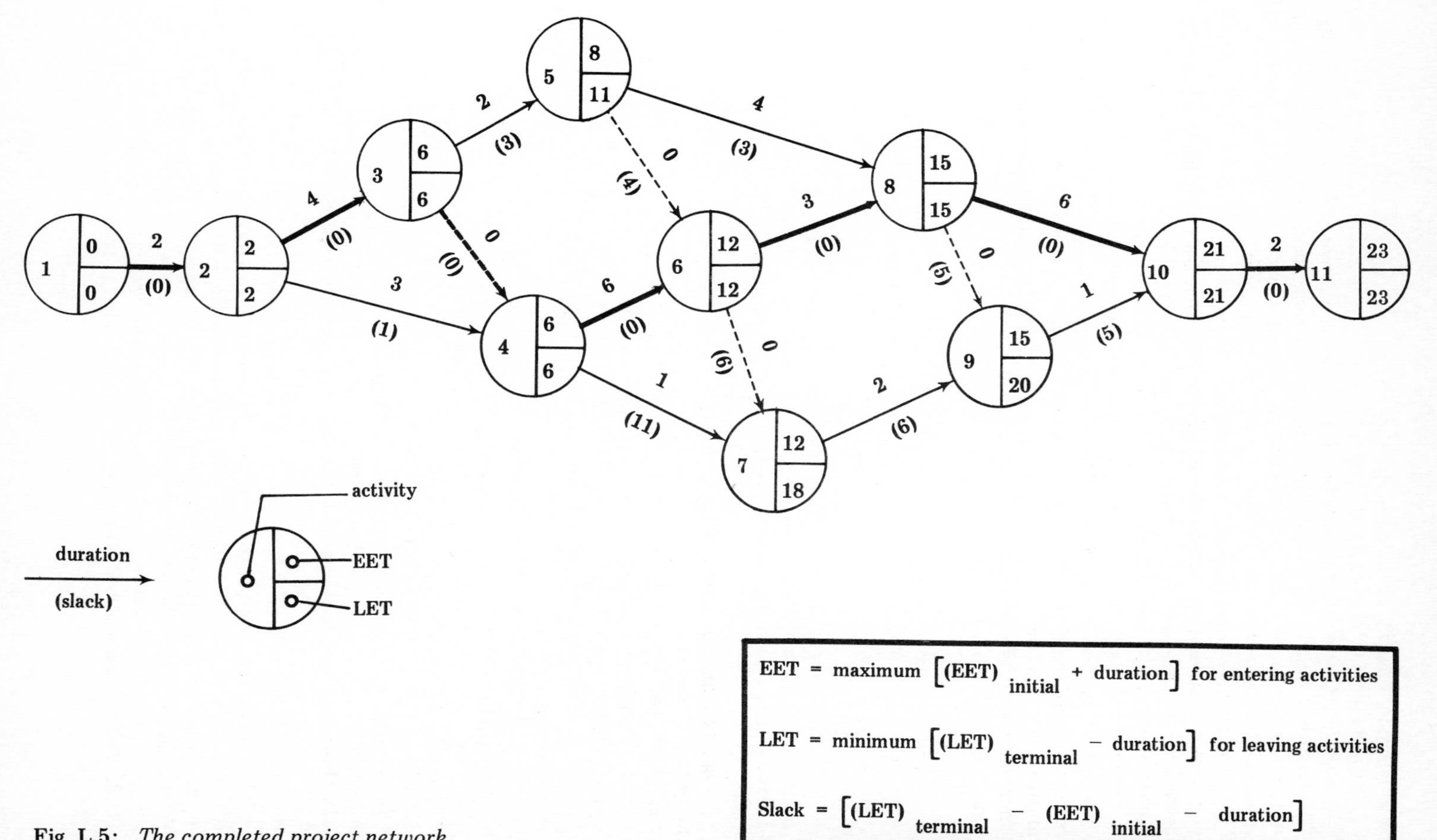

activity
duration
(slack)
EET
LET
EET = maximum [(EET) initial + duration] for entering activities
LET = minimum [(LET) terminal − duration] for leaving activities
Slack = [(LET) terminal − (EET) initial − duration]

Fig. L.5: *The completed project network.*

For example, event 2 has only one entering activity, 1-2 (see Fig. L.5). Logically, what is the earliest time this event can begin? When activity 1-2 is completed. The earliest event time for event 1 is (by definition) zero. The duration of activity 1-2 is estimated at two hours. Activity 1-2 cannot begin before time zero, and will consume two hours; thus it cannot possibly end before time two. Event 2 marks the end of activity 1-2; thus the EET for event 2 must be two hours after the project begins.

Let's consider another example of the earliest event time calculation; we'll focus on event 7. There are two entering activities: 4-7 and 6-7. Prior calculations have shown that event 4 cannot begin before hour six; since activity 4-7 takes one hour, it cannot possibly be finished before hour seven. We also know that event 6 cannot begin before hour twelve, and that activity 6-7, a dummy, requires zero hours; thus it cannot possibly be finished before hour twelve. Event 7 represents the completion of activities 4-7 and 6-7. The former might be completed by hour seven, but the other can't be finished until twelve hours have passed. Event 7 represents the completion of *both*; thus its earliest event time must be twelve, the higher number.

The computed earliest event time for each activity has been added to the project network of Fig. L.5; look to the upper right of the event circles. Take the time to confirm each calculation.

The latest event time represents the latest time an event can possibly begin without affecting the schedule for the project. By convention, the LET of the last or terminal event is equal to its earliest event time; thus both the EET and LET are twenty-three for event 11 (Fig. L.5). For all other events, the latest event time is computed by using the following rules:

1. Consider all activities leaving an event.

2. Subtract each activity's duration from the LET of its terminal event.

3. Select the smallest LET obtained.

For example, consider event 10. It has one leaving activity, 10-11. The latest event time for event 11 is twenty-three hours; the duration of activity 10-11 is two hours; thus the LET for event 10 is twenty-one hours.

The computed latest event times are shown at the lower right of each event in Fig. L.5. Let's verify the calculations for event 8. There are two leaving activities: 8-9 and 8-10. Consider 8-9 first. The previously calculated LET for event 9 is twenty hours. The duration of activity 8-9, a dummy activity, is zero; thus the computed LET for event 8 is twenty hours. Now consider activity 8-10. The LET for event 10 is twenty-one hours; the duration of activity 8-10 is six hours; thus the computed LET for event 8 is fifteen hours. Which number is correct, twenty or fifteen? The third rule says to select the smaller figure; thus fifteen hours is correct. The student should verify the other latest event times on Fig. L.5. Remember to start at the right and work backwards.

The Critical Path

Note that on several of the events shown on Fig. L.5, the earliest and latest event times are equal. These events define the *critical path*, which is marked by a double line. Events along this line must begin on time, and activities that lie on the critical path must require no more than the estimated duration, or the project will not be completed on time. When PERT is used, the critical path becomes the primary focus of management control; monitoring the critical events or milestones provides an early warning if the estimates are inaccurate. When CPM is used, the critical path defines those activities into which additional resources should be poured, as only by shortening the critical path can the project completion time be improved.

Slack Time

Activities not on the critical path can—to a point—start late or exceed their estimated durations without affecting the project's estimated completion time; this property is called *slack* or *float*. The total slack time for an activity is computed by subtracting its duration and the earliest event time of its initial event from the latest event time of its terminal event:

$$\text{Total slack} = \left[(\text{LET})_{\text{terminal}} - (\text{EET})_{\text{initial}} - \text{duration} \right]$$

Using Fig. L.5 as a reference, the slack times shown in Fig. L.6 can be computed. Slack is also shown enclosed in parentheses beneath the activity arrows of Fig. L.5.

Fig. L.6: *Computed slack times for the barracks painting project network.*

Event		LET (terminal)	EET (initial)	Duration	Slack
2-4	Paint side 1	6	2	3	1
3-5	Scrape side 3	11	6	2	3
4-7	Razor side 1	18	6	1	11
5-6	Dummy	12	8	0	4
5-8	Scrape side 4	15	8	4	3
6-7	Dummy	18	12	0	6
7-9	Razor side 2	20	12	2	6
8-9	Dummy	20	15	0	5
9-10	Razor side 3	21	15	1	5

How can an analyst or a manager use slack? One possibility is in scheduling. For example, consider the three activities 4-7, 7-9, and 9-10. All have considerable slack, and all involve razoring. Is it possible that scraping and razoring could be done by the same five soldiers without affecting the project schedule? If so, the job would require ten rather than fifteen soldiers, a significant improvement. Slack is also useful in management control. For any given activity, the slack shows the maximum time the schedule can slip without affecting the project's expected total completion time.

PERT AND CPM

The project network is the foundation of both PERT and CPM, and the basic concepts of critical path and slack are common to both. PERT is used when the activity times are uncertain; the techique serves as an early warning system, alerting the analyst and management to the possible impact as the actual times for the various activities become known. Industry developed CPM to help solve scheduling problems when the activity times are known more precisely. The emphasis is on cost minimization: How can we best allocate limited resources to complete the project in a reasonable time and at minimum cost? The user of CPM would certainly consider cutting resources from fifteen to ten soldiers while maintaining the projected schedule. The PERT user, while interested in cost control, would be more concerned with the accuracy of the estimates for such early activities as 1-2, 2-3, and 3-4, and might postpone the release of the extra soldiers until the successful, on-time completion of event 4 demonstrated the accuracy of the estimated durations.

The project network is, however, the key to both techniques. The analyst can benefit from preparing a project network, even if neither PERT nor CPM is used to its full extent. For example, consider the simple Gantt chart prepared earlier in the chapter (Fig. L.2). It showed an estimated project completion time of twenty-two hours, while the project network of Fig. L.5 estimated twenty-three. Why the difference? Although it is not obvious on the Gantt chart, a study of the project network should reveal that it takes longer to scrape side two than it does to paint side one. See if you can find the relevant activities. The advantage of a project network over a Gantt chart is that the project network explicitly defines the relationships between events and activities, while the Gantt chart merely implies them. Simply, a project network is an easy to prepare, easy to read, highly accurate tool for scheduling and managing a project.

REFERENCES

1. Hartman, W., Matthes and Proeme (1968). *Management Information Systems Handbook*. New York: McGraw-Hill Book Company.

2. PERT Coordinating Group (1963). *PERT: Guide for Management Use*. Washington, D.C.: US Government Printing Office, publication number 0-6980452.

3. Wiest, Jerome D., and Levy (1969). *A Management Guide to PERT/CPM*. Englewood Cliffs, New Jersey: Prentice-Hall, Inc.

Module M

File Design and Space Estimates

OUTLINE

SECONDARY STORAGE

In this module, we will consider problems associated with storing files on such secondary device as magnetic tape, cassette, magnetic drum, magnetic disk, and diskette. Tape and cassette are limited to the sequential storage of data, and are used primarily for backup on non-hobby computer systems. Drum is a high speed, low capacity medium used mainly for special purpose applications such as paging on a virtual memory system. For the average analyst, disk and diskette are the most commonly used secondary storage devices, and thus we will concentrate on them.

DEFINING THE DATA STRUCTURES

During system design, physical files were identified. As part of detailed design, the analyst must specify the contents of these files. The first step is laying out the data structure or structures. Each file represents one or more data stores on the data flow diagram, and each store, in turn, defines one or more data elements. Detailed descriptions of these elements can be found in the data dictionary. Given this information, the analyst can compile a list of data elements that compose a single logical record in the file. The file is built around the resulting structure.

For example, consider the year-to-date file from the payroll application covered in Part I of the text. The data elements that must be stored for each employee are shown in Fig. M.1. The year-to-date statistics are seven digit numbers with a maximum value of 99,999.99; the comma and the decimal point are not stored, but implied. The total length of the structure is 71 characters. The year-to-date file will consist of a series of these logical records, one for each employee.

At this point, the analyst might code the data structures in the source language to be used by the programmers.

Fig. M.1: *The data structure of a year-to-date payroll logical record.*

Element name	Length	Format
Social security number	9	character
Name	20	character
Year-to-date gross pay	7	99999V99
Year-to-date federal tax	7	99999V99
Year-to-date state tax	7	99999V99
Year-to-date local tax	7	99999V99
Year-to-date FICA tax	7	99999V99
Year-to-date net pay	7	99999V99

SELECTING A LOGICAL ORGANIZATION

The analyst's next task is to select a set of rules for organizing the file. One possibility is to use a *sequential organization*. As a sequential file is created, records are added one after another in chain-like fashion. Later, when these data are processed, they are read in the same fixed sequence. Records can be stored in time order, or sorted and stored by some key. The basic idea is that a sequential file must be processed in a fixed order; the only way to access record number 500 is to first access records 1 through 499. Thus, the speed with which a given record can be accessed is a function of its physical location on the file.

The analyst can also select a *direct* or *random organization* for the data. When a direct access file is used, each logical record is assigned a key that corresponds in some way to the physical location of the data on disk. Given this key, it is possible to store or retrieve a record without regard for its position on the file; for example, record number 500 can be accessed just as quickly and just as easily as any other record.

A third alternative is an *indexed file*. As the data are stored, an index is created linking each record's logical key to its physical address on disk; the index is then stored independently. Later, when the data must be retrieved, the index is read into main memory and searched by the logical key, the physical address of the record is extracted from the index, and the record is read directly. Often, the index is maintained in logical key order; if so, by following the index, key by key, it is possible to process such files sequentially.

Not every element of a given data structure must be stored on the same file. For example, consider a student grade history record. It probably contains the student's identification number, name, department of major, and other general information. Following these basic elements will be the course code, description, credit hours, and grade for *each* course the student has taken. How many such entries will be found? The answer depends on the student. A first semester freshman may have none. A second semester senior may have fifty, or even more. One option is storing each student's grade history as a variable length record. An alternative is a series of linked records (Fig. M.2). Each student has a master record containing basic data. Buried in this record is a pointer to the first of the grade records. The first grade record points, in turn, to the second, which points to the third, and so on, until the end of the chain is reached. Splitting a data structure between two or more files should be considered whenever a repeating substructure (such as the student course records) is present, or when a record must be accessed by several different keys; for example, a school might want to access student data by student number, department of major, or local address.

The wave of the future is *data base management*. The task of creating a data base is beyond the scope of this book; the author strongly recommends that any future systems analyst take a good data base course. Assuming that the data base exists, the analyst's main task will be defining the data structures. many of the physical details associated with file design can be ignored in a data base environment. In most organizations, a data base administrator is responsible for creating, maintaining, and controlling access to the data base; once the basic structures have been defined, the analyst should consult with this individual.

Fig. M.2: *Linked records.*

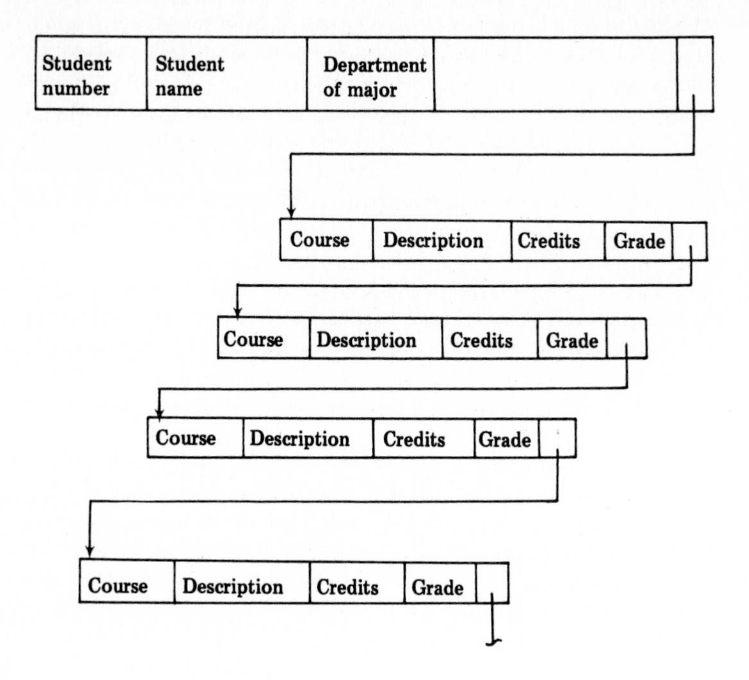

Selection Criteria

Which organization should the analyst select? The answer depends on the application, as each technique has its strengths and weaknesses. Generally, however, the analyst's objective should be to select the data organization that allows the system's functions to be performed at the lowest possible cost. In making this judgement, several factors, including physical storage costs, programming costs, execution costs, and usage costs, must be carefully evaluated and balanced; the concern must be with the *total cost*, not with any single element of that total.

Balancing cost factors is not always easy. For example, when disk is used, moving the read/write mechanism is a major component of execution cost. With well planned sequential files, head movement can be minimized, which suggests sequential access. Often, programmers feel more comfortable with sequential applications, meaning that programming costs will probably be lower than they might be if direct access were chosen. Even physical storage costs tend to favor sequential files, as the larger blocking factors possible on such files mean that the data will occupy less physical space. However, sequential files almost always require that data be processed in batch mode, and the delays inherent in waiting for the computer output can represent a significant user cost that far outweighs the apparent economies of a sequential file. Analysts have been grappling with the problem of balancing these cost factors for years, and have developed certain criteria that seem to suggest particular organizations.

Activity is a measure of the percentage of records in a file that are actually accessed each time the file is processed. It is a function of the required *turnaround* time. Some applications, such as payroll, can tolerate several days between collecting and processing data. Others, such as updating checking accounts, must be run at least daily. Still others require turnaround measured in hours, minutes, seconds, or even fractions of a second. The turnaround required by an application limits the amount of time during which transactions can be collected. If enough transactions occur during this interval, the file will be active; if not, it will be inactive. Highly active files should be organized sequentially; less active files call for a direct organization.

Turnaround is the time between the submission of a job and the return of the results. When this interval becomes very small, a few seconds or less, we stop thinking in terms of turnaround, and begin to concentrate on *response time*. If response time is an important factor in an application, it is likely that the system will be designed to process individual transactions in a random or unpredictable order. Sequential access simply will not do for such applications; direct or indexed access is essential.

Localized activity occurs when a significant percentage of a file's activity is concentrated in a relative handful of records. For example, an airline reservation file might contain data on numerous flights scheduled months in the future, but most customers will be interested in those few flights scheduled to leave within the next several days. When localized activty is expected, direct access is a good choice.

Another important factor is *flexibility*, which favors the indexed technique. When an indexed organization is used, records can be processed either sequentially or directly. Also, multiple indexes can be created; for example, student data might be indexed by student indentification number, department of major, and local address, and subsequently accessed by any of these keys. The indexed technique should not be used when a file is volatile, however. *Volatility* is a measure of the rate of additions to and deletions from the file; for example, a file listing current hit records would be quite volatile. On an indexed file, both the index and the data must be maintained; on sequential or direct files, only the data are involved.

File size is another concern. On disk, for example, direct access is achieved by moving an access arm over the track containing the desired record, and then reading the data as it rotates by. With a small file that occupies only a track or two, there is little advantage to be gained from direct access; it is almost as efficient to search the file sequentially. If sequential I/O instructions are easier for the programmers to code and debug, this factor may well offset the apparent advantages of direct access. In particular, consider using a sequential organization to hold table files. In contrast, consider how a massive, rarely used file might be stored. A sequential file on magnetic tape (or even microfiche) may well be the best answer.

Another important factor is the *status quo*. A new or unfamiliar file organization means that the programmers will have to be trained before they can use it. Additionally, implementing the system will probably take longer than if a more familiar organization were used. In selecting the new organization, the analyst should consider these extra costs. Similar arguments might be advanced for the existing hardware. The point is not to avoid change, but to recognize that change can be expensive.

Few files exist in a vacuum; most coexist with other, related files, and must be organized in a way that is compatible. Be careful to look beyond the single application, however. For example, consider payroll. The activity criterion might suggest a sequential organization, and in fact this might be the best choice for the transaction and year-to-date files. What about the personnel data, however? They are processed by the payroll application, but are needed by other systems as well. Perhaps the best choice is an indexed organization. The ultimate expression of concern for data *integration* is a data base.

These criteria are merely guidelines, not absolutes. They are the result of years of observation, trial, and error, and are in fact closer to folk wisdom than to science. Beginners often go overboard—the activity is high, and therefore the file must be organized sequentially. Don't forget that selecting a file organization is aimed at reducing the sum of several different costs, including the cost of storage, execution, programming, and usage. Selecting a file organization is perhaps the most important decision the analyst can make during detailed design. Take your time, and do it right.

PHYSICAL DATA STORAGE ON DISK

How are data physically stored on disk? That depends on the disk. Some systems divide the surface into a series of fixed length *sectors*. On many large computer systems a *fixed block architecture* is used; logically, there is little difference between the fixed block and sectored approaches, as both store data in fixed length units. Other disk units are *track addressed*. A track is one of the concentric circles around which data are stored on a disk surface. With a track addressed disk, space can be used in any way the programmer or analyst wishes. Track addressed disk is somewhat more flexible, allowing the programmer to fit the physical data structure to the requirements of the application. The sectored and fixed block approaches tend to be a bit easier to use.

When using a track addressed disk, the analyst can choose between a *count/data* or *count/key/data* format. Let's consider the count/data format first (Fig. M.3). For each record, two elements are stored: a count and the data. The count simply indicates the record's position relative to the beginning of the track; the first record is 00, the second is 01, and so on. A gap of unused space separates the count from the data. Imagine that a programmer wants record 04 on track 42. First, the access mechanism is moved to track 42. Next, the track is searched for count 04; when it is found, the

Fig. M.3: *The count/data format.*

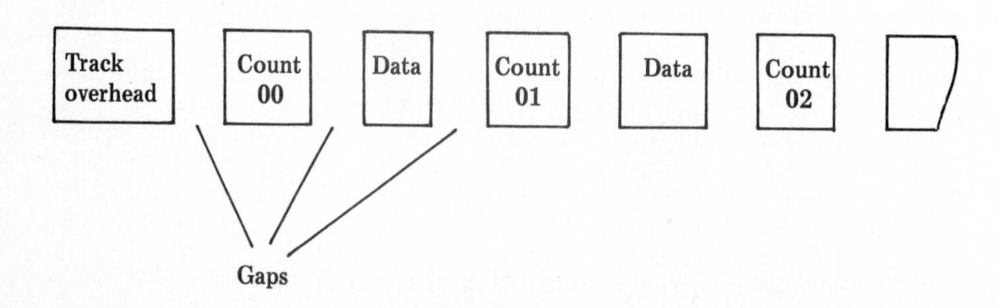

data following it is transferred into the computer. The gap between the count and the data gives the disk's control unit enough time to read the count, check it, and then decide if the data should be accessed or ignored.

Sequential files are stored using the count/data format, as are direct access files that rely on relative record location; for example, COBOL's relative organization. If the programmer chooses to create and maintain an index listing the logical keys and their associated relative record numbers, an indexed file can be physically stored in count/data format, too.

The count/key/data structure (Fig. M.4) is an alternative. The count is still the relative position of the record on a track. The key is a logical entity that uniquely identifies the record, for example a social security number or a student identification number. The first step in accessing a record is still to move the read/write mechanism over the proper track. However, the track is then searched by key, rather than by count.

IBM's indexed sequential access method uses the count/key/data format. Many COBOL programmers are familiar with a direct data structure in which a relative track address and a logical key are provided to the system; this technique also uses count/key/data.

LOGICAL AND PHYSICAL DATA

A *physical record* is the unit of data transferred between main memory and an external device. A *logical record* is the unit of data needed by a single iteration of a program. Code is written to process logical records. Hardware moves physical records. While it is certainly possible that a logical record will be stored as a single physical record, it is not necessary. For example, on a track addressed disk, logical records can be blocked to form larger physical records (Fig. M.5). Why? Remember the gaps of unused space separating the physical records on the disk surface? Ten logical records stored without blocking call for ten sets of gaps. Ten logical records blocked and stored as a single

Fig. M.4: *The count/key/data format.*

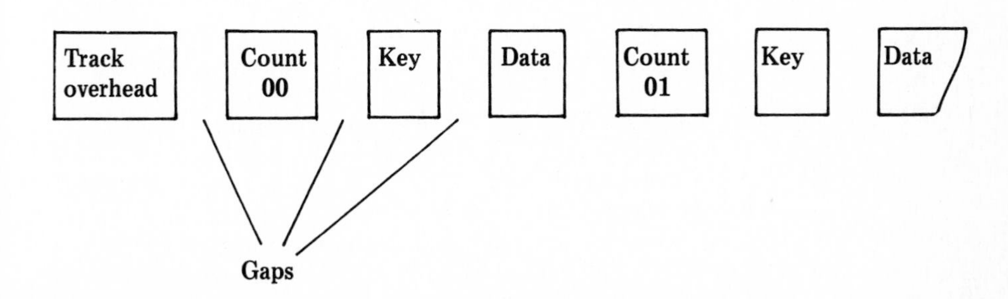

Fig. M.5: *Blocking.*

10 records, unblocked:

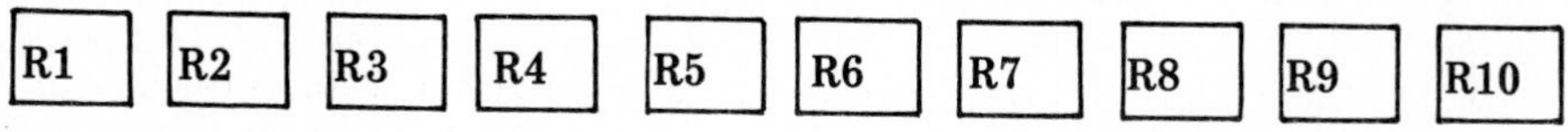

R1 R2 R3 R4 R5 R6 R7 R8 R9 R10

10 records blocked:

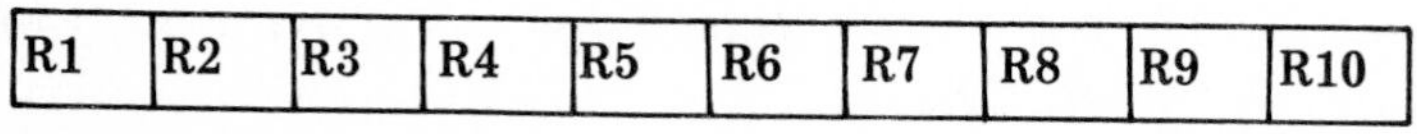

R1	R2	R3	R4	R5	R6	R7	R8	R9	R10

physical record require only one gap. With less wasted space, more data can be stored on a single track. Blocking several logical records to form a single physical record improves hardware efficiency.

On a sectored disk, a sector is the physical unit transferred between main memory and the external device. A common sector size on microcomputer systems is 256 characters or bytes. Storing one 71-character logical record in each sector would waste space. On many systems three such logical records are stored in each sector, thus utilizing 213 of the available 256 storage locations. What happens if a logical record is too big to fit in a sector? Often, a record can be *spanned* over two or more sectors. With blocking, reading a single *physical* record provides two or more *logical* records. With spanned records, accessing a single *logical* record involves reading two or more *physical* records.

The analyst must describe both the logical and the physical data. Programmers work with logical data, and program specifications must be developed during detailed design. It is the physical data that will actually be stored, however. Disk space is normally shared by many files; it is a scarce resource that must be carefully rationed. How much space should be assigned to a file? Is there enough space on a device to hold the necessary data? The analyst must answer these questions.

ESTIMATING SPACE REQUIREMENTS

Let's begin with a sectored disk or diskette. Earlier, the logical record length of a year-to-date record was determined to be 71 characters. If a sector can hold 256 characters, three complete records can be stored in each sector. How many records must be stored? In Chapter 2, employment at THE PRINT SHOP was estimated at 50 people. At three records per sector, how many sectors are needed to hold 50 records? Divide three into fifty: the answer is 16.67. Sixteen sectors are not enough; seventeen will be needed.

What is the capacity of a diskette? That depends on the system. Some small, single-sided 5.25 inch diskettes are designed to hold 160K characters per volume—that's about 640 sectors. Volumes with twice or even four times that capacity can be purchased. Hard disk drives, even for a microcomputer, are rated in megabytes; for example, a 1.6 megabyte disk would have 6400 sectors. The analyst should be familiar with the specifications of disk units used in his or her computer center.

In general, when a disk that stores sectors or fixed blocks is used, the amount of space needed to hold a file can be computed by the following rules:

1. Divide the sector size (or fixed block size) by the logical record length to get the number of records that will fit in a single sector. Ignore fractions, as only complete logical records are stored.

2. Divide the integer quotient from step 1 into the total number of logical records that must be stored in the file. Round the quotient up to the nearest whole number (for example, 15.25 becomes 16), since partial sectors cannot be assigned.

The result is the number of sectors needed by the file. The rules for estimating space on a fixed block architecture system are identical; the sector size is the only thing that changes.

What happens when large records must be spanned over two or more sectors? The rules change a bit, but the intent is still the same. Start by computing the number of sectors required for each logical record, and then multiply by the number of records in the file.

Estimating space requirements on a track addressed system is a bit more complex. In addition to the logical record length and the number of records in the file, yet another variable must be considered: the physical record length or block size. For example, let's use the 71 character logical record of Fig. M.1, and assume that 5000 such records must be stored. Initially, we'll assume that the data are not blocked; in other words, the physical record length is 71 characters.

The direct access device to be used in this example is an IBM 3330. The capacity of a track on an IBM 3330 is 13,030 characters or bytes. Using the count/data format, each record will consume a certain amount of overhead (for the count field and the gaps) plus the record length. The formula for computing the effective space required by each record is:

$$\text{space/record} = 135 + \text{data length}$$

With a data length of 71, each record will need the equivalent of 206 characters. Dividing 206 into the capacity of a track (13,030) yields a quotient of 63 plus a remainder; thus 63 records can be stored on each track. A total of 5000 records must be stored; how many tracks will be needed? Once again, divide; 5000 divided by 63 is

79 plus a fraction, meaning that more than 79 tracks will be needed, so 80 tracks should be assigned to the year-to-date file.

One point in these calculations might be a bit confusing. When computing the number of records per track, we truncated, essentially discarding the fractional part of the answer. Later, when computing the number of tracks, we rounded the answer up to the next larger integer. That seems inconsistent, but it isn't. The rule is that every physical record must be stored as a single entity. The only way we can use the fractional part of a track is by storing a piece of a record, and that is illegal. We compute the total number of tracks by dividing the number of records in the file by the number of *complete* records that can be stored on a single track. A remainder indicates that not all the records could be stored if only the quotient were used; in other words, the product of the quotient and the records per track would be less than the total number of records in the file. By rounding up, we are certain that enough space will be allocated to hold all the records.

What if the data are blocked? Theoretically, less space should be needed, since fewer interrecord gaps are stored. Let's see if the theory works. We'll try a *blocking factor* of ten, meaning that ten logical records will be grouped to form one physical record. The physical record length is 710 characters. Plug the new block length into the formula described above: 135 + 710 is 845. The capacity of a track is still 13,030 bytes, so 15 physical records will fit on each. Remember, however, that a physical record holds 10 logical records. That's 150 logical records per track. Only 34 tracks are needed to hold all the data. By blocking the data, the required space has been reduced from 80 to 34 tracks, and that's significant.

In general, increasing the blocking factor reduces the amount of space required to hold the data. Figure M.6 shows the IBM 3330 space required by the year-to-date file using a variety of blocking factors. Initially, the impact of additional blocking is dramatic, but note how the curve soon begins to level out. The analyst should pick a blocking factor that lies near the beginning of the horizontal portion of the curve; in this example, 14 might make sense.

Note that the curve of Fig. M.6 is not quite smooth; for example, as the blocking factor goes from 9, to 10, to 11, the number of tracks goes from 33, to 34, and back to 33 again. Why? This phenomenon results from using integer arithmetic in computing records per track. For example, imagine that the logical record length were 638 bytes. A blocking factor of ten yields a physical record of 6380 bytes; using the formula above, we would add 135 to get 6515 positions per physical record. The track capacity is 13,030 bytes, so exactly two physical records (twenty logical records) could be stored on each track. Increase the blocking factor to eleven. The physical record length would now be 7018 bytes, but only one such record would fit on a 13,030 byte track. In this example, increasing the blocking factor from ten to eleven reduced the number of logical records per track from twenty to eleven. The analyst should be aware that this happens—more is not always better. A good idea is to write a quick BASIC program to compute space requirements for a variety of blocking factors.

What if keys are used? Another gap (more overhead) and the key length must be added to the overhead and record length described above, and thus the formula (for an IBM 3330) becomes:

Blocking Factor	Number of Tracks
1	80
2	54
3	46
4	41
5	38
6	37
7	34
8	33
9	33
10	34
11	33
12	33
13	33
14	30
15	31
16	32
17	30
18	31
19	30
20	32
21	30
22	29
23	31
24	30
25	29

Fig. M.6: *Disk space as a function of the blocking factor.*

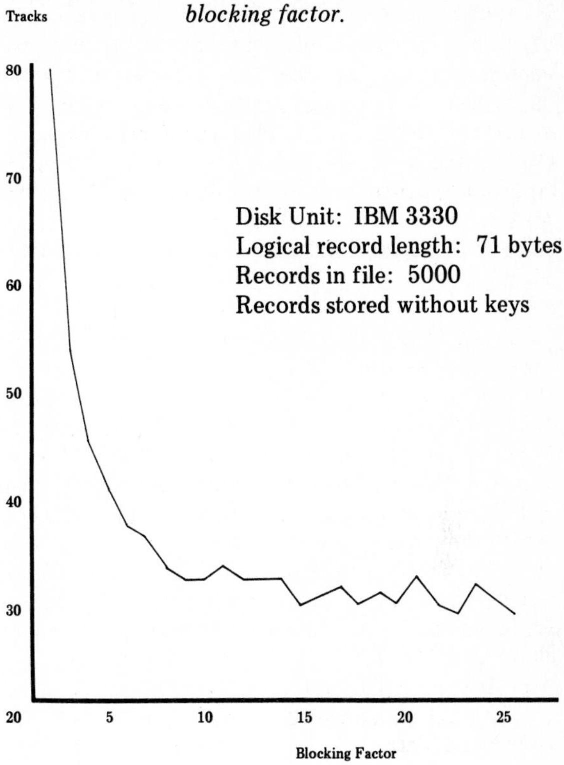

Tracks

Disk Unit: IBM 3330
Logical record length: 71 bytes
Records in file: 5000
Records stored without keys

Blocking Factor

$$\text{space/record} = 135 + 56 + \text{key length} + \text{data length}$$

As an exercise, repeat the calculations using this new formula.

The analyst should consider other factors, too. For example, when an indexed file is used, the index or indexes must be stored, and space must be allocated to store them. Overflow space is also easy to overlook. Many direct access techniques generate *synonyms*, two or more records that randomize to the same physical disk address. It is impossible to store two or more records in the same space; when synonyms occur, the subsequent records must be placed on an overflow area. If overflow is a possibility, the analyst must allocate space for it.

Is there anything the analyst can do to squeeze the same amount of data into even less space? Why not shrink some of the fields? For example, computational fields such as the year-to-date gross pay and the other year-to-date statistics can be stored in numeric rather than in character or display form. If each of the seven-digit numbers in Fig. M.1 were stored as a four-byte packed decimal or BCD field, the logical record length would drop from seventy-one to fifty-three bytes. Recompute the space requirements for the year-to-date file using this new logical record length.

Codes can sometimes be used, too. For example, in an accounts receivable file, a one-character code might be used instead of a four- or five-digit credit limit, thus saving three or four bytes per record. Be careful when using codes, however. Outside of the system that uses them, they are meaningless. It is possible to create so much confusion by using a code that the apparent savings are lost to increased programming and user costs. Don't forget that the objective of file design is minimizing the total cost of the file, not just the physical storage cost.

In the examples presented above, the IBM 3330 and a 256-byte sector were used as illustrations. Other disk units may differ. The analyst should identify the proper formulas or sector size for his or her own installation.

REFERENCES

1. Davis, William S. (1983). *Operating Systems: a Systematic View.* Reading, Massachusetts: Addison-Wesley Publishing Company.

2. IBM Corporation (1974). *Introduction to IBM Direct-Access Storage Devices and Organization Methods.* White Plains, New York: IBM Corporation. Publication number GC20-1649.

3. Lewis, T.G. and Smith (1976). *Applying Data Structures.* Boston: Houghton Mifflin Company.

Forms Design and Report Design

OUTLINE

COMMUNICATING WITH PEOPLE

The basic function of a computer is to process data into information. Most systems involve people; ultimately, they provide the data and use the information. In this module, we will consider three basic media for communicating with people: printed reports, data collection forms, and CRT screens.

A printed report is an output medium. It provides a permanent, inexpensive, portable copy of large volumes of data, but the information is static, and it can quickly become outdated. An alternative is displaying information on a CRT terminal. CRTs are fast, and the information displayed is current. Unfortunately, the display is also temporary, not very portable, and limited in volume. Clearly, the analyst faces a tradeoff in choosing between reports and displays.

Data collection forms are used to capture input data. Forms are inexpensive and, if properly designed, can be used by almost anyone. They are, however, difficult to process. They must be handled and/or transcribed by human clerks, and this is both expensive and error-prone. Using such techniques as OCR, mark sense, and MICR, machine-readable forms can be designed, but these techniques limit the flexibility of the forms. For many applications, a CRT terminal is an alternative. Screens are flexible—they can be used to collect almost any kind of data. Once data are entered through a terminal, they can be stored in machine-readable form; no additional handling is required. However, CRT terminals are relatively expensive, and users require at least a minimum level of training. Once again, we face a tradeoff.

Once a medium is selected, how does an analyst design an appropriate report, data entry form, or CRT screen? While certainly not complete, the material below suggests a number of guidelines.

REPORT DESIGN

The basic work document for designing a report is a printer spacing chart (Fig. N.1). The chart is simply a grid, with each block representing one print position. Using it, the analyst describes, line by line and print position by print position, the structure of a report. Eventually, the analyst or a programmer converts each line on the printer spacing chart to an appropriate set of source language constants or to a data structure.

The report outlined in Fig. N.1 contains headers, detail lines, and summary lines. Page or topic headers identify the report; they are printed at the top of each page or following a major control break. The report date should always be included, and page numbers are helpful, too. Column headers identify the data elements listed in a report, and must be coordinated with the detailed data; more about this later. Headers should clearly describe the report and its contents to the target audience. For example, technical people tend to use acronyms and abbreviations which, while quite reasonable on a technical report, would be inappropriate on one designed for management or the general public. User groups have their own shorthands, too; for example, YTD may be an acceptable abbreviation for "year-to-date" if the target audience is payroll clerks or accountants, but other groups might not assume that meaning.

Fig. N.1: *A mock-up of a typical report on a printer spacing chart.*

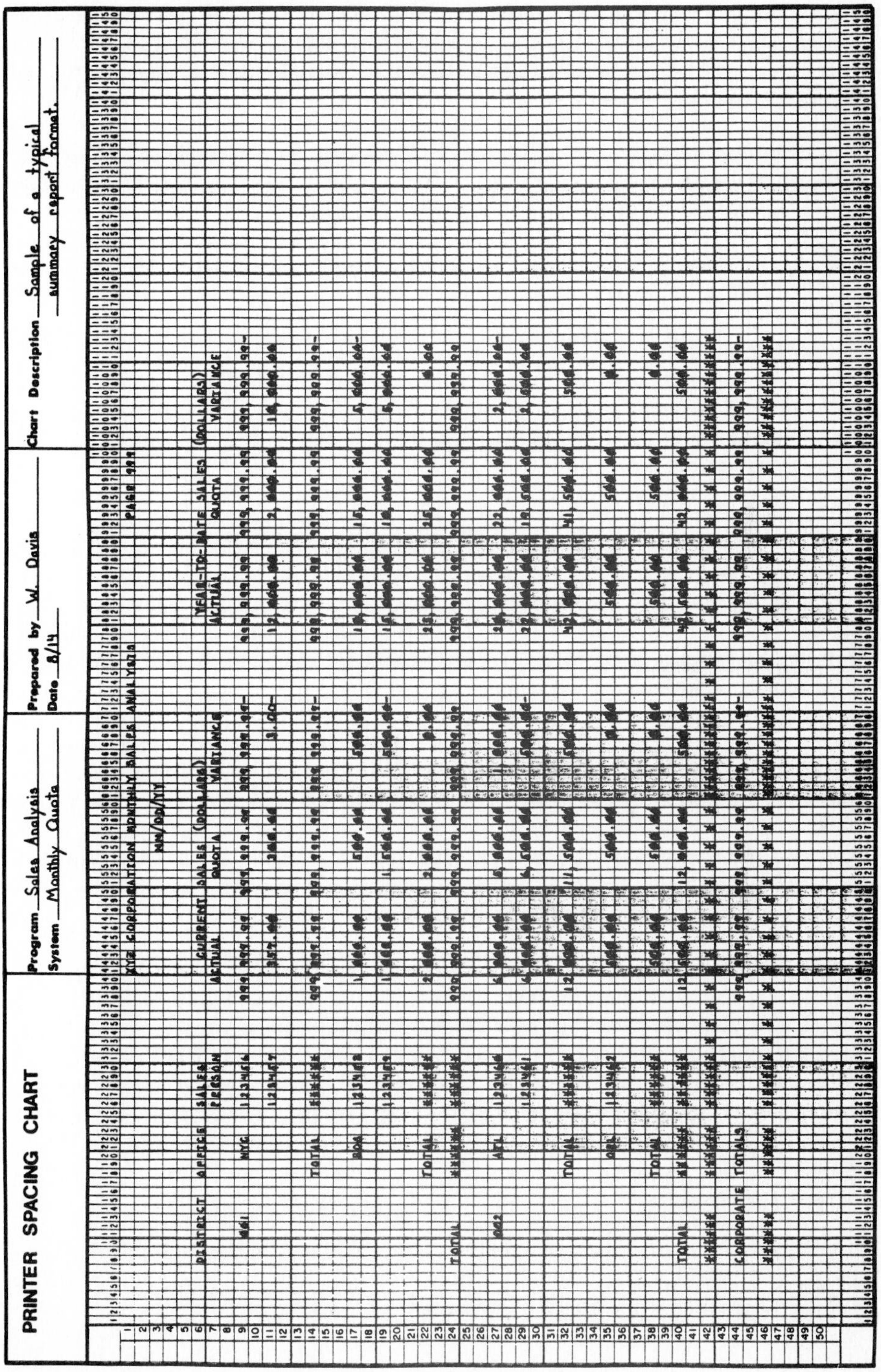

```
PRINTER SPACING CHART
Program  Sales Analysis        Prepared by  W. Davis     Chart Description  Sample of a typical
System   Monthly Quota         Date  8/14                                   summary report format.

 1                                XYZ CORPORATION MONTHLY SALES ANALYSIS                      PAGE 999
 3                                         MM/DD/YY
 6          DISTRICT  OFFICE  SALES         CURRENT SALES (DOLLARS)              YEAR-TO-DATE SALES (DOLLARS)
 7                            PERSON    ACTUAL      QUOTA      VARIANCE       ACTUAL      QUOTA      VARIANCE
 9              001     NYC   123456  999,999.99  999,999.99  999,999.99-  999,999.99  999,999.99  999,999.99-
11                            123457      357.[illegible]      3[illegible]      3.00-   12,[illegible]   2,[illegible]   10,[illegible]
14                    TOTAL   XXXXXX  999,999.99  999,999.99  999,999.99-  999,999.99  999,999.99  999,999.99-
17                      BOA   123458  1,[illegible]       500.[illegible]    500.[illegible]   1[illegible]   15,[illegible]   5,[illegible]-
19                            123459  1,[illegible]     1,500.[illegible]    500.[illegible]-  1[illegible]   10,[illegible]   5,[illegible]
22                    TOTAL   XXXXXX  2,[illegible]     2,[illegible]        0.[illegible]   25,[illegible]   25,[illegible]   0.00
24          TOTAL     XXXXXX  XXXXXX  999,999.99  999,999.99  999,999.99   999,999.99  999,999.99  999,999.99
27              002     ATL   123460  6,[illegible]     5,[illegible]     1,[illegible]   2[illegible]   22,[illegible]   2,[illegible]-
29                            123461  6,[illegible]     6,500.[illegible]    500.[illegible]-  2[illegible]   19,500.[illegible]   2,500.[illegible]
32                    TOTAL   XXXXXX  12,[illegible]   11,500.[illegible]    500.[illegible]   4[illegible]   41,500.[illegible]   500.00
35                      DBL   123462     500.[illegible]     500.[illegible]      0.[illegible]     500.[illegible]     500.[illegible]     0.00
38                    TOTAL   XXXXXX     500.[illegible]     500.[illegible]      0.[illegible]     500.[illegible]     500.[illegible]     0.00
40          TOTAL     XXXXXX  XXXXXX  12,[illegible]   12,[illegible]      500.[illegible]   4[illegible]   42,[illegible]   500.00
42          XXXXXX    XXXXXX  XXXXXX  X X X X X X X X X X X X X X X X X X X X X X X X X X X X X X X X X X X
44          CORPORATE TOTALS          999,999.99  999,999.99  999,999.99-  999,999.99  999,999.99  999,999.99-
46          XXXXXX    XXXXXX  XXXXXX  X X X X X X X X X X X X X X X X X X X X X X X X X X X X X X X X X X X
```

As the name implies, a detail line provides detailed data. The process of designing a report begins here. First, list all the data elements or fields to be included in the report. Group related items; for example, Fig. N.1 shows three clear groupings: identification, current sales, and year-to-date sales. Next, define the number of print positions needed for each element by writing its maximum and/or minimum value; include appropriate punctuation and a sign. For non-numeric items, use the field length.

The column headers are planned next. Write each header beside the maximum/minimum value of the associated data element. If the header is wider than the data, then it defines the column width; for example, in Fig. N.1, the first report column, DISTRICT, is eight positions wide because the header requires eight positions. Otherwise, the data element determines the column width; for example, consider the fourth column, ACTUAL.

Is there enough space to print the complete report? Add the maximum column widths. Remember that there must be at least one space between columns, so count the number of columns, add one (for the left margin), and add this count to the sum of the column widths. Is the total less than or equal to the printer's line width (common widths are 80, 120, and 132 positions)? If not, it may be necessary to delete fields or change headers.

Given the list of fields, their associated headers, and the maximum column widths, the report format can be layed out. Spaces should be left between columns. How many? There is no easy answer. The objective is to produce a report that is easy to read. The only way to determine if a report layout is good is to lay it out and look at it; if it looks good, it probably is. A few suggestions can be made, however. Related items should be grouped; for example, on Fig. N.1, extra space separates the three logical groupings described above. Wider margins should be left on both sides of the paper, and at the top and bottom; people expect margins around printed material. Finally, key fields should be strategically placed; for example, the most significant field on the report outlined in Fig. N.1 is probably the variance (the difference between actual sales and the quota), so this field is placed near the right margin, where negative signs are visually obvious. Key data elements might be further highlighted; for example, the words "BEHIND QUOTA" might be printed to the right of a detail line associated with a sales person who is significantly behind quota.

Avoid printing repetitive data. For example, consider the district identification code in Fig. N.1. Within a district are several sales offices; within each office are several sales people. Should the district identification code be printed on each detail line? No; it should appear only on the first line for a district. Why? Consider the partial report illustrated in Fig. N.2. Which version is easier to read? Note how the version on the right clearly shows which offices belong to which district. Repeating the same descriptive data generates needless clutter, and interferes with the clarity of a report.

What about the summary lines? They should be offset or highlighted in some way. Note how, in Fig. N.1, we have printed the word "TOTAL" in the appropriate column and skipped an extra line before and after each summary line. Key totals might be boxed, and asterisks are commonly used to visually mark key summary lines.

Fig. N.2: *A partial report illustrating the value of suppressing repetitive information.*

a. With repetitive data displayed.

DISTRICT	OFFICE
001	NYC
001	BOS
001	PHL
001	WAS
002	ATL
002	ORL
002	DAL
002	HOU
003	CHI
003	STL
003	MSP
003	KC
004	LA
004	SF
004	SEA
004	ANC

b. With repetitive data suppressed.

DISTRICT	OFFICE
001	NYC
	BOS
	PHL
	WAS
002	ATL
	ORL
	DAL
	HOU
003	CHI
	ATL
	MSP
	KC
004	LA
	SF
	SEA
	ANC

The objective is to create a report that clearly conveys the necessary information. There is an acid test—can the intended audience read it? There is only one way to find out—ask. A printer spacing chart should be used to prepare a mock-up of a report, and this mock-up should be reviewed by the intended audience. If the people who must use a report have trouble reading it, the fault is probably not theirs, and the report should be redesigned.

FORMS DESIGN

A report provides output information to people; a form is designed to collect input data from people. To an extent, a form can be viewed as a report with blank spaces to be filled in; thus, many of the suggestions made under report design apply to forms design, too. Once collected, however, the data will have to be entered into the computer, and the analyst must take this eventual end use into account.

Some applications call for free-form data entry, while others demand that the data be rigidly structured. A key variable is the expected data volume. Designing a form is independent of the volume; however, the costs of collecting, entering, and processing the data increase as the volume increases. Unstructured forms are easy to design; to cite an extreme example, a tape recorder might be an acceptable free-form medium. Unfortunately, free-form data are difficult to enter, and even more difficult to process. It takes time to design a rigidly structured form, but once the form has been designed, data collection, data entry, and processing are greatly simplified. The more rigid the form, the more time it takes to design it. The more free-form the data, the longer it takes to extract and enter them. With relatively little data, it might make sense to use free-form data collection techniques and spend extra time on data entry. With volumes of data, time spent on form design can save a great deal of data entry time.

Often, a standard form—punched cards, test score sheets, MICR documents, OCR forms, UPC forms—can be adapted to a new application. Existing forms are not always suitable, however, and new ones must be created. While a detailed discussion of forms design is beyond the scope of this book, a few suggestions can be offered. One common technique is providing a series of boxes or blanks to be filled in, with headers or labels identifying the data elements to be entered. Another option is a list of possible choices for ticking or circling. Many professional printers distribute sample forms; a quick perusal might suggest an approach. Finally, when in doubt, seek expert advice. Many large organizations employ graphic designers. A smaller firm might contact a local printer; most are more than willing to help in exchange for some business.

How might the data be entered? Some forms—OCR, mark sense, bar-codes—can be read electronically. Others must be transcribed by data entry clerks. Traditionally, data were recorded on coding sheets and then given to a keypunch operator. Today, a more common option is to enter the data through a CRT terminal; often, a mock-up of the form is displayed, and a clerk simply fills in the blanks. What function is performed by the clerk? The data are copied. Copying data is expensive and error-prone. Why not use the CRT terminal as a data *collection* device? We'll consider this option in the next several paragraphs.

SCREEN DESIGN

A CRT terminal can be used for input, output, or both. Some are designed to read and write single lines. On input, each line holds one logical record. Often, such terminals are used much like keypunches, with data simply copied to the screen from coding forms. Conversational data entry is an option. A control program begins by displaying a prompt asking the clerk to enter a specific data element. In response, a value is entered and stored, and the prompt for the next data element displayed. On output, single line terminals are similar to printers; each line represents one line of a report. However, as a screen fills, old information begins to scroll off the top as new information enters from the bottom. This can make a screen difficult to read, and forces the user to work at the screen's pace. Consequently, large reports are rarely output through a screen.

Other CRT terminals perform full screen I/O. Using this approach, the contents of a full screen are defined in a buffer and then displayed. A visible marker called a cursor indicates the position on the screen where the next character typed by the user will appear. Buttons to move the cursor up, down, right, or left are generally available, allowing the user to position the cursor (and thus type) anywhere on the screen. Using full screen I/O, a dummy version of a form can be displayed, and data entered by filling in the blanks.

Look at a screen. Count the number of characters displayed on a single line. Now, count the number of lines. These two numbers define a rectangular grid. For example, the draft of this book was prepared on a personal microcomputer system. The screen displayes 25 lines of 80 characters each, and thus can be represented by an 80 by 25 grid. To lay out a screen, start with a sheet of graph paper, and block out a grid pattern equal to the capacity of your CRT. By printing one character per box, labels, headers, and data elements can be positioned, and the contents of the screen planned; the process is similar to defining a report format.

Several techniques can be used to highlight key data elements or to distinguish names, labels, and data. A simple approach is to use upper case for names and labels, and lower case for data. With more sophisticated terminals, different colors or reverse video can be used. Blinking characters or fields seem almost to jump out from the screen, and boxes or arrows are almost as effective.

Computer graphics is a developing technology. Space does not permit a detailed explanation of graphic techniques; the interested student might consider the book by Foley and Van Dam referenced at the end of this module. Touch sensitive screens and graphic input are other technologies of the near future. A good course in computer graphics is strongly recommended.

REFERENCES

1. Fitzgerald, Jerry, Fitzgerald, and Stallings (1981). *Fundamentals of Systems Analysis*, Second Edition. New York: John Wiley & Sons.

2. Foley, James D. and Van Dam (1982). *Fundamentals of Interactive Computer Graphics.* Reading, Massachusetss: Addison-Wesley Publishing Company.

3. Gore, Marvin and Stubbe (1979). *Elements of Systems Analysis for Business Data Processing.* Dubuque, Iowa: Wm. C. Brown.

4. Parkin, Andrew (1980). *Systems Analysis.* Cambridge, Massachusetts: Winthrop Publishers, Inc.

Decision Tables and Decision Trees

OUTLINE

PROCESSES WITH MULTIPLE NESTED DECISIONS

Algorithms involving multiple nested decisions are difficult to describe using structured English (Module H), pseudocode (Module I), or logic flowcharts (Module J). In this module, two alternative tools, decision tables and decision trees, are introduced. We'll use a simple example to illustrate both: Assume that the basketball coach has asked us to look through the student records and produce a list of all full-time male students who are at least 6 feet 5 inches tall and who weigh at least 180 pounds. First, we'll consider decision tables.

DECISION TABLES

A sample decision table is shown as Fig. O.1. It is divided into four sections: a condition stub at the upper left, a condition entry at the upper right, an action stub at the lower left, and an action entry at the lower right. The questions are listed in the condition stub; note that each requires a yes/no response. The associated actions are listed in the action stub. The responses (Y or N) are recorded in the condition entry, while the appropriate action is indicated in the action entry.

Perhaps the easiest way to understand a decision table is to read one. Let's begin with the first question: Is the student male? There are two possible answers: yes (Y) or no (N). What if the answer is yes; can we make a decision? Not yet; three more tests must be passed. What if the answer is no? If the student is not male, she is not a candidate for the basketball team, and thus can be rejected. Move down the column containing the N, and note the X on the action entry line following "Reject the student."

FIG. O.1: *A decision table for the basketball problem.*

Condition Stub:	Condition Entry:				
Is the student male?	Y	N			
Is the student taking at least 12 credit hours?	Y		N		
Is the student at least 77 inches tall?	Y			N	
Does the student weigh at least 180 pounds?	Y				N
Action Stub:	**Action Entry:**				
List the student's name and address	X				
Reject the student.		X	X	X	X

Move on to the second question: Is the student taking at least 12 credit hours? Again, there are two possible answers: yes or no. Note how the answers are recorded on Fig. O.1. The second Y is directly under the first one; this implies that the answers to both questions must be yes before the action identified by an X in that column's action entry can be taken. Is a yes response to both questions sufficient? No; two more tests remain. What if the answer to this second question is no, however? The student can be rejected. Why isn't the second N aligned under the first one? Because any single N, by itself, is enough to reject the student.

Read the rest of the table. It clearly shows that the student's name and address will be listed only if the answers to all four questions are yes, but that the student will be rejected if the answer to any one is no. When an algorithm involves more than two or three nested decisions, a decision table gives a clear and concise picture of the logic.

DECISION TREES

As an alternative, the analyst might use a decision science or management science tool called a decision tree. Let's consider the general structure of this tool, and then discuss how it might be applied to systems analysis and design. Imagine that you have an opportunity to purchase, for $50,000, exclusive rights to market a new product, magnetic soap. If the product catches on, you stand to make a great deal of money. On the other hand, magnetic soap might flop, and you could lose your entire investment.

The decision tree of Fig. O.2 is a graphic representation of this problem. The tree starts (on the left) with an *act fork*, represented by a small box. Emanating from it are two branches; they represent your possible decisions; buy or don't buy the rights. Move along the "buy" branch. The dot represents an *event fork*. Coming from it are two more branches representing the consequences of this decision—perhaps the product will be successful; perhaps it won't. Each branch terminates in an *outcome*.

Fig. O.2: *A simple decision tree.*

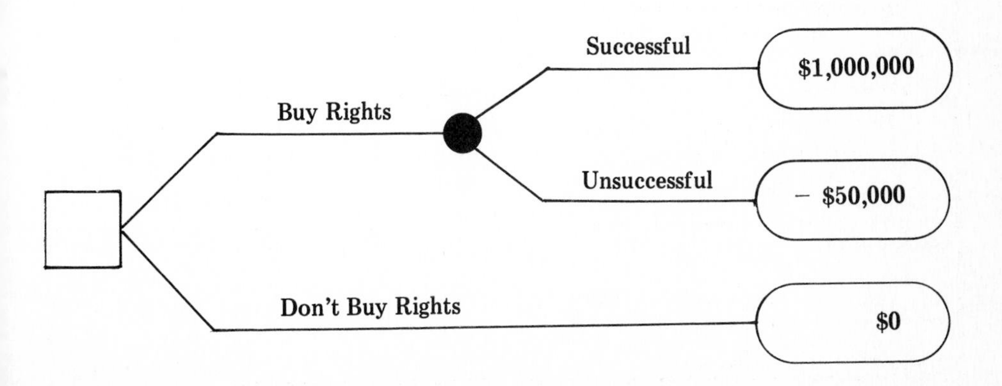

Fig. O.3: *A decision tree for the basketball problem.*

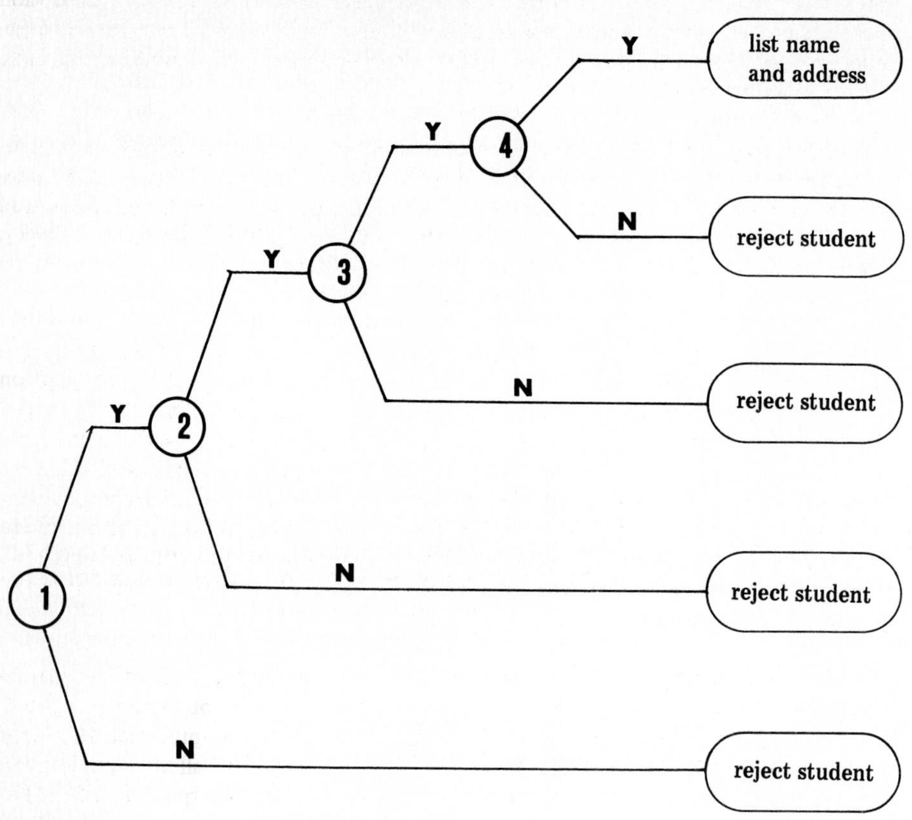

Decisions:

1. Is the student male?

2. Is the student taking at least 12 hours?

3. Is the student at least 77 inches tall?

4. Does the student weigh at least 180 pounds?

If the product is successful, you stand to make a million dollars. If it fails, you lose $50,000. Go back to the act fork. Before the decision is made, you have an option not to buy the rights. Clearly, this decision will cost you nothing. There are no consequences associated with this choice, either; your outcome is zero whether magnetic soap catches on or not.

The decision tree is a graphic representation of the decision, events, and consequences associated with a problem. Once the tree is drawn, probabilities can be associated with each branch, and expected values of the outcomes computed. Interested students may find decision trees excellent for illustrating alternative system strategies; the text referenced at the end of this module is a good source. We are, however, more interested in using decision trees to graphically represent nested decision logic.

Let's return to the basketball problem. The algorithm consists of a series of four nested questions or decisions; Fig. O.3 is a decision tree for this logic. Each question or decision is represented as a circle. Begin with the first one: Is the student male? There are two possible responses: yes or no. If the answer is no, the student is rejected; a yes answer, on the other hand, leads to another question. Again, there are two possible answers: yes and no. Again, a no leads to an outcome: reject the student, while a yes response leads to yet another question. Follow each branch on the tree to its logical outcome. Can you see how a student's name and address are listed only if all four questions are answered affirmatively? Decision trees are easy to construct, and graphically illustrate decision logic.

As we mentioned earlier, once a decision tree is drawn, probabilities can be associated with each branch, and the expected values of the outcomes computed. The systems analyst can take advantage of this basic idea to improve the efficiency of an algorithm without affecting its clarity. Consider the basketball problem. Imagine that you have one hundred student records to check. The first question asks if the student is male. How many students are eliminated by this first question? Probably, about half; we would expect about fifty of the one hundred students to be male. Consequently, the second criterion, credit hours, is checked for fifty students. How many are taking twelve hours or more? On some campuses, most of them; on others, very few; let's assume forty of the fifty are full-time. Student height is thus checked for forty students; how many exceed the height limit? Very few; perhaps three or four are checked for weight. Consider the total number of tests performed—100, plus 50, plus 40, plus 3 or 4 is over 190 decisions.

Change the sequence of the tests. Which of the four questions is likely to eliminate the most students? Relatively few will be 77 inches tall or taller; thus height is a good first screen. After one hundred records are checked for student height, how many will remain? Perhaps five or six. Thus, only five or six records are checked for the other criteria; consequently, the total number of tests is reduced. In general, when designing a nested decision algorithm, the most discriminating tests should be performed first. The result is improved program efficiency at little or no cost.

REFERENCES

1. Brown, Rex V., Kahr, and Peterson (1974). *Decision Analysis for the Manager.* New York: Holt, Rinehart and Winston.

Index